Electronic Techniques

SECOND EDITION

Electronic Techniques:
SHOP PRACTICES
AND CONSTRUCTION

ROBERT S. VILLANUCCI

ALEXANDER W. AVTGIS

WILLIAM F. MEGOW

Wentworth Institute of Technology
Boston, Massachusetts

Prentice-Hall Inc., Englewood Cliffs, N.J. 07632

Library of Congress Cataloging in Publication Data

Villanucci, Robert S   date–
  Electronic techniques.

  Bibliography: p.
  Includes index.
  1. Electronic apparatus and appliances—Design
and construction.   I. Avtgis, Alexander W.,
date–   joint author.   II. Megow, William F.,
date–   joint authors.   III. Title.
TK7870.V497   1981   621.381   80–17892
ISBN 0-13-252486-4

Editorial/production supervision and interior design by
Steven Bobker and Mary Carnis
Cover Design by Edsal Enterprises
Manufacturing Buyer: Joyce Levatino

Printed in the United States of America

10 9 8 7 6 5 4 3

Prentice-Hall International, Inc., *London*
Prentice-Hall of Australia Pty. Limited, *Sydney*
Prentice-Hall of Canada, Ltd., *Toronto*
Prentice-Hall of India Private Limited, *New Delhi*
Prentice-Hall of Japan, Inc., *Tokyo*
Prentice-Hall of Southeast Asia Pte. Ltd., *Singapore*
Whitehall Books Limited, *Wellington, New Zealand*

To

Annette, Gloria, and Helen

Contents

Preface

The second edition of *Electronic Techniques: Shop Practices and Construction* was prepared in order to further reduce the gap which exists between industry and an educational environment in the layout and fabrication of modern electronic packaging with its almost exclusive reliance on printed circuits. Owing to the wide acceptance of the first edition of *Electronic Techniques*, much of the material has been retained. New material added reflects current industrial practices in the fabrication of printed circuits. The additional subjects include *dry-film laminating*, *tin-lead electroplating techniques* and *double-sided printed circuit layout* and *fabrication*.

In addition, the *projects* section of the text has been up-dated to include projects which reflect high interest by the students, thus further motivating them to higher performance.

Advances in electronic packaging have introduced dramatic innovations in recent years. The development of new devices for high-density packaging in conjunction with modern printed circuit technology has brought about more sophisticated construction and packaging techniques that require today's technician to develop special skills. It is the purpose of this second edition to continue to present a practical and realistic approach for developing these skills in planning, laying out, and constructing electronic equipment.

This text has been designed to fulfill the needs of any electronic shop program. The material presented is directed toward educational institutions such as technical-vocational schools, technical institutes, and junior colleges, as well as industrial and military training programs.

The training of a skilled craftsman requires not only clear and detailed explanations, but also visual and graphic aids. For this reason, over 500 figures and drawings have been included to illustrate fundamental techniques used in electronic design, construction, and packaging. In addition, many problems have been included to aid in the thorough understanding of these basic concepts.

Although the design and construction techniques are presented in a logical sequence, individual chapters may be studied independently. No prerequisite knowledge of electronic circuits is necessary to understand the material presented, although a deeper insight may be realized if the technician has a fundamental background in electronics.

A high-fidelity stereo amplifier system has been selected as the motivating teaching vehicle and appears pictorially throughout each chapter. Its function is to generate interest and to present in a logical sequence the fabrication techniques necessary to construct any package or system. It is not necessary to build the amplifier system in order to derive maximum benefit from the text. This approach allows complete flexibility to permit an instructor to select other included projects. He may, in addition, use any of his own choosing.

Chapter 1 is a discussion of the factors that must be considered in packaging any electronic system. Chapter 2 introduces the high-fidelity stereo amplifier system. In Chapter 3, preliminary considerations of the package are converted to detailed engineering drawings and sketches. The layout, fabrication, and finishing of sheet metal chassis elements are covered in Chapters 4 through 9. Chapters 10 through 15 provide detailed information on designing and fabricating single-sided printed circuits by the print-and-etch technique. In these chapters, all pertinent information on available printed circuit materials and techniques are presented to serve as working guides that will allow the technician to become proficient in single-sided printed circuit technology. These chapters are sufficiently detailed to provide the necessary information for establishing a safe and economical printed circuit prototype laboratory. Chapters 16 and 17 present material on double-sided printed circuit layout and fabrication using the print-plate-and-etch technique including the use of dry-film photo-resist. Finally, assembly, wiring and harnessing techniques are covered in Chapters 18 through 20 in order to combine the printed circuit boards and the chassis elements into a completed system. Chapter 21 contains a selection of projects to (1) further develop student interest and (2) provide practice in developing the skills discussed in this text. New projects included in the second edition employ devices and circuitry which are consistent with modern technology.

We express our gratitude to Mr. Charles M. Thomson for his valuable advice and criticisms of the original manuscript and to Messrs. Robert F. Coughlin and Frederick F. Driscoll for their many helpful comments and suggestions. We also appreciate the assistance given us by the staff of the Air Force Contracts Section at Wentworth Institute of Technology. We thank the manufacturers who have furnished photographs for the text. Our sincere thanks also are extended to Mr. Thomas E. Naylor, Jr. for the processing of photographs for both editions and to Mrs. Pauline Campbell, Mrs. Mary Ellen Hatfield and Ms. Michele Reilly for their excellent typing of the manuscript.

Last, we are grateful to our families who were able to maintain their sense of humor throughout the preparation of the second edition.

Robert S. Villanucci
Alexander W. Avtgis
William F. Megow

1

Planning and Designing Electronic Packaging

1.0 INTRODUCTION

A high-quality piece of electronic equipment can result only from thorough, careful planning and application of the fundamental mechanical and electronic design factors. Of these factors, *reliability* is one of the most important. A highly reliable piece of equipment will result if all the design factors that govern packaging are considered during the preliminary planning stages.

This chapter examines the numerous, common packaging problems involved in the preliminary planning stage of constructing electronic equipment. Consideration is given to *space* and *weight requirements*, *component* and *hardware selection*, *chassis materials* and *configurations*, *component* and *hardware positioning*, *maintenance*, and *human engineering*. Specific problems are related to the packaging of a *stereo amplifier* that serves as an example packaging problem throughout this text. Each step of the packaging phase is applied, in sequence, to this amplifier, which illustrates most of the common packaging problems.

1

1.1 SPACE AND WEIGHT REQUIREMENTS

Once the need for a system has been determined, the mechanical and electronic design engineers will stipulate the many specifications for the completed unit in order to ensure that it will be compatible with its predicted environment. For example, packaging for *vehicular* installation often introduces complex dimensional and weight restrictions. If space is at a premium and size and weight must be kept to a minimum, the system must be packaged in subassemblies or adapted to a modular design that will fit into the various-shaped spaces that are characteristic of vehicles such as ships and air or space craft (Fig. 1.1a). *Stationary* equipment, such as that used in homes and laboratories, is normally subject to fewer severe environmental constraints and is designed around chassis, cabinets, racks, panels, and consoles that are commercially available. Modern stationary equipment makes extensive use of single- and double-sided printed circuit boards (Fig. 1.1b).

FIGURE 1.1. (a) Packaging of subassemblies for space research. Courtesy of the Air Force Cambridge Research Laboratories, Hanscom Field, Bedford, Massachusetts. (b) Microprocessor trainer.

(a)   (b)

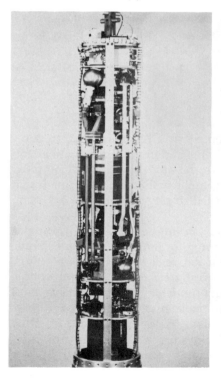
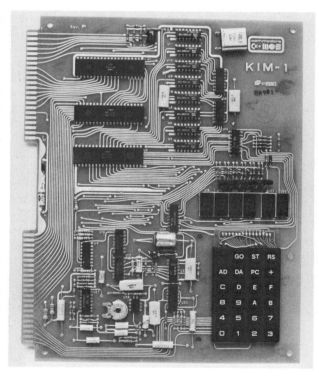

1.2 COMPONENT AND HARDWARE SELECTION

Once the component values of the system have been designed, the circuits are tested in experimental or *breadboard* form. It is generally the responsibility of the design engineer to determine the final selection of component values and ratings. Technicians must also be acquainted with the factors that govern these selections in order to enable them to choose the best available components for the specific applications. The final choice of components and hardware will be somewhat dependent on *cost* and *availability*, but the prime concern should be reliability. *Size*, *shape*, and the *finished appearance* of the unit also dictate part selection.

All electronic equipment should be as compact as possible, without reaching the point at which extreme component density prohibits assembly and service. Very often, the size of the unit is predetermined by its application, and the technician is compelled to work within established limits. The variation in size and mounting characteristics among commonly available components of the same electrical ratings makes it possible to satisfy even extreme dimensional restrictions. If, for example, the height of a unit employing a transformer is a limiting factor, the technician could choose a transformer that is longer and wider, thus taking advantage of a reduced height to minimize the overall vertical dimension of the package (Fig. 1.2).

The first criterion used in selecting components is, of course, that they meet the circuit's electrical demands with high reliability. Manufacturers' rated tolerance must be within design values. The closer the tolerance to the rated value of the component, the more it costs. Since cost is always a consideration, a technician should never choose com-

FIGURE 1.2. Variety of transformer styles provides for packaging flexibility.

ponents with closer tolerances than are necessary for the particular application. For example, if a resistor value is designed for 220 ohms with 10% tolerance, it certainly would be uneconomical to select a 1% resistor, since it may not appreciably improve circuit performance or reliability.

Voltage, current, power levels, frequency, and *continuous* or *intermittent operation* are the major circuit parameters that govern the selection of both the component value and type. In prototype construction the maximum electrical ratings (*stress levels*) of the components must be *derated* to ensure reliability. Derating provides a sizable safety factor so that components will operate well within their rated stress limits. In general, resistors are operated about 50% below their power stress levels. As an example, if a circuit resistor is required to dissipate ¼ watt, the resistor actually chosen would have a ½-watt rating. On the other hand, a limit must be placed on over-derating or compactness will be compromised. Using a 2-watt resistor where a ¼-watt resistor will operate within derating values or an electrolytic capacitor with a working voltage of 100 volts where 10 volts would be more than adequate would sacrifice size, weight, and expense without improving circuit performance.

Environmental factors and *human engineering* also come into play in the determination of which component or style of hardware to use. To illustrate this, hermetically sealed components would be necessary for protection against extremely humid conditions. From the standpoint of safety, external accessibility, and appearance, an extractor post fuse holder would be selected over a fuse block (Fig. 1.3).

It is important that the technician realize and appreciate that in many cases, component and hardware selection is one of compromise. This compromise comes about through the combination of availability, cost, size, and tolerance. Regardless of the choice of components, in the final analysis the system must function within the established operating specifications. Otherwise, it is of little value, even though it may be well packaged.

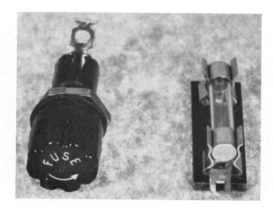

FIGURE 1.3. Extractor post fuse holder and fuse block.

Hardware items are sometimes a matter of personal choice. However, the size of such items must be realistic and fit the general scheme of the system. Miniature and subminiature hardware should not be selected when high component density is not required. In many instances, the selection of hardware such as switches, indicator lamps, connectors, jacks, and plugs is determined by those styles and types that are conventionally used in similar applications. If the unit being designed is to be used in conjunction with other units in an existing system, much of the hardware would be selected on the basis of mechanical and electrical compatibility and similar physical appearance. As an example, an existing system whose interconnections are terminated in *Cannon* connectors would require a special cable if a newly designed addition to the system used *Jones*-type connectors. Again, if a system makes use of *pc* (printed circuit) board connectors, additional boards used in the same system must be of a similar type (Fig. 1.4).

With the ever-widening use of integrated circuits, packaging has become more complex and a heavy reliance on printed circuitry has resulted. *Printed circuits* provide the only realistic solution to the problems associated with mounting transistors and the intricate wiring that is characteristic of integrated circuits. In the preliminary planning stage of construction, the technician is faced with the problem of using partial or total printed circuitry in a design. The solution to this problem often dictates the selection of the most suitable component and hardware types and styles. Although complex, printed circuits have reduced many packag-

FIGURE 1.4. Typical printed circuit board connectors.

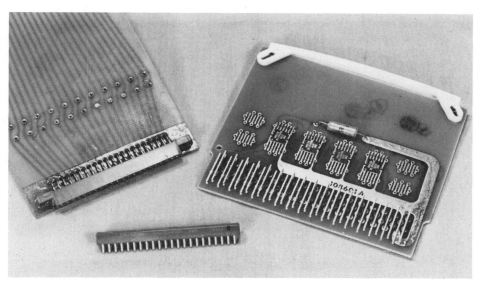

FIGURE 1.5. Screw-adjustable potentiometer for printed circuit applications.

ing problems involving size, weight, and production uniformity. Since such a large percentage of components are designed for use with printed circuit boards, the technician will encounter little problem in finding components available.

Some circuits require variable components, which are used only for initial adjustments for optimum circuit performance. Once these adjustments are made, ready access to the component is not necessary. These components are normally chassis-mounted or adapted to printed circuit design, as shown in Fig. 1.5.

1.3 CHASSIS MATERIALS AND CONFIGURATIONS

Sound mechanical design is an essential element of electronic equipment packaging. Components must be provided with a sturdy mounting base that will hold them in a fixed position so that they may be electrically connected. The metal frame on which the components are normally mounted is called the *chassis*. The overall dimensions of the chassis are determined by the final component layout of the circuit. The chassis configuration is selected through an evaluation of the following factors: *circuit function*, *size*, *area restrictions*, *rigidity*, *component accessibility*, and *available manufacturing* and *fabricating facilities*.

The two basic types of chassis are the *box* and the *flat plate* configurations. Other chassis shapes are variations of these. Figure 1.6 illustrates several common commercially available chassis that are widely used in industrial and consumer product applications.

FIGURE 1.6. Commercial chassis configurations. (a) U-shaped. (b) Standard box. (c) Small utility box. (d) Front panel and enclosure. (e) Portable equipment case. Courtesy of Bud Radio, Inc., Willoughby, Ohio.

Factors that determine the style of chassis to use in a particular application are: *weight of the heaviest components; use of controls and indicators* such as switches, potentiometers, meters, and lamps; *circuit and operator protection; portability;* and *appearance.* If heavy items are to be mounted, a simple U-shaped chassis, as shown in Fig. 1.6a, may not provide enough strength, especially if aluminum is the chassis material. A more sturdy design, such as the box chassis in Fig. 1.6b, should be considered. Here, tabs on the long sides of the chassis are folded behind the short sides and allow for the corners to be mechanically secured by sheet metal screws, rivets, or spot welds. Small self-contained circuits are packaged in chassis-enclosure combinations, as shown in Fig. 1.6c. Some circuit applications, particularly test equipment, require extensive use of meters, knobs, and indicator lamps and in addition must be portable. A reasonable selection for this application is the chassis and enclosure configurations shown in Fig. 1.6d or e, which provide a front panel for mounting controls and indicators, and, in some cases, handles for improved portability. Figure 1.7 illustrates a typical portable package.

Chassis are generally fabricated from sheet metal but may also be cast. The most common types of materials used for chassis fabrication are *aluminum, aluminum alloys, low-carbon cold or hot rolled steel,* and

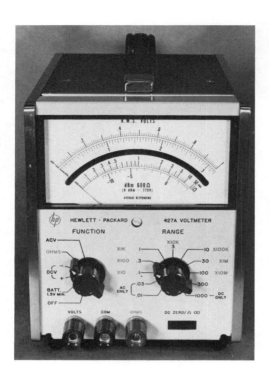

FIGURE 1.7. Handles improve portability of test equipment.

molded plastics. Special applications call for chassis constructed of *copper*, *brass*, or *magnesium*. The choice of materials for chassis depends on *strength*, *weight*, *environmental conditions*, *finish*, *electrical and thermal requirements*, *cost*, and *special circuit demands* such as shielding. Aluminum chassis are the most readily available commercially in a variety of height, width, and depth dimensions with *gauges* (metal thicknesses) ranging from Nos. 14 to 20 in a wide selection of alloys and *tempers* (metal hardness). (See Appendix I for a table of metal gauges and their decimal equivalent thicknesses for aluminum and steel.) Aluminum used in chassis construction is light, strong, easy to work, and relatively inexpensive. It also requires no protective finish, since it does not normally corrode. Steel chassis need to be finished with a protective coating to prevent corrosion and to improve the appearance. (Additional information on chassis finishing is found in Chapter 9.)

Although steel chassis are more difficult to work than aluminum and require additional care in finishing, they are more rugged than a comparable gauge in aluminum. A steel chassis would be preferred if a circuit, such as a power supply, requires the use of several heavy components such as power transformers and filter chokes (Fig. 1.8). If aluminum were used for this application, a heavy gauge would be necessary.

If circuit design characteristics include frequencies such as *VHF* [very high frequencies — 30 MHz (megahertz) to 300 MHz] and *UHF* (ultra high frequencies — 300 to 3000 MHz), copper or copper-plated

FIGURE 1.8. Steel chassis used to support heavy components.

chassis are used because their conductivity is higher than that of aluminum or steel. This characteristic is especially desirable if the chassis is used as part of the electrical circuit.

Once the chassis configuration and material have been determined, the next step is to establish the size of the finished chassis. Factors that determine the size are the *number* and the *dimensions* of the components and the *component density* and *positioning*.

1.4 COMPONENT AND HARDWARE PLACEMENT

The packaging of an electronic system involves the positioning of parts and components to determine their optimum density and location, finished chassis size, and a balanced layout. Both electrical and mechanical factors are taken into consideration when packaging. Logical sequence dictates that mechanical criteria are applied first.

Large, heavy components such as transformers and filter chokes should be placed near the corners or edges of the chassis, where the greatest support exists. If several heavy components are being used, they should be evenly distributed over the entire chassis. Power transformers are normally positioned to the rear of the chassis to avoid the ac line cord running near or through critical circuit areas. Fuses, cable connectors, and infrequently used binding posts and jacks for external antennas, or remote speaker connections also should be positioned to the rear (Fig. 1.9a). Control knobs, switches, frequently used binding posts

(a)

(b)

FIGURE 1.9. External hardware and component placement. (a) Typical rear-mounted hardware. (b) Typical front-mounted components and hardware.

or jacks, meters, dials, and indicator lamps are generally located to the front of the chassis (Fig. 1.9b). These items are not mounted directly to the chassis but to a front panel or cabinet. Electrical connections between the panel-mounted and chassis-mounted components are then made directly or to intermediate terminal points. Panel-mounted hardware and components should be positioned to provide balance in their appearance and to ensure ease of operation, unobstructed visibility, and a neat and orderly wired package.

Whenever possible, speakers should be front-positioned so that the sound is directed toward the operator. At times, however, when tone quality is important, a larger speaker is required and it may be necessary, because of physical size, to position it elsewhere. Sacrificing position for tone quality is not an uncommon practice.

Electrical criteria often impose restrictions on component placement. Common circuit problems that must be considered in packaging to ensure proper circuit performance are *thermal radiation* from components such as resistors and power transistors; *ac hum* introduced into an audio circuit; and *distributed lead capacitance and inductance*, which can drastically affect audio circuits and RF (radio-frequency) circuits. These circuit problems are discussed in the following paragraphs.

Low-power transistors and integrated circuits are heat-sensitive devices and must be positioned a reasonable distance from any component that generates considerable heat. One component that gives off a large amount of heat is the *power transistor*, because it is designed to operate at near-maximum ratings. In most applications, normal air convection does not provide for adequate heat dissipation and *heat sinks* are necessary. Heat sinks are metal configurations that, when brought into physical contact with the transistor case, absorb and dissipate a large portion of the generated heat. The chassis itself can be used as a heat sink, but power transistors are manufactured with the collector electrically connected to the case. If this creates an electrical problem when mounted directly to the chassis, a *mica washer* is used. This mica washer electrically insulates the case from the chassis without causing severe adverse effects to the thermal conductivity between the case and the chassis. This mounting arrangement is shown in Fig. 1.10. Silicon grease is also used on both sides of the mica washer to increase thermal conductivity. If chassis sinking alone would not provide adequate heat dissipation, fin-type sinks like those shown in Fig. 1.11a are employed. If low-power devices require heat sinking, the type shown in Fig. 1.11b, which is not chassis-mounted, is often used. The proper size and configuration of sink required is predetermined during the circuit design stage. The problem of providing adequate space for the sink is left to the technician.

What is commonly referred to as *ac hum*, or 60-hertz hum, is a condition common to most electronic circuits employing power transformers. A 60-hertz signal, usually originating at the power transformer, can be induced into the input stages of the circuit. This induced signal is

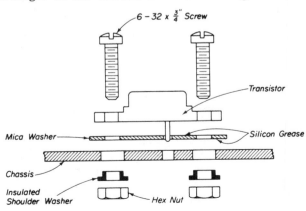

FIGURE 1.10. Correct mounting arrangement of a power transistor to a metal chassis.

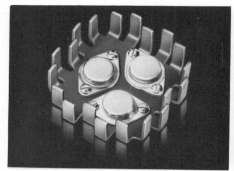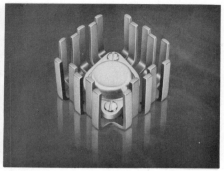

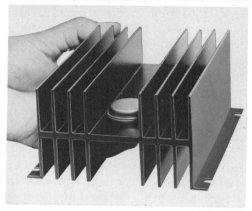

(a)

(b)

FIGURE 1.11. Typical heat-sink configurations. (a) Chassis-mountable heat sinks for medium and large power devices. (b) Heat-sink style for low-power devices. Courtesy of Thermalloy, International Electronic Research Corporation, and Wakefield Engineering Company, Inc.

then amplified along with the desired signal, resulting in a steady hum. Shielding the signal lead or, by proper placement, making the input lead very short, will minimize this problem. It is advisable, however, to make it a common practice to locate power transformers, ac leads, or any source of electromagnetic radiation away from the signal input leads.

If circuitry employing RF frequencies is to be packaged, special attention must be given to the lead lengths required by the component and hardware positions. Because of distributed lead capacitance and inductance, which become a problem at these higher frequencies, wires used for electrical connections must be kept as short as possible. This is especially true at the input stages, as in the case of a receiver's local oscillator section, wherein the leads are extremely susceptible to noise pickup. Figure 1.12 shows an example of a compact RF package with the components positioned so as to keep the leads as short as possible.

If antenna positioning is involved in the circuit layout, it is important to locate the antenna input as close to the input circuit as possible to avoid stray pickup. In addition, if the antenna is designed to be totally

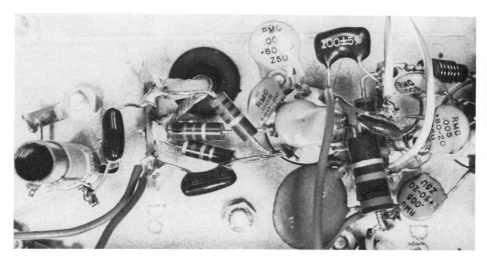

FIGURE 1.12. Typical RF package design.

contained within the unit, it should be mounted on a nonmetallic surface and kept as far away from metal elements as possible. This avoids the problem of picking up undesired frequencies.

Component and hardware placement is by far the most involved and time-consuming phase of planning a package. Performance and reliability of the system will depend heavily upon this phase of design.

1.5 MAINTENANCE

Maintainability is a measure of the ease, accuracy, and time required to restore a circuit or system to its normal mode of operation once a failure or abnormal function is detected. Maintenance covers two general procedures. First is the repair of failures and defects. Second is *preventive maintenance*, which deals with normal or routine checks and servicing of equipment. The object of preventive maintenance is to keep equipment operating trouble-free and at optimum performance. Certain precautions should be taken in the design of any package to minimize unnecessary maintenance. Whenever leads must pass through a hole in a metal chassis or bend around the edge of a bracket, either a grommet or clamp should be used to prevent abrasion of the lead insulation, which would cause a short circuit (Fig. 1.13).

High ambient temperatures within enclosures can eventually cause components to overheat and fail. To ensure cool, trouble-free operation, approximately 6 cubic inches of space per watt of power to be dissipated is necessary if normal air convection is to be relied upon for cooling a unit. Vents or holes must be provided, preferably at the top of the enclosure surrounding the circuit. If units are stacked in a rack, those units that dissipate the most heat should be located in the uppermost

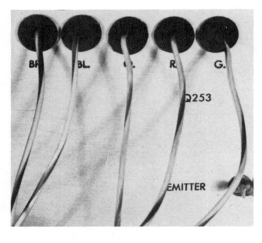

FIGURE 1.13. Grommets prevent abrasion of lead insulation.

position to prevent the heating of components in other units. Whenever possible, liquid cooling and blowers should be avoided. Liquid cooling systems are subject to failure, and blowers could cause air circulation that will cool certain components but overheat others. Air filters are often required when blowers are used, thus adding another maintenance dimension to the unit. Liquid cooling and blowers also add bulk and weight to the unit.

Enclosures, in addition to allowing circuits to function without overheating, must also be tight enough to isolate the components and wiring from dust or other contaminants that could accumulate in the circuit and cause failures. If large holes have to be formed in an enclosure for speakers, the entry of dust can be minimized by using a *speaker cloth*. For additional protection of the speaker cone, a grid or wire screen can be used with the cloth (Fig. 1.14).

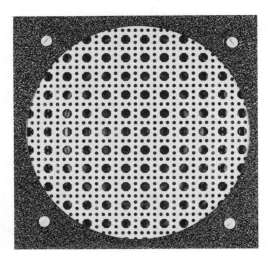

FIGURE 1.14. Metal grid to prevent speaker cone damage.

FIGURE 1.15. Power cord attachment design necessitates its removal for safety before servicing.

Well-designed enclosures minimize maintenance problems through the various types of circuit protection they afford. They can, however, present problems in terms of accessibility. Ease in maintenance is directly related to component accessibility. By locating access plates, doors, and openings near those components that are most subject to maintenance needs, repair and service can be achieved easily, quickly, and accurately. Safety precautions should be built into the design when providing access to electronic circuits. Safety devices may have to be incorporated with the removable elements to avoid the possibility of electrical shock when working on the circuit through any access opening (Fig. 1.15).

Providing for maintenance in packaging design is a compromise involving three factors: *sufficient environmental protection*, *accessibility*, and *safety*.

1.6 HUMAN ENGINEERING

In electronic packaging design, human limitations associated with the use of electronic equipment must be considered. These human limitations involve such things as vision, arm reach, and hand manipulation, which must be considered in order to select the most suitable controls and visual displays. The ease and accuracy with which the unit is operated depends on the proper selection and placement of these components.

Meters should not be so small as to make the reading of major and minor scales difficult. The type of scale used should fit the function of the unit. For example, if only circuit conditions are to be monitored rather than the finite values of electrical parameters, the use of smaller meters having fewer scale divisions is desirable. If, on the other hand, accurate measurements are required, a larger meter with numerous minor scale divisions is preferable.

Glare must be avoided when continuous dial and scale readings are necessary. If reflected light in the unit's environment is excessive, glare shields such as the one shown in Fig. 1.6d, should be employed. In addition, the panel should be finished in a color possessing low reflective characteristics (see Chapter 9).

Control knobs and switches should be mounted within easy and comfortable reach of the operator and should not be crowded. If they are, accurate individual adjustments will be difficult to make and other settings may be disturbed. Keeping with accepted conventions in mounting panel elements is very important. For example, toggle switches that are to be operated in a vertical manner should be "on" when the lever is in the upward position. Potentiometer adjustments in a clockwise direction should cause an increase in the magnitude of the electric quantity. In rack- or console-mounted equipment, all visual displays and indicators, such as meters and indicating lamps, should be mounted as close as possible to the eye level of the operator.

Stability is essential, especially if the unit is portable and relatively light. Some controls, especially rotary switches, require substantial torque to actuate and may cause the unit to become unsteady while in use. This problem can be reduced or overcome by either mechanically securing the unit to a mounting surface or attaching rubber pads to the bottom corners. The pads not only provide greater friction but also offer protection to the working surface, in addition to serving as electrical insulators for the unit.

The consideration of human limitations and capabilities, then, must be of concern to the technician designing a package. Many manufacturers employ human engineering consultants to render advice about the placement of control elements. A unit may function properly, but it will not be acceptable if it is difficult to control, read, or adjust.

2

High-Fidelity Stereo Amplifier System

2.0 INTRODUCTION

A stereo amplifier, employing many of the latest semiconductor devices, is used as a means of illustrating the practical applications of packaging techniques discussed in this book. In this way, familiarity is gained with the application of the latest design considerations, conventions, and practical procedures for sound electronic packaging.

Since a basic understanding of the electronic circuitry involved is essential in order to fully interpret and apply the necessary procedures for packaging, a brief discussion of the circuit's electronic operation and special characteristics is presented here.

The block diagram of the amplifier is shown in Fig. 2.1. This type of diagram identifies the function and interrelationship of each circuit. Figure 2.2 shows a detailed schematic diagram of one channel of the amplifier and the power supply. Only the left channel is shown because the right channel is an exact duplicate.

Incoming audio signals from an AM (amplitude modulation) tuner, FM (frequency modulation) tuner, or magnetic phonograph cartridge must be increased in power from their relatively low output levels (milli-volt range) to a level sufficient to drive a low-impedance speaker, such as

17

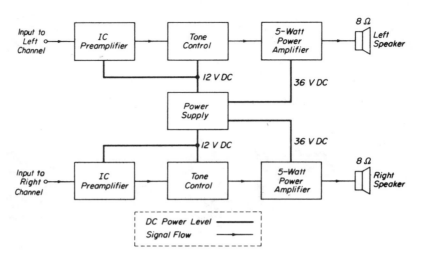

FIGURE 2.1. Block diagram of the stereo amplifier. Subassemblies and their interrelationships are shown by signal flow paths and DC power levels.

the 8-ohm speaker shown in Fig. 2.1. Not only must the incoming signal be increased in level, but the output must also be a faithful reproduction of the input. For this reason, relatively sophisticated circuitry is necessary. Although Fig. 2.1 appears to be complex, this entire diagram represents a complete system composed of seven individual but related circuits. Therefore, to most readily grasp a basic understanding of the operation of the system, the function of each circuit must be investigated. This will serve as an essential guide to the technician for isolating all the important packaging problems. Notice that this system consists of *two channels*. Each channel is terminated with its own speaker and all circuits in each channel are energized from one common power supply.

The first circuit that the signal encounters in either channel is the *preamplifier*. The purpose of this circuit is to amplify a signal level of only a few millivolts to a voltage level on the order of 500 millivolts. This increased level or amplified signal is now large enough to drive the power amplifier once it passes through the tone control circuit. The preamplifier consists of a single active device or integrated circuit (IC) with its support components (resistors and capacitors) through which the first increase of input signal magnitude is achieved.

The increased signal of each preamplifier output is next applied to the inputs of the *tone control circuit*. The function of each tone control or *frequency shaping* network is to allow the operator to select and compensate for individual listening preferences in program material. The human ear tends to detect higher frequencies, at the same output level, more readily than lower frequencies. Consequently, the operator may find it desirable to increase or decrease the low-frequency output level without

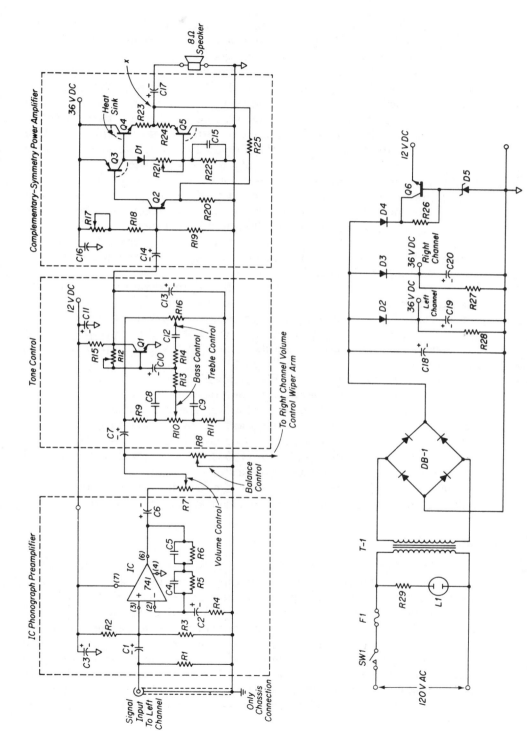

FIGURE 2.2. Schematic diagram of one channel of the stereo amplifier and power supply.

19

altering the high-frequency level. This technique, called *base boost* or *low-frequency cut*, respectively, is possible with the tone control. *Treble* control capability is also provided with the tone control circuit by boosting or cutting high frequencies. Each channel's tone control circuit consists of a high-frequency, high-gain *silicon transistor* with the associated frequency-selective circuitry performing the tone shaping from the transistor's input to output. Control is achieved through two potentiometers, with the base and treble control independently adjusted. These tone control potentiometers can also be adjusted for *flat control* (neither cut nor boost) wherein the gain is unity for all frequencies. In this mode of operation, the output level of the tone control is equal to the output level of the preamplifier and is sufficient to drive the power amplifiers to their full output. The *complementary-symmetry power amplifiers* represent the final stage of signal amplification. The function of the power amplifiers is to provide the necessary voltage and current to drive the speakers while maintaining the proper impedance match for maximum power transfer to these speakers.

Although both channels of the amplifier are operated independently of each other to reproduce stereophonic program material, they receive power from a common supply. This *power supply* must provide *two* levels of dc output. The lower level, 12 volts, is used to energize the preamplifiers and tone control circuits. The higher voltage level, 36 volts, is used to operate the power amplifiers.

The schematic diagram of Fig. 2.2 will now be discussed in detail, not only to increase the technician's familiarization with each individual circuit, but also as an aid to understanding specific packaging problems related to these circuits. The operation of the power supply and one channel only will be described since, as mentioned previously, the second channel is an exact duplicate. However, *both* channels will be constructed throughout the text to form a complete stereo system. Since this system is designed to faithfully reproduce audio frequencies (20 hertz to 20 kilohertz), higher-frequency packaging problems are not considered.

2.1. PREAMPLIFIER

The preamplifier circuit shown in Fig. 2.2 utilizes a 741 high-gain, wideband *integrated circuit operational amplifier* (op-amp) with a relatively high input impedance. The input signal is restricted to a maximum value of 10 millivolts (rms). If this signal level is exceeded, the power amplifier will be overdriven, thus distorting the signal and reducing the bandwidth of frequencies to be faithfully reproduced and amplified. The input signal is applied to the *noninverting input terminal* of the op-amp through a *phono jack*. The use of this type of jack is essential so that *shielded wire* can be used between the input source and the preamplifier. All precautions should be taken to eliminate undesirable pickup or noise from the input of the preamplifier because this noise will be amplified along with the signal. Shielded cable protects against pickup by shorting undesired signals to *ground*. Resistors R_2 and R_3 provide the proper input bias current for the op-amp inputs. The input impedance for the preamplifier circuit is R_1, R_2, and R_3 in parallel, or approximately 47 kilohms, which is suitable for magnetic cartridge impedance matching. Since magnetic cartridges are nonlinear in their response, one function of the preamplifier is to compensate for this characteristic. This is accomplished by the equalization circuit R_5, R_6, C_4, and C_5. Voltage gain is controlled by the ac feedback network, consisting of the equalizing circuit together with R_4 and C_2. Resistors R_5 and R_6 also supply dc feedback to the inverting terminals of the op-amp to stabilize the operating point.

Because aluminum is an excellent electrical conductor, *common*, or ground, points are often connected to the chassis. This technique is called *chassis ground*. A problem may arise in the use of chassis ground that should be pointed out. If the terminal points are connected to many widely separated locations on the chassis, the small amount of resistance in the aluminum may be just enough to set up a *potential difference* among these points. This condition is known as a *ground loop*. The path of this loop may include the input of the amplifier and would therefore introduce an undesirable signal between output and input. To prevent the possibility of a ground loop in this amplifier system, the circuit should be connected to the chassis at only *one* point. Usually, a point as close as possible to the amplifier input jacks is chosen for this purpose.

The output of the preamplifier, through coupling capacitor C_6, is coupled to volume control R_7, which adjusts the signal to the tone control input. R_8 is the balance control. Each side of this potentiometer is connected to the wiper terminals of the volume controls. The wiper of the balance control is connected to the common side of the circuit. (Of course, balance control is eliminated if only one channel is to be used.) R_8, then, controls the relative signal level from each channel being delivered to the tone control circuit.

2.2 TONE CONTROL

Coupling to the tone control is provided through C_7. Transistor Q_1 is biased at its low noise point by R_{15} and the adjustable local feedback resistor R_{12}. Base boost and base cut control is accomplished by R_9, R_{10}, R_{11}, C_8, and C_9. With the wiper of R_{10} in the uppermost position (boost), represented by a *clockwise* rotation of the control shaft of the potentiometer, the gain of the amplifier at low frequencies is the ratio of R_9 to R_{10}. As the frequency increases, capacitor C_9 bypasses R_{10}, and at approximately 600 hertz the gain of the amplifier approaches unity. The treble control consists of C_{12} and R_{16}. High-frequency gain is approximately equal to the ratio of the bottom part of R_{16} to the top part of R_{16} as determined by the wiper section. The tone control circuit is decoupled through C_{11}.

Wiring of this circuit is critical in terms of reducing pickup of undesired signals. Consequently, all lead lengths should be as short as possible.

2.3 POWER AMPLIFIER

The output of the tone control is coupled to the complementary-symmetry power amplifier through C_{14}. The input impedance to this amplifier is approximately 20 kilohms. This circuit requires 0.5 volt rms to deliver a true power output of 5 watts. Resistors R_{17}, R_{18}, and R_{19} provide bias to the transistor Q_2. R_{17} should be adjusted to establish point x (the common point of R_{23}, R_{24}, R_{25}, and C_{17}) at a potential of approximately 18 volts ($\frac{1}{2}$ V_{cc}) with no signal applied. Transistor Q_3 is the driver for the complementary output transistors Q_4 and Q_5. The driver must be able to dissipate approximately 1 watt and will require heat sinking for safe operation. Resistor R_{25} provides both dc and ac negative feedback, to allow for operating point stability and gain stability. Operating-point temperature compensation can be achieved by mounting diode D_1 on the common heat sink provided for Q_4 and Q_5 so that it encounters the same thermal conditions as these transistors. R_{21} can be adjusted to minimize crossover distortion. Capacitor C_{17} isolates the speaker from the dc current of the power supply.

Since Q_4 and Q_5 must dissipate 25% of the rated output power under maximum load conditions, they will require heat sinking. An aluminum plate $\frac{3}{16}$ inch thick by 2 inches by 3 inches should be used as the minimum-size heat sink.

Because the circuit for this power amplifier makes no provision for short-circuit protection at the output, care must be exercised to prevent the output terminals from being shorted when the speakers are connected.

2.4 POWER SUPPLY

The dc power to the stereo amplifier system is supplied by transformer T_1, conventional bridge rectifier DB_1, and filter capacitor C_{18}. Diodes D_2 and D_3 are used for channel separation. Capacitors C_{19} and C_{20} provide additional filtering for the 36 volts for the power amplifiers, which is obtained at the cathodes of the diodes D_2 and D_3. The 12 volts necessary to power the preamplifier and tone control circuits is obtained through the use of an emitter follower regulator that establishes its reference 12 volts from zener diode D_5. No diodes are necessary for separation when a regulator is used. Therefore, both channels may be driven from the emitter of Q_6. D_4 is used to separate the regulator from both channels.

2.5 AMPLIFIER PACKAGE DESIGN

Normally, a semiconductor amplifier has little space or weight restrictions. Therefore, the first phase in the design of this amplifier package will be the selection of components and hardware. Resistors and capacitors are selected on the basis of the circuit's voltage, current, and power requirements. The power levels in circuits of this type are generally very low. For this reason, as well as for their small physical size, $\frac{1}{2}$-watt resistors will be used throughout the design *with the exception* of R_{22}, which must be capable of dissipating at least 2 watts, and R_{23} and R_{24}, which must dissipate at least 1 watt. All these resistors will have a tolerance of 10%. *Electrolytics* will be used for coupling and decoupling capacitors as well as in the power supply filter circuit, owing to the larger capacitive values required for these applications. Electrolytic capacitors are selected over other available types because they provide the large values of capacitance required with the smallest physical size. The balance of the capacitors in the system will be the *disc type*. These have the advantage of possessing an extremely high leakage resistance. In addition, their small physical size makes them easily adaptable to printed circuit design. The *working voltage* for the disc capacitors, based on circuit requirements, will be 100 volts and for the electrolytics, 50 volts.

The transformer (T_1) selected is a 25.2 volts at 2.8 amperes power transformer, even though the circuit is not expected to draw more than 1.6 amperes.

Diodes, transistors, and integrated circuits are selected in the same manner, with the voltage, current, and power requirements of the circuit being satisfied. Cost and availability serve as secondary factors in determining the choice of these components.

A parts list will initially be prepared. This list must be complete, including not only the components but also the associated hardware and materials to complete the package. Figure 2.3 shows the complete list of components and parts for the amplifier. As much information as possible should appear on the parts list, such as types, ratings, and tolerances.

The type of aluminum that will be used for the chassis, front panel, and enclosure elements will be *16 gauge, No. H–1100*. This choice is made on the basis of the desirable characteristics of this material — easily worked, sturdy, lightweight, conductive, and readily available. The heaviest component (transformer T_1) weighs less than 32 ounces and is easily supported by this material. In addition, aluminum lends itself readily to the finishing techniques discussed in Chapter 9.

The box-type chassis, shown in Fig. 1.6b, will be selected for this design. Even though the heaviest component in the system weighs less than 2 pounds, the box chassis possesses advantages other than strength. It is most suitable for mounting printed circuit boards (which must be used in this type of circuitry), wherein large areas of metal can be removed from the top surface without seriously weakening the overall chassis.

Resistors*

R1 — 47 kΩ
R2, R3 — 1.5 MΩ
R4 — 820 Ω
R5 — 56 kΩ
R6 — 820 kΩ
R9, R11, R13, R14, R19 — 22 kΩ
R15 — 10 kΩ
R18 — 270 kΩ
R20 — 150 Ω
R22 — 390 Ω, 2 W
R23, R24 — 0.47 Ω, 1 W
R25, — 3.3 kΩ
R26 — 1 kΩ
R27, R28 — 27 kΩ
R29 — 180 kΩ
*All resistors are $\frac{1}{2}$ W, 10 %
unless otherwise specified

Potentiometers

R7 — 10 kΩ, 2 W audio taper
R8 — 5 MΩ, 2 W linear taper
R10, R16 — 250 kΩ 2 W linear taper
R12 — 1 MΩ, $\frac{1}{8}$ W trim pot
R17 — 500 kΩ, $\frac{1}{8}$ W trim pot
R21 — 50 Ω, $\frac{1}{8}$ W trim pot

Capacitors*

C1 — 2 μF
C2 — 10 μF
C3, C11, C16 — 0.1 μF disc
C4 — 0.0015 μF disc
C5 — 0.005 μF disc
C6, C7, C10, C13, C14 — 5 μF
C8, C9 — 0.01 μF disc
C12 — 500 pF
C15 — 0.02 μF disc
C17, C19, C20 — 1000 μF
C18 — 1500 μF
* All capacitors are electrolytics
with a 50 WVDC rating unless
otherwise specified
* All disc capacitors are rated
at 100 WVDC

Devices

IC — 741 linear op-amp
   (case style TO-5)
Q1, Q2 — 40408
Q3 — 40406
Q4 — MJE 521
Q5 — MJE 371
Q6 — 2N3053
D1 — 1N3754
D2, D3, D4 — 1N4002

D5 — 1N963 Zener diode 12 V
DB-1 — Motorola MDA 960-2
   diode bridge (100 P.I.V.
   at 2.5 ADC)

Components

T1 — Stancor P8388 transformer
   (25.2 V at 2.8A)
L1 — NE — 51 neon lamp
F1 — 1$\frac{6}{10}$ A, type 3AG, slow-blow
SW1 — single-pole, 2-position wafer switch

Hardware and Materials

Chassis material — 16 gauge, H-1100
   aluminum
Cartridge-type fuse holder
Strain relief
40 inch molded, two prong, AC line cord
4 — $\frac{3}{8}$ inch grommets
1 — $\frac{1}{4}$ inch grommet
1 — 2 input phono jack
2 — 2 screw speaker connectors
1 — pilot lamp holder
5 — control knobs
3 — Wakefield 200 series, heat sinks
44 — 0.045 x $\frac{1}{2}$ inch square wire wrap pins
32 — 2-56 x $\frac{1}{4}$ inch machine screws and nuts
14 — 4-40 x $\frac{1}{4}$ inch machine screws and nuts
4 — 8-32 x 1$\frac{3}{8}$ inch machine screws and nuts
4 — #6 sheet metal screws

FIGURE 2.3. Parts list for stereo amplifier.

A preliminary sketch of the amplifier component layout is next undertaken. This sketch is made simultaneously with the technique of positioning the components on a similar mounting surface to that selected for the final package. A preliminary sketch of the amplifier chassis is shown in Fig. 2.4. This type of pictorial is necessary to fully appreciate whether or not the chassis style and shape are completely compatible with both the functional and esthetic aspects of the design.

A starting point in the packaging layout is the placement of the power transformer, which should be mounted as far as possible from the amplifier's inputs to avoid the possibility of introducing 60-hertz hum. For this design, the transformer (T_1) will be mounted on the top right-hand rear corner of the chassis. This will allow the power cord to enter directly into the rear of the chassis. Some form of restraint or clamp is necessary to secure the line cord. This prevents undue strain on terminals or the wires from separating from soldered connections if the cord is roughly handled. The pilot lamp will be mounted on the front panel for ease of observation. This location, however, will require a twisted pair of leads between the lamp base and the primary winding of the transformer. This twisting of the leads cancels the generated magnetic field about the leads, thereby preventing hum. The power supply circuit will be positioned close to T_1 to further minimize lead length carrying 60-hertz frequency. Power for the preamplifier, tone control, and power amplifier circuits should be provided on the side of the power supply printed circuit board closest to these circuits.

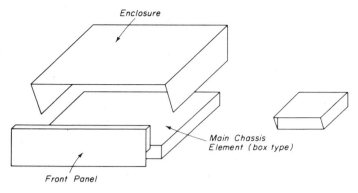

FIGURE 2.4. Preliminary sketch of intended style for the chassis, front panel, and enclosure elements.

Enclosure

Main Chassis Element (box type)

Front Panel

Since the circuit is fused on the primary side of the transformer, the fuse holder will be a cartridge or extractor post type, which is mounted through the rear of the chassis close to the power cord. This type of fuse holder is extremely safe with respect to fuse changing, since its design eliminates the possibility of electrical shock.

All control knobs will be mounted on the front panel and generously spaced for balanced appearance and ease of operation. The position of the potentiometers will influence the placement of the tone control circuit in terms of short lead length. All input and speaker connections will be to the rear of the chassis to keep their leads from interfering with the operation of the controls and to preserve the appearance of the package. Phono jacks will be used for input connections to the unit to take advantage of the shielding of the input leads associated with this type of connector. Insulated screw-type jacks will be used for the speaker connections, although phone jacks also could be used.

The various types of mounting hardware, such as machine screws, nuts, washers, sheet metal screws, and special fasteners, are discussed in Chapter 18. The terminals on the printed circuit boards used for interconnecting wiring will be the *wire-wrap* type. This terminal is readily adaptable to printed circuit work because of its small size. In addition, it is more economical when plug-in modular design is not required, as is the case with this stereo system. (Printed circuit hardware is discussed in Chapter 14.)

Because of the number of components involved and overall circuit complexities, each of the seven circuits in the system will be packaged on an individual pc board. This technique has several advantages over other packaging arrangements, the most important being that each individual circuit in the system can be designed, constructed, and individually tested. Upon completion of these individual circuits (i.e., the power supply, preamplifiers, tone controls, and power amplifiers) simple interconnections among these pc boards will result in a functional amplifier system.

Along with all the inputs, the preamplifiers (the first stage of signal input) will also be positioned to the rear of the chassis. The numerous connections and terminals on the tone control dictate that these pc boards be positioned forward so that they can be connected conveniently to the panel-mounted potentiometers. The output terminals of the power amplifiers will be positioned to the rear of the chassis to minimize lead length to the speaker terminals. Finally, the power supply pc board will be placed as near to the transformer as possible.

Figure 2.5 shows a layout of the chassis with the intended individual board locations in addition to some hardware and component placement. This layout helps in estimating if wiring will be interconnected simply and conveniently.

At this stage, it is not important to consider the exact component positioning on the pc boards. All that is necessary is to determine the *approximate* component layout to gauge the correct overall size of these boards and their *exact* location with respect to the associated chassis-mounted hardware and components whose position is dictated by external connections. For example, input and output coupling capacitors must be mounted to an edge of a board closest to their chassis-mounted terminals. The interconnections among boards will also determine the relative positions of these components (Fig. 2.6).

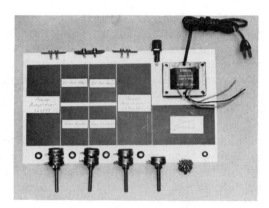

FIGURE 2.5. Chassis layout with intended PC board locations and major hardware and components positioned.

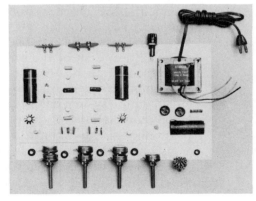

FIGURE 2.6. Major PC board components are positioned to observe if any interconnecting problems will be encountered.

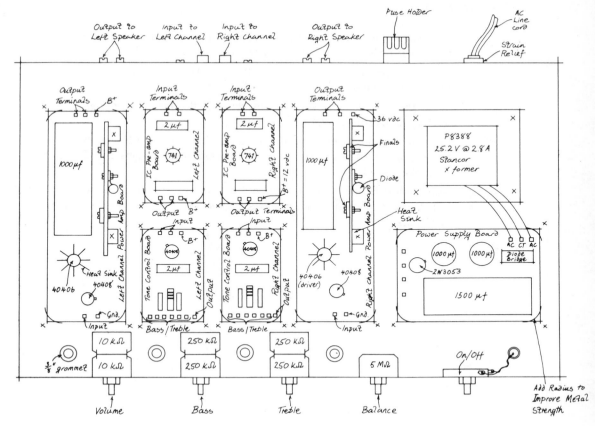

FIGURE 2.7. Sketch of major component positioning for the stereo system.

To maintain a balanced appearance, the control shafts of the base, treble, volume, balance, and on–off controls will be spaced evenly across the front panel.

A sketch of the final placement of the components is completed when the desired location of all parts has been determined and a reasonable balance has been achieved. Figure 2.7 shows the results of major component positioning and mounting-space requirements. From this layout, the optimum-size chassis, panel, and enclosure elements can be designed and fabricated.

To convert the final layout into useful detailed drawings for the actual construction of the unit, accurate measurements of components and their locations must be made. Procedures for determining accurate dimensional information and the application of this information in the preparation of detailed drawings are discussed in Chapter 3.

PROBLEM SET

2.1 Determine the approximate parts position and chassis element configurations for packaging just the power supply section shown in Fig. 2.2

2.2 Give the complete package design for just the two-channel preamplifier and tone control combination taken from Fig. 2.2.

2.3 Show the complete package design for just the power amplifier and power supply sections of Fig. 2.2.

2.4 Estimate the maximum amount of 16-gauge aluminum or 18-gauge steel required to fabricate all the chassis elements selected for Problems 2.1, 2.2, and 2.3 from blank stock.

2.5 Using industrial electronic component catalogs determine the cost per watt of just the power amplifier and power supply sections of the stereo system shown in Fig. 2.2.

Preparing Detailed Drawings

3.0 INTRODUCTION

Once the preliminary plans for a package are completed and sketches of the chassis configuration are prepared, *detailed drawings* must be developed to aid in construction. These drawings entail the final positioning of all components and hardware. Sizes of these components and hardware items are first measured to obtain overall dimensions necessary to select and complete optimum orientation and position. Many of these measurements will also determine the size and location of mounting holes during the fabrication of the chassis. These detailed drawings must include all pertinent information with respect to component assembly and positioning and chassis fabrication and finish. It is important to mention here that someone other than the technician who designed the original package may be assigned the task of construction. For this reason, the drawings must be completely informative, accurate, and perfectly clear.

This chapter contains specific information on techniques of using measuring instruments such as scales, calipers, micrometers, and templates. Also included is relevant information on engineering sketches, basic drafting equipment, drafting standards, dimensioning, and procedural information for developing chassis layout drawings along with assembly and auxiliary views for detail work.

3.1 MEASUREMENTS WITH A SCALE

In the planning stage, approximate positioning of components and hard-ware was estimated by physically placing the components in the desired locations. This procedure is shown in the example package of Fig. 3.1. From this preliminary method, component density can be observed. It is now necessary to take all dimensions from the final component locations and transfer this information to a sketch. These resultant sketches will be used to produce the working drawings.

To obtain most of the required measurements, a simple *steel scale* is commonly employed. Steel scales are available in a variety of styles, including *steel tape rule*, *steel rule*, and *hook rule.* The 6-inch and 12-inch steel rules and the 6-inch hook rule are the types commonly used in electronic construction (Fig. 3.2). The major scale division on these rules is the inch, with half, quarter, eighth, sixteenth, thirty-second, and sixty-fourth subdivisions. Steel scales are also available with decimal graduations having major and minor divisions of 0.1 inch and 0.020 inch, respectively. This rule is desirable when the dimensioning format of the sheet metal drawings is in decimal form. The smaller graduation scales are used where close tolerance measurements are required.

FIGURE 3.1. Overall height, width, and depth dimensions for the chassis are determined directly from the component layout.

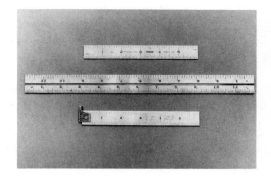

FIGURE 3.2. Steel rules.

Overall length and depth measurements for the chassis determined by the final component layout can be made by alternately positioning and reading the scale along the lines bounding the area previously established in the planning stage. The height of the unit is determined by reading the vertical measurement of the tallest component together with the desired enclosure clearance. This measurement must be added to the height of the base or main chassis to determine the overall height of the finished unit. The measurement techniques just described are shown in Fig. 3.1.

Component measurement for mounting purposes can also be made with a scale. As an example, a method for determining on-center dimensions for mounting holes is shown in Fig. 3.3. Mounting information, however, is usually provided by the manufacturer in the form of a paper template (see Appendix II).

For maximum accuracy when using the 6-inch or 12-inch steel rule, the scale should be positioned perpendicular to the surface being measured. In this way the graduation marks on the scale will be in direct contact with the work and will eliminate the possibility of error due to the thickness of the rule if it is placed flat on the work. In addition, neither end of a rule should be used as the reference or starting edge. For accurate measurements, the work should be aligned with the first major inch graduation mark. This assures that scale-edge deformities will not cause errors.

The hook rule, shown in Fig. 3.4, provides an accurate reference edge along the inside face of the hook. Thus, the zero graduation mark is perfectly aligned with the edge of the work, assuring more exact measurements. It can also be used for rapid settings of inside calipers, as discussed in Sec. 3.2.

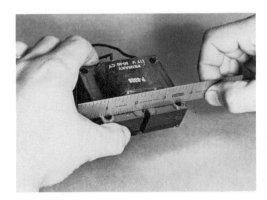

FIGURE 3.3. Method for determining on-center dimensions.

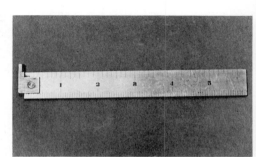

FIGURE 3.4. Hook rule.

3.2 MEASUREMENTS WITH CALIPERS

Although rules are adequate for most measurement applications, *calipers* are necessary if direct measurements are impossible. Calipers are classified as *outside* or *inside*, depending on the shape and position of the legs as well as their specific use. Both types are shown in Fig. 3.5. The main parts of the caliper are the *legs*, *spring crown*, and *adjusting knob*.

The outside caliper is readily adaptable to measuring diameters of round stock, such as the shaft of a potentiometer or switch. This caliper is held by the spring crown cupped in the palm of one hand with the thumb and forefinger free to apply a rotational force to the adjusting knob (Fig. 3.6a). Rotating the adjustment knob will close the legs of the caliper on the surface to be measured. Once the inside tips of the legs contact the surface to be measured, a back-and-forth motion of the caliper along the surface will indicate the correct measurement. The calipers are set properly if there is a *slight friction* as the inside tips of the legs move across the widest portion of the work. Practice is necessary to acquire the correct "feel" for this optimum setting. With the leg spacing set, the measurement can now be determined by placing the caliper along the edge of a scale and reading the distance between the inside of the legs. This procedure is illustrated in Fig. 3.6b.

FIGURE 3.5. Inside and outside calipers. Courtesy of The L.S. Starrett Company, Athol, Massachusetts.

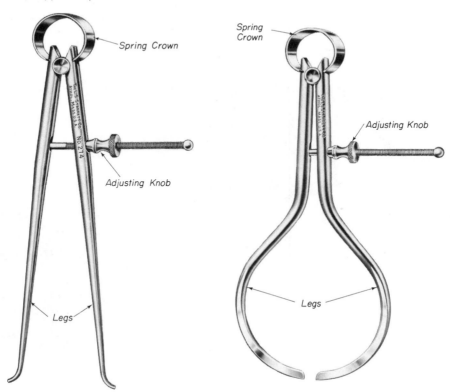

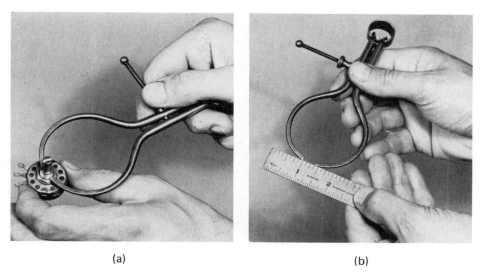

(a)   (b)

FIGURE 3.6. Outside calipers used to measure round stock. (a) Correct method of setting caliper tips. (b) Distance between tips is determined with a steel rule.

Drill gauges can also be used as a quick means of measuring round bodies such as diodes, transistors, and integrated circuits. Figure 3.7 shows the measurement of the outside diameter of a transistor using a drill gauge.

An inside caliper is required to obtain and transfer dimensions of inside distances. The procedure is equivalent to that used with the outside calipers. Figure 3.8 illustrates the use of the inside caliper. If greater accuracy than can be obtained from the basic caliper is required, *dial*

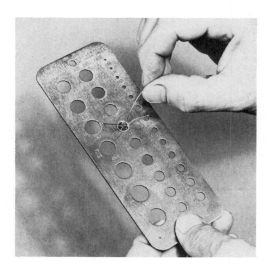

FIGURE 3.7. Drill gauge used to measure the case diameter of a transistor.

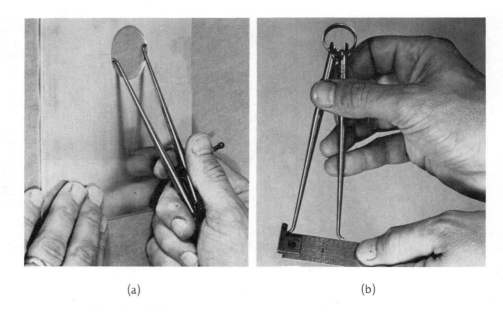

(a)   (b)

FIGURE 3.8. Inside calipers used to measure internal openings. (a) Correct method of setting caliper tips. (b) Distance between tips is best determined with a hook rule.

calipers are used. The 6-inch dial caliper is common in electronic construction. This precision instrument consists of *two pairs of jaws* (one for inside and one for outside measurements), a *dial indicator*, and a *thumb adjusting knob*. This instrument is capable of accuracies from 0.001 to 0.0001 inch (Fig. 3.9). Before using the dial caliper, a zero check should be made to determine if the dial indicator is properly calibrated with the jaws. First, the contacting surfaces of the outside jaws are wiped clean with a soft cloth to remove any foreign matter or

FIGURE 3.9. Dial caliper.

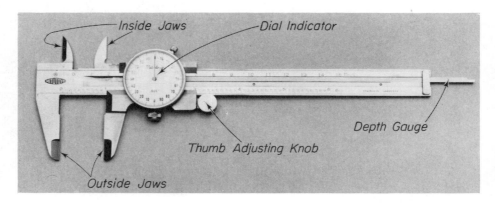

Inside Jaws   Dial Indicator

Depth Gauge

Thumb Adjusting Knob

Outside Jaws

dust. The thumb adjustment screw is then rotated clockwise to close the outside jaws. The dial indicator should read zero when the jaw surfaces make contact. If not, the rack and pinion must be adjusted. The two securing screws at both ends of the rack gear are loosened and the rack is slowly moved until the dial indicator reads exactly zero. The securing screws are tightened and the instrument is set for accurate measurements. Figure 3.10a illustrates the proper method of supporting the dial caliper when making measurements. With the adjusting knob

FIGURE 3.10. Dial caliper used to obtain accurate measurements. (a) Properly held dial caliper. (b) Rocking motion to determine minimum reading.

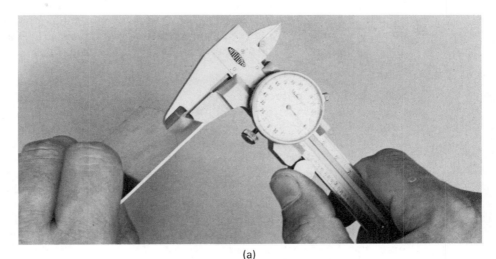

(a)

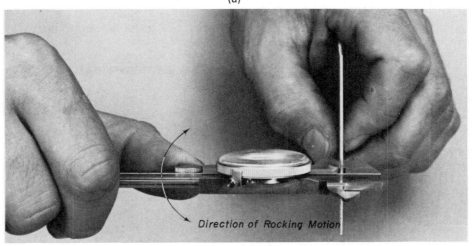

(b)

rotated to close the outside jaws on the work, the correct amount of jaw pressure is obtained through a *slipping action* whereby continued rotation will result in the adjusting knob slipping with no further forward motion imparted to the jaws. This action eliminates damage to the caliper and also measurement error.

All settings should be checked for correct jaw alignment before measurements are read. To ensure accuracy, the jaws of the calipers are slightly rocked over the work. The minimum deflection of the dial pointer will result in the most accurate reading (Fig. 3.10b).

3.3 MEASUREMENTS WITH A MICROMETER

The micrometer, like the dial caliper, is capable of measuring to accuracies of 0.001 inch and estimates to 0.0001 inch. The 1-inch outside micrometer is typical, although larger-capacity micrometers are available in addition to special-purpose micrometers, such as screw-thread measuring instruments.

The 1-inch micrometer is used primarily to measure diameters and thicknesses. This micrometer, shown in Fig. 3.11, consists of a *frame, anvil, spindle, barrel, thimble,* and *ratchet stop.* The proper use of this micrometer is as follows. The little finger is hooked around the frame while the thumb and forefinger of the same hand are free to rotate the thimble (Fig. 3.12a). Clockwise rotation of the thimble will close the spindle on the anvil, which is rigidly supported to the frame. Counterclockwise rotation of the thimble widens the distance between the spindle and the anvil. The work to be measured is held in the other hand (Fig. 3.12b) and placed on the anvil. The thimble is rotated to close the spindle on the work. Once the spindle makes *light contact* with the work, the thumb and forefinger are moved to the ratchet stop. Clockwise rotation

FIGURE 3.11. One-inch micrometer with ratchet stop. Courtesy of The L.S. Starrett Company, Athol, Massachusetts.

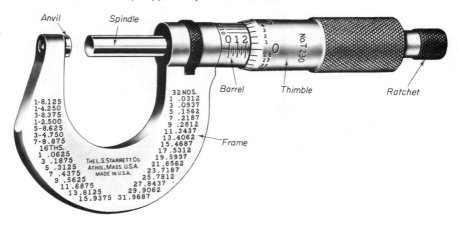

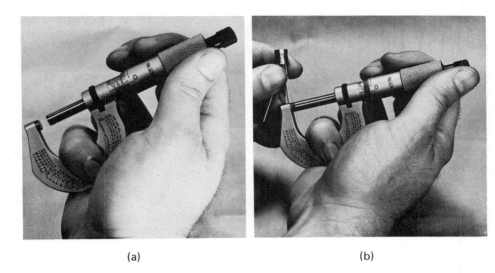

(a)   (b)

FIGURE 3.12. Correct positioning of the 1-inch micrometer. (a) Properly held micrometer. (b) Technique for measuring small work.

of the ratchet stop for two "clicks" will apply the correct amount of pressure to the work. Excessive pressure may damage the finely threaded spindle or mar the machined surfaces of the anvil and spindle, thus causing measurement errors. It is good practice to maintain continuity in measurement by duplicating the applied pressure. The two "clicks" of the ratchet stop will assure this duplication of pressure.

The spindle has a precision of 40 threads per inch. Thus, one complete rotation will move the spindle exactly $\frac{1}{40}$ or 0.025 inch. Each minor graduation mark on the barrel scale is therefore equal to 0.025 inch. Each major numbered division on the barrel scale represents 0.100 inch, requiring four complete turns to move from one major numbered division to the next.

The thimble has 25 graduation lines. Each line represents $\frac{1}{25}$ of $\frac{1}{40}$ of an inch, or 0.001 inch. Therefore, as the thimble is rotated from one division to the next in a counterclockwise direction, the spindle moves 0.001 inch away from the anvil.

To obtain the spindle to anvil opening, three readings are necessary. First, the highest *numbered* graduation shown on the barrel is noted. (Each numbered graduation represents 0.100 inch.) The second reading is the number of minor divisions following the major number on the barrel, each minor division representing 0.025 inch. The third reading is the lowest-numbered graduation line on the thimble that most closely aligns with the horizontal scale line on the barrel. This value is in one-thousandths (0.001) of an inch and is added to the sum of the preceding readings to obtain the final measurement.

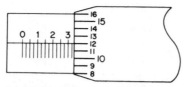

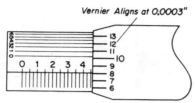

(a) Readings to One-thousandth Inch Using the Barrel and Thimble Scales.

FIGURE 3.13. Illustrations of micrometer scale readings. (a) Readings to the one-thousandth inch using the barrel and thimble scales. (b) Readings to the ten-thousandth inch using barrel, thimble, and vernier scales.

Vernier Aligns at 0.0003"

(b) Readings to Ten-thousandth Inch Using Barrel, Thimble, and Vernier Scales.

Figure 3.13a represents a reading of 0.337 inch obtained as follows:

Reading 1: The largest visible number on the barrel is 3, or 0.300 inch.

Reading 2: The number of smaller graduations after the number 3 is 1, or 1 × 0.025 = 0.025 inch.

Reading 3: The lowest number of the graduation lines on the thimble that most closely aligns with the horizontal scale line on the barrel is 12, or 0.012 inch.

The sum of the three readings is 0.337 inch.

Some micrometers are equipped with a *vernier scale* to extend the measurement accuracy to one ten-thousandths (0.0001) of an inch. This vernier scale is immediately above the barrel scale and its graduation marks extend along the barrel to the thimble scale. The vernier and thimble scales both run horizontally. Each of the 10 graduation lines of the vernier scale, marked 0 to 9, represent $\frac{1}{10}$ of the value of the divisions on the thimble scale — that is, $\frac{1}{10}$ of $\frac{1}{1000}$, which equals $\frac{1}{10,000}$ (0.0001) inch.

To obtain a reading with a micrometer equipped with a vernier scale, the previously discussed three readings are first taken. The fourth reading is obtained by noting which graduation line of the vernier scale aligns with any line on the thimble scale. This reading, in one ten-thousandths (0.0001) of an inch, is added to the sum of the other three readings. Figure 3.13b represents a reading of 0.4583 inch, obtained as follows:

Reading 1: The largest visible number on the barrel is 4, or 0.400 inch.

Reading 2: The number of minor graduations after the number 4 is two, or 2 × 0.025 = 0.050 inch.

Reading 3: The lowest number of the graduation lines on the thimble that most closely aligns itself with the horizontal scale line on the barrel is 8, or 0.008 inch.

Reading 4: The number of the division on the vernier scale that aligns with a division on the thimble is 3, or 0.0003 inch.

The sum of the readings is 0.4583 inch.

3.4 ENGINEERING SKETCHES

A preliminary step in the preparation of the finished drawings is the development of *engineering sketches*. Figure 3.14 shows a sheet metal sketch of the amplifier's front panel. Using quadruled paper eliminates the need of special drafting equipment and at the same time results in a relatively clear picture. Notice that the engineering sketch shown in Figure 3.14 contains not only *dimensional information* but specifications for *finish*, *type of materials*, *labeling*, and *hole positioning*. Also useful are manufacturer's specifications that provide hole-spacing information for mounting holes, pin sizes and configurations, case dimensions, and other pertinent information to aid in layout. Examples of this type of hardware information are given in Appendix III, which includes some of the common solid-state case configurations currently employed in industry.

3.5 DRAFTING EQUIPMENT

To produce good-quality drawings, the technician must have available a minimum selection of drafting equipment and materials and a basic understanding of their function. Tools common to the draftsman are (1)

FIGURE 3.14. Engineering sketch of amplifier's front panel.

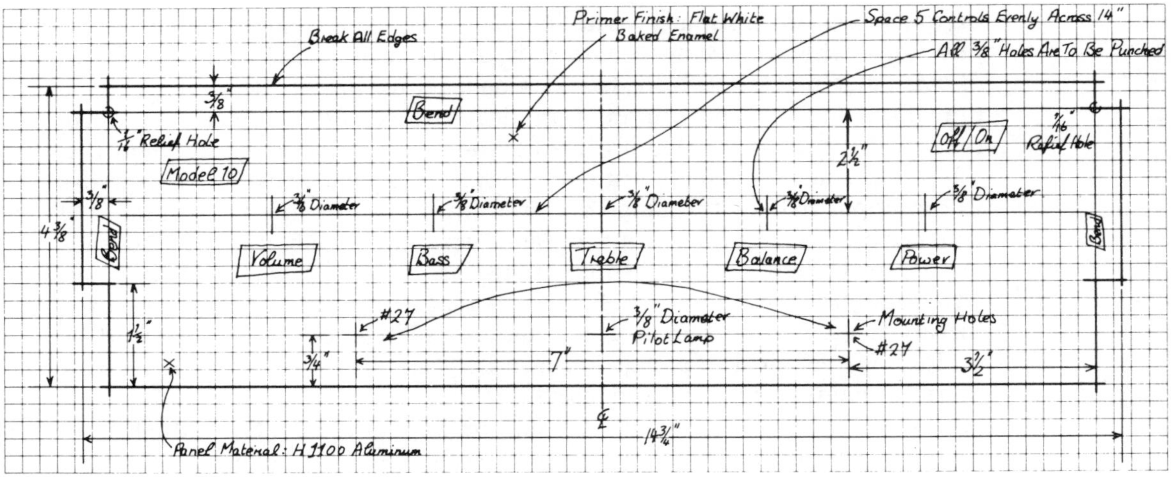

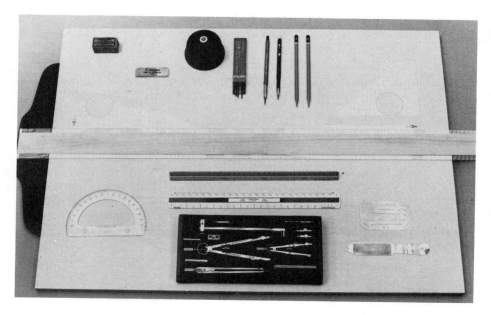

FIGURE 3.15. Basic drafting equipment for electronics.

pencils, (2) *drawing boards*, (3) *T-squares*, (4) *triangles*, (5) *protractors*, (6) *engineering scales*, and (7) *compasses*. Figure 3.15 shows a basic drafting setup.

Pencil lead is available in a variety of grades from extremely *hard lead* for fine light lines to *soft lead* for dark bold lines. Selecting the proper lead grade depends primarily on the application. Table 3.1 lists various lead grades and their applications. Hard leads range from 4H to 9H, the larger number indicating the harder lead. Those leads in the upper half of the range are normally used for diagrams, charts, and graphical computations where extreme accuracy is of prime concern. The lower half of this range represents leads that can be used for line work on engineering drawings, but they have limited application because the lines they make tend to be too light. Medium lead grades are B, HB, F, H, 2H, and 3H. These leads are for general-purpose work in technical drawings. The grades in the upper half of the range are used primarily for line work; grades H and 2H find wide application on pencil tracings for blueprints. The leads in the lower half of the medium range are used for technical sketches, lettering, and freehand work. Soft leads range from 2B to 7B and are used for art work and certain full-size detail drawings. These leads should be avoided in technical drawings because they smudge. Lead should be kept sharp to maintain line-width consistency in the work.

Wooden and *mechanical lead supports* are available. Since the lead in the mechanical or refill type can be interchanged, it is preferred over the wooden type.

TABLE 3.1

Pencil Grades

Grade	Hardness Classification	Application
9H	HARD	For extreme accuracy, charts, diagrams, and
8H	HARD	graphical computations
7H	HARD	
6H	HARD	For construction line work on engineering
5H	HARD	drawings
4H	HARD	
3H	MEDIUM	Line work on machine, engineering drawings
2H	MEDIUM	and pencil tracings for blueprinting
H	MEDIUM	
F	MEDIUM	Technical sketching, free hand work, and
HB	MEDIUM	lettering
B	MEDIUM	
2B	SOFT	Various types of artwork and architectural
3B	SOFT	drawing.
4B	SOFT	Too soft for mechanical drawings
5B	SOFT	
6B	SOFT	
7B	SOFT	

Drafting boards are constructed from thoroughly seasoned, straight-grained soft wood strips. The top or working surface must be both smooth and a true plane. Both side edges must be perfectly straight and parallel so that either one may be used as a working edge with a T-square. The working edges are joined to the main body using tongue-and-groove joints to prevent warping. A drafting board should be checked for warping prior to its use. This can be done by simply placing it on a flat surface and observing if the board rocks. The working edges of the board should also be tested with a T-square that is known to be in alignment. This test is made by placing the bottom edge of the T-square blade along the working edge of the board. If the board is true, the blade will touch along the entire length. This procedure is shown in Fig. 3.16. Any defect can usually be corrected by sanding or planing the working edge. The most convenient size of drafting board is 18 by 24 inches. Although larger boards are available, this size is most suitable for electronic drafting.

The *T-square* is most commonly formed with a *blade* rigidly attached to the *head*. The inner side of the head and the edges of the blade serve as the working edges of the T-square. As such, they should be perfectly straight and at exactly 90 degrees to each other. This angle can be quickly tested by using a triangle, as shown in Fig. 3.17. T-squares are generally made of hardwood, usually the head of black walnut and the blade of maple. Blades are available lined with celluloid, which resists warping and

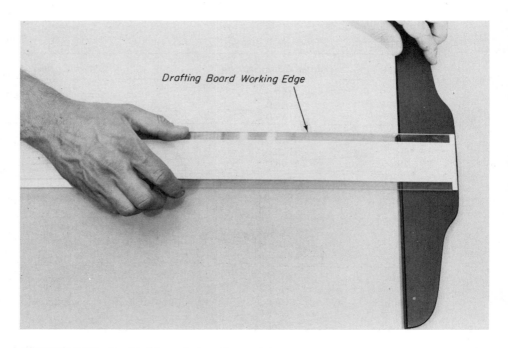
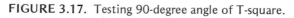

FIGURE 3.16. Drafting board edge-alignment test.

FIGURE 3.17. Testing 90-degree angle of T-square.

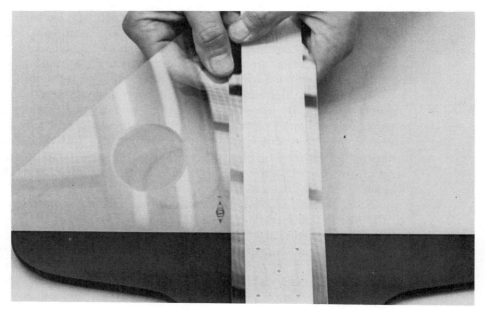

allows lines to be viewed beneath the working edge. Although available in various lengths, the 24-inch T-square is suitable for most applications. The T-square is a fragile instrument and should be handled carefully to avoid loosening the head. When not in use, it should lie flat on a drafting board or be hung from the hole on the end of the blade.

In use, the T-square is placed on the drafting board with the head along the left-hand working edge and the blade extending along the surface of the board. In this position, it is used to draw horizontal lines and serves as a reference for positioning triangles and templates.

Triangles are necessary to simplify the construction of angles. The most commonly used types are the *45-degree* and the 30- by 60-degree triangles (Fig. 3.18a). Triangles made of celluloid are considered to be the best quality and are available in different sizes and thicknesses. Most often, they are clear or transparent, but amber, red, and green shades are not uncommon. The most widely used triangles are 0.08 inch thick. The 45-degree triangle with 6- or 8-inch legs used with a 30- by 60-degree triangle with the longest leg 10 or 12 inches long is a suitable combination for many applications. By proper positioning of these triangles, angles of 30, 45, 60, 75, 90, 105, and 135 degrees are easily measured (Fig. 3.18b). When angles cannot be obtained with these common triangles, the *protractor* is used. For most applications, the clear or shaded celluloid protractor (Fig. 3.19a) is satisfactory. When a high degree of accuracy is required, the *nickel-silver* protractor with movable vernier scale and extension blade should be used (Fig. 3.19b).

Mechanical engineer's scales are the most useful in electronic drafting (Fig. 3.20). These scales provide full-size, one-half, one-fourth, and one-eighth divisions that are satisfactory for most size reductions. *Full-divided*

FIGURE 3.18. Drafting triangles. (a) Standard 45-degree and 30- by 60-degree triangles. (b) Using triangles in combination to increase layout versatility.

(a)

(b)

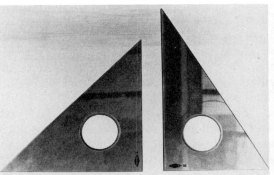

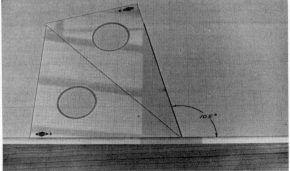

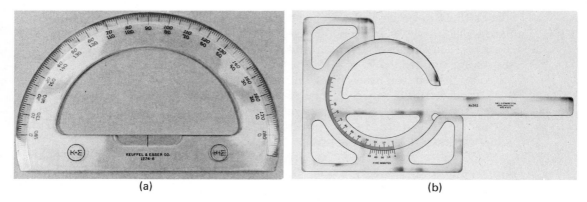

(a)   (b)

FIGURE 3.19. Protractors. Courtesy of The L.S. Starrett Company, Athol, Massachusetts.

FIGURE 3.20. Mechanical engineer's scale.

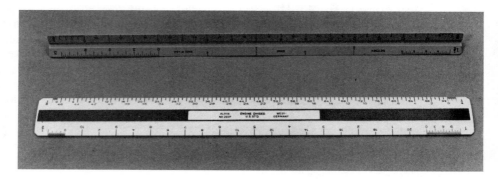

and *open-divided* scales are available, on both of which the main divisions represent 1 inch. On the full-divided scale, the basic units are subdivided throughout the length of the scale. On the open-divided scale, only the end segment of the scale is subdivided. Scales are available in either triangular cross section, combining as many as 11 scales, or the conventional flat type having only two scales. The flat type is easier to use because scales can be selected very quickly. Scales are used to obtain full or reduced dimensional ratios. These dimensions can be laid off directly onto the drawing from the scale or by transferring measurements from the scale to the drawing using dividers. Scales are usually made of boxwood, and some have white celluloid composition edges.

Drawing instrument sets, such as the one shown in Fig. 3.15 containing *dividers, compasses, and inking pens,* are necessary to produce a quality, accurate plate. Occasionally, however, time is more important than precision. The use of *templates,* such as those shown in Fig. 3.21, can aid in producing good drawings with a considerable saving of time

FIGURE 3.21. Drafting and electronic layout templates.

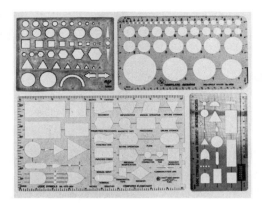

and effort. Templates are used extensively as a quick means of reproducing electronic symbols and other shapes, particularly circles. They consist of various shaped cutouts in a piece of clear or shaded plastic. When these cutouts are outlined with a pencil or pen, the result is a uniform and neatly shaped electronic symbol or figure. Templates were not intended to replace drafting instruments but to provide a faster means of producing sketches and drawings. The quality and accuracy of a plate produced with drafting equipment cannot be duplicated by using templates.

Knowledge of the size and type of paper used in the drawing is helpful. Commonly available drawing paper sizes are: *A-size* (8½ by 11 inches), *B-size* (11 by 17 inches), *C-size* (17 by 22 inches), and *D-size* (22 by 34 inches). Larger sizes are available, of which *K-size* (40 by 50 inches) is normally the largest used in engineering work. The paper size is, in many cases, selected to be compatible with the filing system in use. In all cases, the paper should be of sufficient size to show detail work without crowding, yet not so large as to have excessive unused areas or overly large borders. The preferred paper for electronic drafting is the *HP* (hot-pressed) *cream* or *buff-colored* type. This paper is sufficiently heavy and hard-surfaced to be used for high-grade, accurate detailed work. *CP* (cold-pressed) paper is less expensive and is used for general drawing or sketching. The surface of this type of paper will not hold up to erasing as well as the HP type. When blueprints are to be made from pencil drawings, a white, lightweight bond paper should be used. This type of paper allows blueprints to be produced directly without the need for tracing.

3.6 BASIC LINE CONVENTIONS FOR DRAWINGS

Since the primary objective in engineering drafting is to transmit detailed information pictorially, a basic knowledge of the various types of lines used and their application is essential. The symbolic lines conventionally accepted are shown in Fig. 3.22. All lines should be made dark with the

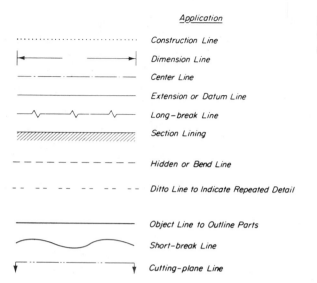

Application

Construction Line

Dimension Line

Center Line

Extension or Datum Line

Long – break Line

Section Lining

Hidden or Bend Line

Ditto Line to Indicate Repeated Detail

Object Line to Outline Parts

Short-break Line

Cutting-plane Line

FIGURE 3.22. Basic line conventions used in engineering drafting.

exception of *construction* lines. In addition, the line width will vary depending on the type of symbol used. Thin lines are used for *dimension, center, extension, long-break,* and *section* symbols. Medium thickness is used for *hidden* and *ditto* lines. Thick lines are reserved for *object, break,* and *cutting-plane* lines.

3.7 DIMENSIONING FOR CHASSIS LAYOUTS

There are several acceptable dimensioning techniques used in chassis layouts, the choice of which will depend on the configuration being dimensioned and the degree of accuracy required. These techniques are *datum-line* or *base-line, center-line, continuous,* and *tabulation* dimensioning.

Datum-line or *base-line dimensioning* uses any two edges that are perpendicular to each other. These reference edges are called *datum lines.* Each dimension made uses one of the datum lines as a reference. This technique eliminates cumulative error, since each dimension is independent of all others. This method of dimensioning is often used for close-tolerance work, especially when mating parts are to be fabricated. Figure 3.23 illustrates base-line dimensioning.

In *pure center-line dimensioning,* the stock is first equally divided vertically and horizontally. The locations of all holes are made from these reference lines. No other lines are used as a reference. In electronic layout, datum-line dimensioning is preferable when locating holes, since edge reference facilitates this layout, as described in Chapter 5.

When holes are to be formed that match a particular pattern on a component, such as the mounting holes on a speaker, a variation of pure center-line dimensioning is preferable. Base-line or pure center-line dimensioning would create a larger tolerance in this application. For this reason,

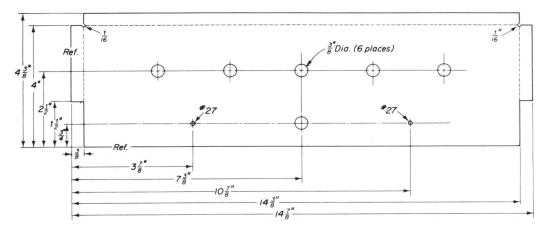

Material : HIIOO aluminum
Gauge : 16
Finish: Spray Painted with Baked Enamel
Color of Finish : Flat White
Special Construction Notes:
(1) Finished Outside Dimensions after Bending (4"x14")
(2) Five ⅜" Diameter Holes are to be Evenly Spaced
across panel

FIGURE 3.23. Base-line dimensioning of the amplifier's front panel.

independent dimensioning of related holes or of hole patterns should be avoided. In this technique the center of the largest or major hole in the pattern is located from datum lines. All other hole centers are dimensioned using these major hole center lines as a reference. Figure 3.24 shows an example of center-line dimensioning.

As drawings become more complex, it is necessary to use both datum-line and center-line dimensioning, especially when the exclusive use of either one would adversely affect clarity. Figure 3.25 shows an example of combining both techniques on one drawing. Notice that the use of both techniques reduces the number of dimension lines crossing. Although it is acceptable practice for dimension lines to cross, dimensions may become difficult to read if this practice is used extensively.

When tolerances are not critical, *continuous dimensioning* may be used. This method is also shown in Fig. 3.25, and is commonly used in combination with center-line dimensioning. Notice that only one arrowhead is used on the dimension line, since the measurements run continuously from a datum line. The longest dimension in the series should be made separately for additional clarity. This technique of dimensioning can cause large tolerances and should be used only when critical dimensions are not required.

When dimensioning complex, high-density hole patterns, hole-center locations may be presented in *tabular* form, with no dimension lines

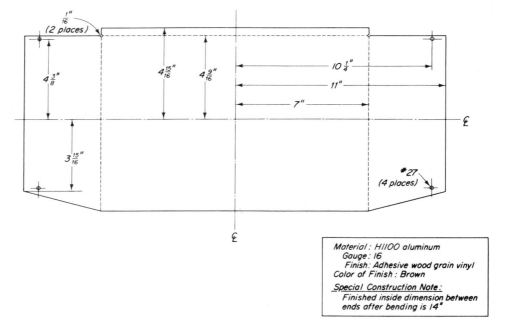

FIGURE 3.24. Center-line dimensioning of the amplifier's enclosure element.

shown on the drawing. Two datum lines are selected and identified with an appropriate designation, such as *X* and *Y*, for the horizontal and vertical datum lines, respectively. All hole centers are then identified literally, the same designation being used for common-size holes. A table identifying each hole, with its size and distance from each datum line, is then prepared and attached to the drawing.

When a high degree of dimensional tolerance is required, measurements should be taken in decimal form. Since this degree of accuracy is not normally required in prototype construction, fractional measurements are used throughout this section. With fractional dimensioning, accuracy to $\frac{1}{64}$ inch is easily obtainable. Maximum allowable variation above and below the nominal value by which the given dimension may differ from the drawing to the actual work appear with the dimension where applicable.

3.8 ASSEMBLY DRAWINGS

After a prototype has been completed, final drawings and photographs are made of all assemblies and subassemblies. These drawings may be of several variations: *two-dimensional top*, *bottom*, and *side views*; *three-dimensional views*; *isometric projections*; and *exploded views*. The designated views must contain all pertinent information concerning the proper

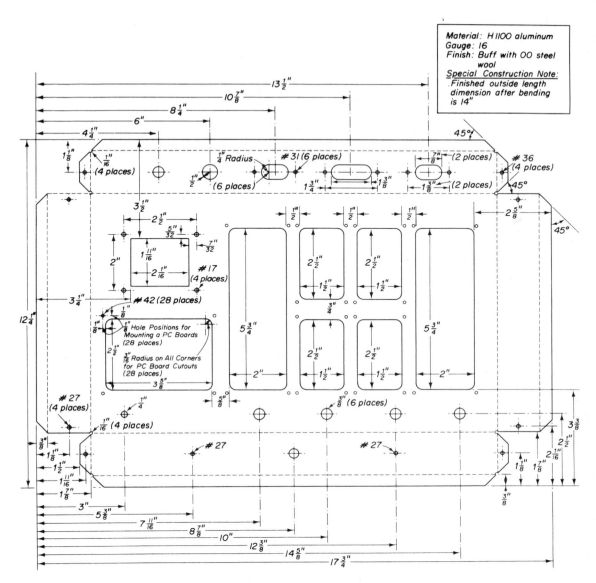

FIGURE 3.25. Amplifier main chassis element layout.

location and orientation of each component and mechanical part. All parts shown in assembly drawings should be identified and labeled consistent with the designations used on the schematic diagram. A common arrangement is a numbering system relating to the parts and accompanying the drawings.

Auxiliary views are frequently necessary, especially when complex or high-density packaging obscures some of the major sections of the pack-

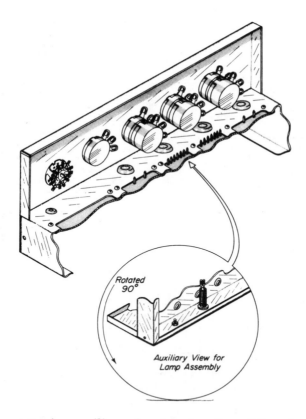

FIGURE 3.26. Front panel assembly drawings.

Rotated 90°

Auxiliary View for Lamp Assembly

age. An auxiliary view is a drawing of the obscured portion of the package after eliminating the obstructions of the primary views. Figure 3.26 shows a partial assembly layout for the amplifier package, including an auxiliary view.

Photographs are particularly advantageous because of their high quality and three-dimensional perspective characteristics. In addition, nomenclature and reference information can be superimposed on photographs. Drawings, however, are absolutely indispensable for fabrication, assembly, and wiring because of the excessive amount of explanatory information necessary.

The next phase of construction after the drawings have been completed is to shear the chassis element into the desired blanks. In the following chapter, detailed procedures for shearing sheet metal are presented. Specifically, the amplifier chassis elements, as shown in the drawings in this chapter, are blanked and prepared for layout.

PROBLEM SET

3.1 Read the measurements shown in Figs. 3.3, 3.6b, and 3.8b.

3.2 Read the micrometer setting shown in Fig. 3.11 and those in Fig. 3.27a and b.

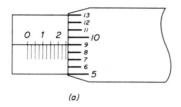

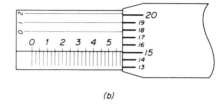

(a)

(b)

FIGURE 3.27.

3.3 Read the vernier caliper settings shown in Fig. 3.28a and b.

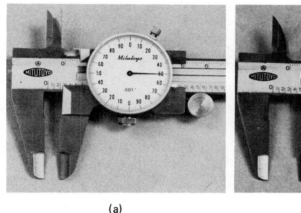

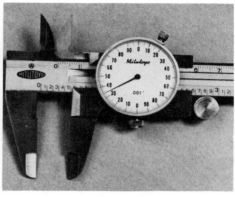

(a)

(b)

FIGURE 3.28.

3.4 Measure the body and lead diameters and body lengths of ¼ watt through 10-watt resistors. Also determine the body and lead diameters of several electrolytic capacitors and the body thickness and lead spacing of various-sized disc capacitors. Tabulate all measurements for future reference.

3.5 Construct the engineering drawings, chassis element layouts, and assembly drawings for the package designed in Problem 2.1.

4

Shearing

4.0 INTRODUCTION

Industrial descriptions of machine tools include sheet metal *shears* in the category of *metal-forming machines*. The action of a shear is primarily to cut pieces of sheet and bar stock (or structural shapes) to desired dimensions. The construction of electronic equipment begins with metal-cutting operations. Selecting and properly using the most appropriate tools for metal cutting is essential for quality workmanship. Bulk sheet stock must be cut into blanks of the desired overall dimensions for the chassis (or enclosure elements) before layout work begins.

Two basic types of shears are typically used in electronic fabrication. These are the *blade shear* and the *notcher*. The vertical drive for shear machines is either manual or power-operated. Some shear machines are also equipped for *reciprocating* vertical motion. Shears are capable of producing a neat, straight cut without edge imperfections.

This chapter considers in detail the alignment and practical operating information for the foot-operated and hand-operated *squaring shear* and the *hand notcher*. Also included are illustrations of the amplifier chassis elements in the construction project to demonstrate the versatility of these shears.

4.1 FOOT-OPERATED SQUARING SHEAR

The foot-operated squaring shear, shown in Fig. 4.1, is used for (1) *cutting* large pieces of sheet metal into *blanks*, (2) *trimming* stock to obtain a finish cut or reference edge, and (3) *squaring*. The basic parts of this shear, shown in Fig. 4.2, are two *cutting blades, bed, foot treadle*, and four *alignment gauges*.

Accurate and professional chassis layout begins with blank stock that is not only cut to the correct overall dimensions, but also is cut squarely. Before the 90-degree alignment of the shear blades with the side gauges is tested, the following inspections of the shear should be made: (1) uniform hold-down bar pressure, (2) proper shear blade clearance, and (3) even tension when the foot treadle is depressed. Each of these tests is described below.

A test piece of scrap metal of uniform thickness is used to test for uniform hold-down bar pressure. With the test piece under the hold-down bar and the foot treadle fully depressed, proper adjustment is made via two adjusting screws located on the front side of the hold-down bar. The screws are set so that the work is held firmly, with care taken so that excessive pressure does not deform the metal.

Proper adjustment of shear blade clearance between the upper and lower blades is a uniform gap, typically from three to five thousandths of an inch. Adjustment screws for varying clearance to this distance are located on the front end of the shear frame.

FIGURE 4.1. Foot-operated squaring shear. Proper positioning of hands and feet is illustrated along with the correct implementation of the back stop.

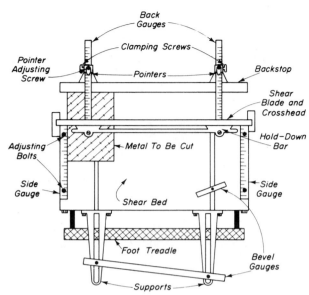

To ensure a clean cut, there must be uniform translation of motion between the foot treadle and the upper cutting blade. This tension adjustment is made by turning the two *turnbuckles* that link the *foot treadle* and the *crosshead.*

When the foregoing tests and adjustments are completed, the 90-degree alignment of the blades and the side gauges may be tested. To determine this alignment, a *carpenter's square* is placed on the shear bed, with the longer leg flush with the lower cutting blade. The shorter leg of the square is placed against one of the side gauges. For square cuts, the side gauges must be perpendicular to the cutting edge and flush against the square. Any necessary correction, shown in Fig. 4.3a, can be made by loosening the two *securing bolts* holding the side gauge in place and positioning the gauge so that it rests flush against the leg of the square. When

FIGURE 4.3. Side gauge alignment test. (a) Adjustment is necessary as detected by use of carpenter's square. (b) Proper side gauge position.

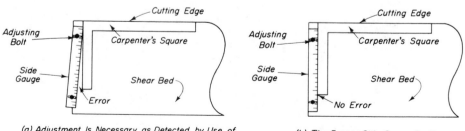

(a) Adjustment Is Necessary as Detected by Use of Carpenter's Square.

(b) The Proper Side Gauge Position.

the securing bolts are tightened, the shear is properly aligned for square cuts. Correct alignment is shown in Fig. 4.3b. *Both* the gauges on the shear bed should be checked for alignment before starting any cutting operations. Once the side gauges are properly adjusted, stock placed against them for cutting will have a right angle between the edge of the stock against the gauge and the cut edge. Although either side gauge may be used, the gauge on that side of the shear where the cutting action *first* occurs is used when cutting small pieces of stock.

Many shears are equipped with *back gauges* and a *stop* for *back shearing*, as shown in Fig. 4.2. Although front shearing is more accurate when using the side gauges, the back gauges are extremely convenient when a large quantity of stock is to be cut to the same dimensions. The back gauges are calibrated in sixteenths of an inch measured from the rear edge of the lower blade. The gauges also serve as the supporting arms of the back stop. To ensure that the back gauge pointers are properly set, a long piece of stock of a known uniform width is inserted between the back stop and the rear edge of the lower blade. The pointers should be readjusted if they do not read the correct dimensions.

The desired depth of cut is easily set by moving the stop along the back gauges to the required dimension using the clamping screws to secure the back stop in place. Metal to be cut is placed on the shear bed and moved under the upper blade until it contacts the back stop. With the metal held firmly in position, pressure is applied to the foot treadle. The hold-down bar will contact the metal first, securing it firmly against the shear bed. Additional pressure applied to the treadle engages and depresses the upper blade until the metal is cut. Removing the pressure from the treadle releases the blade and then the hold-down bar, in sequence. To aid in shearing large pieces of stock, *extension arms* are available. These arms bolt to the front edge of the shear bed and support the material.

Careful attention to all the adjustments mentioned in this section results in a neatly cut edge free of waves and burrs. The *cutting capacity* of the shears should *never be exceeded*. The cutting capacity of the type of shear shown in Fig. 4.1 is approximately 16 gauge for mild steel and 12 gauge for the softer, nonferrous metals (aluminum, brass, and copper), with a maximum cutting length of 36 inches.

Several safety precautions must be observed when using a foot-operated squaring shear. Of course, fingers must be kept clear of the hold-down bar and cutting blades at all times. In addition, the operator's feet must be positioned so that no injury occurs during depression of the foot treadle.

4.2 SQUARING IRREGULAR STOCK

If it is necessary to begin cutting operations by squaring an irregularly shaped piece of stock, a properly adjusted foot shear is used (Sec. 4.1). One straight edge must be cut initially if one does not already exist on the

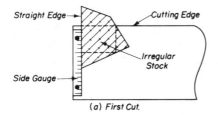

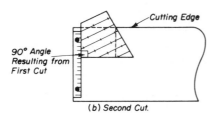

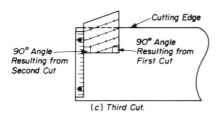

FIGURE 4.4. Sequential procedure necessary to completely square a piece of irregular stock by alternate shearing and counterclockwise rotation. (a) First cut. (b) Second cut. (c) Third cut.

stock. This straight edge is placed against the side gauge and another cut is made, as shown in Fig. 4.4a. If the left-side gauge is used as illustrated, the stock is rotated *counterclockwise* and a second cut is made (Fig. 4.4b). Again the stock is rotated in a *counterclockwise* direction and the final cut is made (Fig. 4.4c). If the *right*-hand side gauge is used, the operation is similar except that the stock is rotated in a *clockwise* direction.

4.3 HAND-OPERATED SQUARING SHEAR

For work on smaller pieces of stock, the *hand shear* is usually preferable since it is more accurate and easier to operate than the foot shear. The hand shear consists of a *table*, a *side gauge*, *back gauges*, and two *cutting blades*. A *hand lever* causes two *eccentric cams* to move the *crosshead*. This imparts the vertical motion to the *upper blade*.

Use of the hand shear is basically the same as that of the foot shear. Hand shear tables vary in size from 6 inches to 36 inches and the shearing capacities are similar to those of foot shears. Sheet metal is cut by holding the stock firmly against the side gauge or back gauge stop with one hand and pulling the shear handle forward with the other. Figure 4.5 shows the front-panel blank of the amplifier being cut on a hand shear.

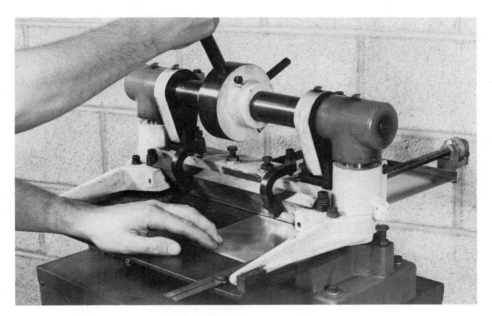

FIGURE 4.5. Using the hand-operated shear to trim chassis blanks. Metal to be cut is positioned against the right-side gauge and held firmly with the left hand. The micrometer back gauge is also shown.

The hand shear should also be checked for correct adjustment prior to use. The adjustment procedures are basically the same as those for the foot shear (Sec. 4.1). If the shear table is not wide enough to accommodate a carpenter's square, a piece of sheet metal that has been previously squared serves the purpose.

The *side gauge* on the shear table is calibrated from the *cutting edge* of the *lower* shear blade. This gauge should be checked for accuracy before any cutting operations by placing a piece of square stock of known dimension on the shear table with one edge flush with the lower blade and the other against the side gauge. The known length of stock should exactly measure the dimension shown on the gauge. If adjustment is necessary, the securing screws on the gauge are loosened and the gauge is moved until the correct calibration mark exactly aligns itself with the edge of the stock, as shown in Fig. 4.6. With the side gauge properly adjusted, measurements can be taken directly from drawings and stock can be cut without the need for layout marks scribed into the work. The degree of accuracy of any cut is directly related to the accuracy to which the side gauge is adjusted. The side gauge is normally calibrated in sixteenths of an inch. If the drawing requires a closer tolerance than permitted by this calibration, lines scribed into the stock may be used to determine the correct cutting position. For an even greater degree of dimensional accuracy, *micrometer back gauges* are available with precise settings as low as 0.001 inch.

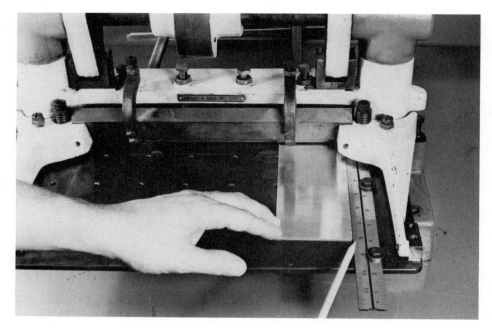

FIGURE 4.6. Side gauge adjustment is tested with a known length of stock.

4.4 HAND NOTCHER

In the construction of electronic equipment, it is often necessary to shear *notches* of various sizes and shapes into a piece of stock. The *hand notcher* is an extremely useful tool for this purpose. The basic parts of the notcher are the *operating handle, table, dies,* and *blades.* The operating handle is attached to an *eccentric cam.* A *V-ram* and a *vertical block,* to which the blades are attached, moves vertically as the hand lever is operated.

The notcher is used for shearing short lengths of stock, cutting 90-degree notches, and making various angular trim cuts. The two cutting blades are mounted at right angles to each other and positioned over a pair of dies that are secured to the table. The edges of the blades and dies are at 45-degree angles to the sides of the table. Figure 4.7 shows a hand notcher and illustrates its use.

The notcher is operated by sliding the metal to be sheared under the notcher blades and pulling the handle forward. This results in a cut notch having a 90-degree angle. The two table scales should be checked with a combination square to ensure accurate results. The combination square is placed on the notcher table, as shown in Fig. 4.8. The scale edge of the combination square is aligned with one blade of the notcher. The zero graduation mark, which is perpendicular to the cutting edge, should

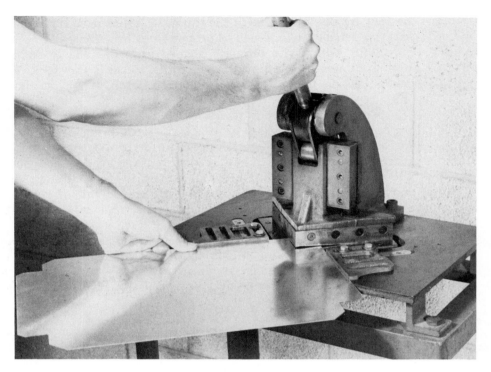

FIGURE 4.7. Metal to be cut is held firmly against the notcher table and stops with one hand, while the handle is operated with the other. The stops assist in positioning metal quickly for duplicating work.

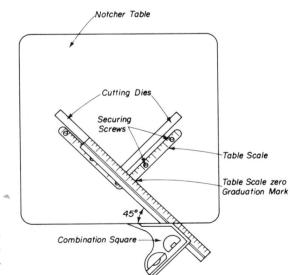

Notcher Table

Cutting Dies

Securing Screws

Table Scale

Table Scale zero Graduation Mark

45°

Combination Square

FIGURE 4.8. Testing for the proper alignment of a table scale and cutting die using a combination square.

be split by the scale edge of the square. If adjustment is required, the two *securing screws* are loosened to properly position the scale.

With both scales properly aligned, it is possible to cut notches of desired dimensions without making scribe marks in the stock. The stock is positioned on the scales for the correct depth of cut, as shown in Fig. 4.9.

For notches having angles *less than* 90 degrees, a *bevel square* or *protractor head* provides a means of transferring drawing dimensions directly to the work. The bevel square or protractor head is set to the desired angle of cut. The base of the square or head is then placed against the bevel blade or rule. While maintaining the desired angle, the work is moved along the table scale for the required depth of cut, as shown in Fig. 4.10. For notches having angles *greater than* 90 degrees, it is necessary to perform *two* cuts.

When quantity duplication of work is performed, it is tedious to use the table scales alone or in combination with a square to align the stock. Stops are available, similar to those shown in Fig. 4.7, which are easily bolted to the notcher table over both scales. Each stop is slotted for ease of adjustment.

FIGURE 4.9. Accurate notches can be made without marking the metal by using the table scales to position the stock for the desired depth of cut.

FIGURE 4.10. Proper use of protractor head to simplify cutting angles other than 90-degrees.

Extreme care must be taken to keep fingers away from notcher blades. Unlike the hand- and foot-operated shears, which provide some degree of protection to the operator, the notcher is *not* equipped with a hold-down bar.

The correct cutting position of the notcher blades and dies must be continually maintained; otherwise, both will be damaged.

The maximum capacity of the hand notcher is approximately the same as for the foot shear shown in Fig. 4.1 (16 gauge for mild steel and 12 gauge for the softer metals), with a cutting length of 6 inches.

4.5 HYDRAULIC AND MECHANICAL POWER-DRIVEN SHEARS

Hydraulic or mechanical power-driven shears have a wider range of metal capacities than those of the manually operated type. Power shear capacities range from $\frac{1}{8}$ inch by 24 inches to 1 inch by 30 feet mild steel with cutting pressures up to 8000 tons. Power shears for even thicker and wider stock are also available for special applications.

Hydraulic shears offer some advantage over the mechanically powered motor-driven models. The hydraulic drive provides both variable speed and variable blade pressure. This allows wider adjustments to obtain optimum cutting action for any specific type of stock material.

Power shears have distinct advantages over manually operated models if quantity cutting is a major concern. These shears may be set for single or multiple shearing operations, thereby increasing cutting rates and lowering costs.

4.6 SHEARING APPLICATION

The three sheet metal amplifier components may be sheared to the desired dimensions by following the procedures described above. These components are shown in Fig. 4.11. Subsequent chapters continue the logical sequence in fabricating the amplifier chassis, front panel, and enclosure.

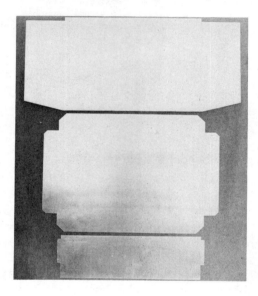

FIGURE 4.11. Three chassis elements for the amplifier after basic shearing operations are completed. These components are now ready for layout.

PROBLEM SET

4.1 Test the 90-degree alignment of the side gauges to the cutting blades, side gauge 1-inch graduation distance from cutting blades, and parallel alignment of back stop to cutting blades for identical back gauge settings on a foot-operated shear.

4.2 Check the accuracy of the 45-degree angle between the cutting blades and front edge of a sheet metal notcher. Also inspect the alignment of the scale zero graduation marks and the leading ends of the cutting blades.

4.3 Using a shear, notcher, combination square, and protractor, fabricate a drill-point sharpening gauge to the dimensions specified in Fig. 4.12 from 16-gauge aluminum.

FIGURE 4.12.

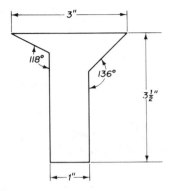

4.4 Shear and notch a piece of 16-gauge aluminum to the dimensions given in Fig. 4.13 for a soldering iron stand. The layout is completed in Problem 5.2 and bending is accomplished in Problem 8.2.

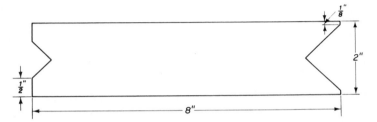

FIGURE 4.13.

4.5 Fabricate a piece of 16-gauge aluminum into the configuration shown in Fig. 4.14 for a transistor heat sink. The heat-sink layout will be completed in Problem 5.3, drilling and deburring in Problem 6.2, and bending in Problem 8.3.

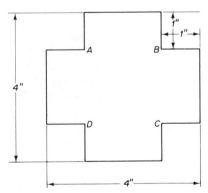

FIGURE 4.14.

5

Chassis Layout Techniques

5.0 INTRODUCTION

In the preceding chapter, stock was sheared into *blanks* for the sheet metal elements of the amplifier. All the information relative to hole size, hole position, bend lines, and so on, resulting from the planning stage will now be transferred from the chassis layout drawings to the work. Techniques and necessary tools for prototype development are considered in this chapter and attention is given to methods for producing accurate layout. Selecting and using the proper tools is stressed to guide the technician in developing greater skills to minimize the inherent errors associated with this type of work.

Specific information is provided in this chapter on *layout dye*, *layout tables*, and procedures using the *scribe* and *scale*. Also included are discussions on the proper use of the *hermaphrodite caliper*, *combination square*, *punches*, and *dividers*. The progressive layout of the amplifier's three sheet metal elements are shown to provide a better understanding of layout tool operation.

5.1 LAYOUT DYE

Layout dye is a convenient aid in sheet metal layout work. Commercial dyes are generally *blue* or *yellow*, the selection depending on the application. Blue dye is preferable for chassis layouts since it provides more contrast with the light metal surface. Layout dyes are fast-drying and are easily removed. For this reason, care must be exercised to prevent oil or solvents from contacting the surface during the layout operations.

The surface to be dyed must be free from dirt and oil in order for the dye to adhere. Any common detergent will serve as a cleaning agent. A light coat of dye can be easily applied with a small nylon brush or cotton swab. The dye should be applied to the side of the metal that will *not* be visible after the metal is bent to the desired form. After approximately 10 minutes' drying time, the surface is ready for working. Light pressure applied with a pointed object on the dye results in a smooth, clean line where the dye is removed. The resulting contrast between the dark color of the dye and the exposed underlying metal facilitates rapid and accurate location and identification of layout marks for punching, drilling, cutting, and bending operations. If the applied pressure is excessive, the surface will be gouged. Enough pressure must be applied, however, to ensure the removal of a suitable amount of dye to achieve the desired contrast.

Removal of the dye, after the chassis fabrication operations have been completed, requires the application of a suitable solvent, such as wood alcohol, and light rubbing with a cloth or cotton swab.

5.2 LAYOUT TABLE

For precision chassis layout work, a *layout table* will aid in producing good-quality results. Layout tables are commonly made of cast iron and are generally referred to as *bench plates* or *surface plates*. These plates are machined on the top and sides with the edges perpendicular to the top surface and at right angles to each other. This level surface can serve as an excellent means of testing for flatness. Any rocking motion of a surface placed on the bench plate indicates a variance from true flatness.

Figure 5.1 shows a typical layout table. In scribing layout lines on a chassis blank, the following procedure is followed. The blank is placed in one of the corners of the layout table with two of its adjacent edges exactly flush with two adjacent edges of the table. *Parallel clamps* firmly secure the metal to the table. Excessive clamp pressure should be avoided to prevent damage to the work. Usually, finger-tight pressure is sufficient. It is often necessary to reposition the clamps when they obstruct the layout operations. With the metal thus properly positioned, the edges of the layout table serve as a convenient reference for measuring, tool positioning, and alignment.

FIGURE 5.1. Dyed sheet metal clamped to layout table.

5.3 SCRIBE AND SCALE

The *scribe* serves as the instrument to mark construction lines on the metal surface. The scribe, as shown in Fig. 5.2, is a sharp-pointed implement that makes a fine groove in the dye covering the metal. The handle is knurled to reduce the possibility of slipping. The tip has a taper of approximately 15 degrees, resulting in an extremely fine point, to facilitate a high degree of accuracy in scribing lines. Many scribes have *tungsten carbide* tips and seldom require sharpening. Tips made from hardened tool steel need to be sharpened occasionally. Care must be exercised in this sharpening process to maintain the original taper over the entire length of the tip to avoid *rounding*. Rounding prevents the tip from getting as close as possible to the scale edge when laying out measurements.

Dimensional information from the chassis layout drawings is transferred to the blank stock with the aid of a scale. The end of the scale should never be used as a measuring reference unless it is established that the dimension from each end to the first major graduation is accurate. It is common practice to use one of the major numbered graduations as the reference. This procedure is shown in Fig. 5.3. A small *light* scratch is made on the metal with the point of the scribe to the desired measurement on the scale. To position a hole, *two* perpendicular dimensions transferred from the drawing must be constructed on the work. The point of intersection of these two lines represents the center of the hole.

FIGURE 5.2. Layout scribe.

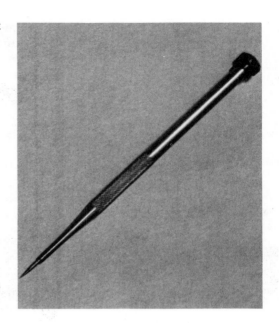

FIGURE 5.3. Transferring dimensional information to the sheet metal with scale and scribe.

To minimize errors, all measurements should be made from a reference edge of the metal. Measuring from a previously marked point could compound errors. It may be necessary, then, to add or subtract measurements from the drawings before transferring them to the metal.

When scribing lines such as bend lines along the entire length of the metal, two *location marks*, spaced as far apart as possible, are first made. The edge of the scale is then placed along these marks. The scribe, with the point along the scale edge and tilted for close contact with this edge, is drawn *lightly* across the metal. The resultant line drawn on the dye should completely include or cover the location marks to ensure accuracy. If a correction has to be made, a new line must be drawn. To avoid later confusion brought about by a double line, the incorrect line should be blotted out with dye.

5.4 HERMAPHRODITE CALIPERS

A simple way of producing scribed lines parallel to the edge of the metal blank is to use the *hermaphrodite caliper*. This instrument is shown in Fig. 5.4. It consists of an *adjusting knob* and *two legs*, one having a *scribe point* and the other a *hooked end* for alignment along the metal edge. These legs are joined at the top with a *spring crown* that provides tension. To set the desired width of scribe, a scale is first placed on the layout table. The scribe tip of the caliper is then set on a convenient reference graduation (Fig. 5.5). The adjusting knob is rotated until the face of the hooked end exactly aligns with the desired measurement of the scale. A scribed line is made on the dye by holding the hooked end perpendicular to the edge of the blank and resting it on the surface of the layout table. To avoid error in measurement, the top of the hooked end should not drop below the top surface of the work when the bottom edge of the hook is resting on the layout table. This problem arises when working with stock exceeding $1/16$ inch in thickness. This type of measurement error is shown in Fig. 5.6. Notice how the tip is made to follow toward the outer edge of the work. This problem can be overcome by first setting the caliper to the desired dimension and then placing the top of the hook even with the top surface. A small location mark is made in the work surface. The hooked end is then lowered to the surface of the work table and the adjusting knob is reset to allow the scribe tip to return to the location mark. Although the width of the caliper spread is now larger than the original measurement, the scribed line will be in the correct location on the work.

Care must be taken to maintain the correct position when using the caliper and also to apply the proper amount of pressure. The caliper should be tilted at approximately 45 degrees toward the operator (Fig. 5.7). While maintaining this 45-degree angle, a constant downward pressure is applied with the index finger to the scribe leg as the caliper is drawn across the work from the farthest end. Although the hermaphrodite

FIGURE 5.4. Hermaphrodite calipers for sheet metal layout.

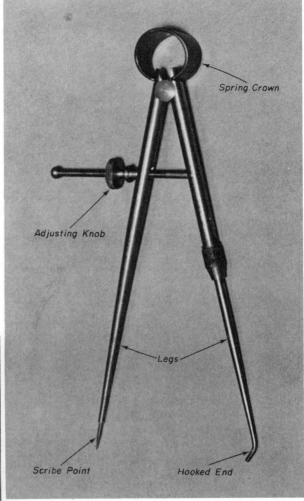

Spring Crown

Adjusting Knob

Legs

Scribe Point

Hooked End

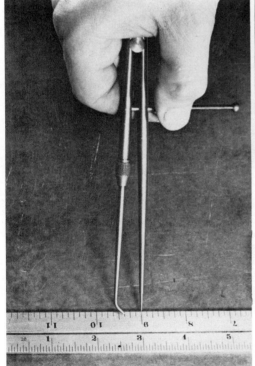

FIGURE 5.5. Setting the hermaphrodite caliper.

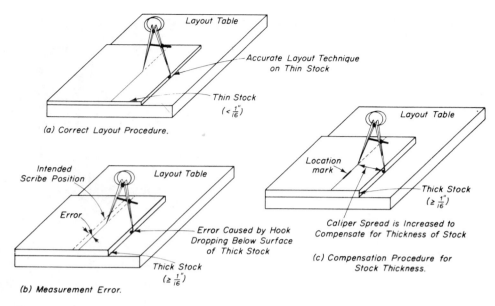

FIGURE 5.6. Layout procedures for hermaphrodite caliper.

FIGURE 5.7. Correct pressure and position are important when using the hermaphrodite caliper.

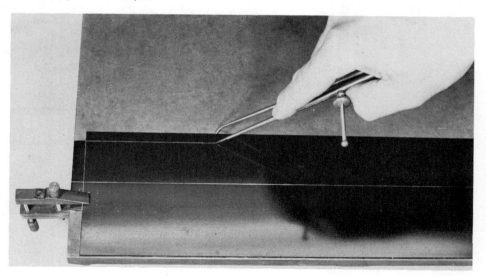

caliper legs are rigidly positioned, lateral pressure can cause the tip distance to change as the legs tend to flex. If an error in marking is made, it can be blotted out with dye and another line drawn. An effort should be made to draw the scribe tip across the work only once. Otherwise, a double line could result, inviting error. Since alignment and pressure must be maintained simultaneously for optimum results, experience will provide the necessary "feel" for this instrument.

5.5 THE COMBINATION SQUARE

The combination square, shown in Fig. 5.8, is an extremely useful and versatile tool in layout work. It consists of four parts: the *grooved blade*, a *squaring* and *mitre head*, a *bevel protractor head*, and a *centering head*. The blade is a steel rule normally graduated on all four edges, providing two $1/16$-inch scales, one $1/8$-inch scale, and one $1/32$-inch scale. This blade is grooved to accept a *binding pin* operated by a *thumb screw* associated with each of the three removable heads. This binding pin clamps the blade in place. To mark a 90-degree line with respect to a reference edge of the work, the blade is inserted into the squaring head. With the thumb screw loosened, sufficient length of blade is exposed to traverse the intended layout. After securing the blade, the 90-degree inside face of the squaring head is placed along a reference edge of the layout table and blank, as shown in Fig. 5.9. With the head and blade in this position, a

FIGURE 5.8. Combination square. Courtesy of The L.S. Starrett Company, Athol, Massachusetts.

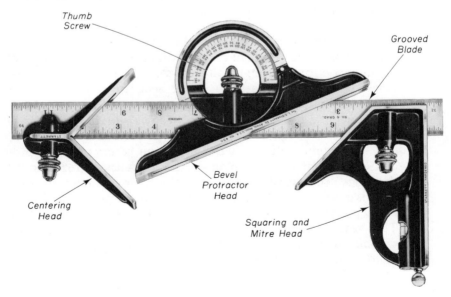

Thumb Screw

Grooved Blade

Centering Head

Bevel Protractor Head

Squaring and Mitre Head

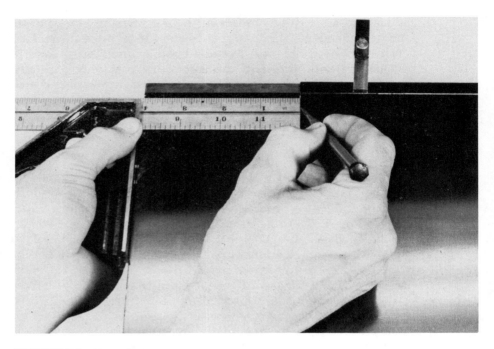

FIGURE 5.9. Use of the squaring head allows for rapid transfer of dimensional information.

scribed line along either edge of the blade will be perpendicular to the reference edge. A line parallel to the reference edge can be drawn by firmly holding a scribe on the work against the *end* of the blade while moving the head along the reference edge. A caution should be pointed out. As mentioned previously, the end of a scale may not be accurate for dimensioning. If checked and found to be accurate, direct positioning of the blade for scribing parallel lines can be made using the scale's graduations. If, however, the scale is not accurate at the end, a location mark to the desired dimension is first constructed. The end of the rule is then positioned to this mark and the head is slid to the reference edge and secured.

The combination square, using the squaring head, provides a rapid and convenient means of transferring dimensions for hole locations from drawings. The measurement from the reference edge to the center of the hole is first set on the blade. With the head moved to cover the approximate position of the hole center, a line is scribed against the blade end. The head is then moved to a second reference edge with the correct dimension set between blade and head and another line is scribed perpendicular to the first. The intersection of the scribed lines locates the center of the hole.

To scribe 45-degree angle lines, the 45-degree angle edge of the mitre head is used. Placing the inside face of the head against the reference edge, the blade makes a 45-degree angle with respect to that edge (Fig. 5.10). Lines may be scribed at this angle using either edge of the blade.

For angles other than 45 or 90 degrees, the bevel protractor head is used. This head is adjustable through 180 degrees, calibrated from zero degrees center scale to 90 degrees at either scale end (Fig. 5.11). With the edges of the blade parallel to the face of the protractor head when placed against the reference edge of the blank, the scale reading is zero degrees and is aligned with the index mark on the frame of the head. As the blade is rotated, the angle made may be read directly as it aligns with the index mark.

When the center of a piece of round stock is to be located, the centering head is used. The round stock is held firmly against the back side of the scale and lowered snugly between the arms of the V-shaped centering head (Fig. 5.12). The scale now cuts the center of the stock and a line is scribed. By rotating the stock approximately one-quarter turn, while in the same position, another scribed line is drawn. The intersection of these lines locates the center of the stock.

Before using the combination square with any of the heads, care must be taken to assure that the blade is secured to the head by tightening the thumb screw. Otherwise, angular and scalar errors may occur.

FIGURE 5.10. Mitre head is used to aid in scribing lines at 45-degree angles.

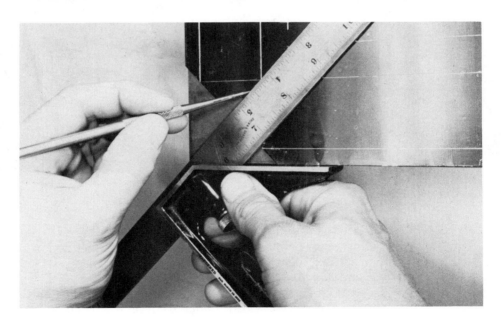

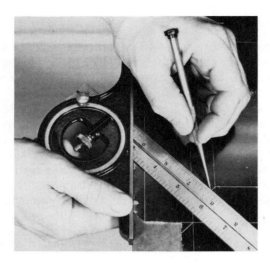

FIGURE 5.11. Bevel protractor head and blade.

FIGURE 5.12. Locating the center of round stock with the aid of a centering head. Courtesy of The L. S. Starrett Company, Athol, Massachusetts.

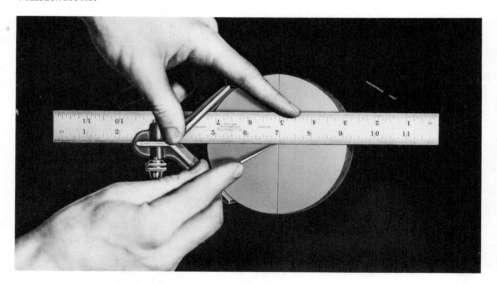

5.6 PUNCHES

To facilitate the drilling process, as discussed in the next chapter, a conical indentation must be made in the metal at the intersection of the location marks of all hole centers. These indentations are made with the use of a *hammer* and *punch*.

Punches are classified as *prick* and *center*, depending on the angles of the point (Fig. 5.13). Both punches have hardened steel points and knurled handles for ease in gripping. The prick punch has a point ground

FIGURE 5.13.
Punches.

Prick   Center

to 30 degrees and it serves to introduce a small, round indentation primarily for location purposes. The center punch has a point ground to 90 degrees. Its function is to provide a larger-diameter indentation in the metal at the proper angle to accept the narrowest part of the drill bit web for alignment or centering. This indentation also prevents the drill bit from "walking" from the center position of the hole during the drilling operation.

With the punch held perpendicular to the work and the point at the center of the intersecting lines, it is *lightly* tapped with a hammer (Fig. 5.14). An 8-ounce ball peen hammer is sufficient for use in punching. It is important to place the work on a hard, flat surface for complete support during the punching operation. Use of a heavy hammer or excessive striking force will damage the work as well as the punch. It is advisable to test for the correct striking force on a sample piece of stock, applying only one strike with the hammer. Any error in the location of the indentation will result in a corresponding error in the location of the hole. To correct an off-center error in punching, the tip of the center

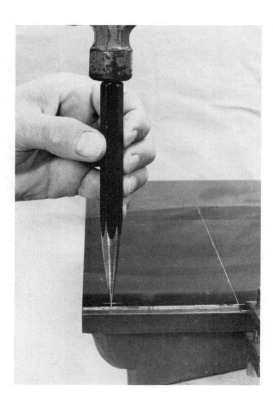

FIGURE 5.14. Correct perpendicular positioning of prick punch.

punch is placed in the indentation at an angle such that when struck the hammer will tend to drive the point *toward* the intended intersection. Figure 5.15 illustrates this technique.

Automatic punches are available with prick or center punch tips (Fig. 5.16). These punches eliminate the need for a hammer because they contain an adjustable spring-loaded driving mechanism in the handle that forces the tip into the work when pressure is applied to the handle. The amount of striking force required for the proper indentation is determined by the type and hardness of the metal used. The pressure is adjusted by turning the upper portion of the punch clockwise to compress the spring for increased pressure or counterclockwise to expand the spring for reduced pressure. To use the automatic punch, it is positioned perpendicular to the work surface with the tip exactly at the point of intersection of the location marks. A steady downward pressure is applied to the handle until the spring is released, driving the punch tip into the work. If the indentation is made slightly off center, the punch can be tilted in the original mark toward the center point and activated as many times as necessary to correct the error. As with most hand tools, practice on sample stock with the automatic punch will help in selecting the optimum setting.

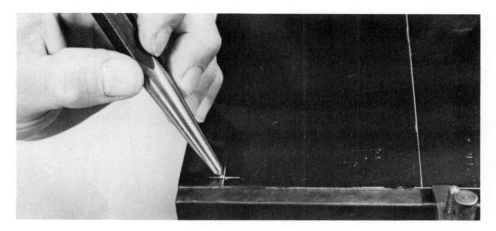

FIGURE 5.15. Error correction center-punching technique.

FIGURE 5.16. Automatic punch. Courtesy of The L. S. Starrett Company, Athol, Massachusetts.

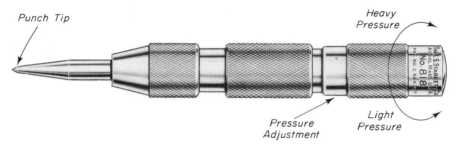

5.7 DIVIDERS

Dividers serve several purposes in layout work. In addition to locating centers of evenly spaced holes and dividing lengths into equally divided parts, they also are used to scribe circles about hole location marks to ensure accuracy of drilling or chassis-punching operations. The common divider, as shown in Fig. 5.17, consists of *two legs* with scribe points, *spring crown*, and *adjusting knob*. The adjusting knob is used to set the divider in a way similar to the way the calipers were set in Chapter 3.

To set the divider for scribing a circle, one tip is placed at a convenient graduation mark on a scale, such as the 1-inch mark. The adjusting knob is then rotated until the other tip is on the graduation mark corresponding to the radius of the circle. To position the divider for use, one tip is placed in the indentation previously made in the punching operation with the other tip contacting the work surface. While main-

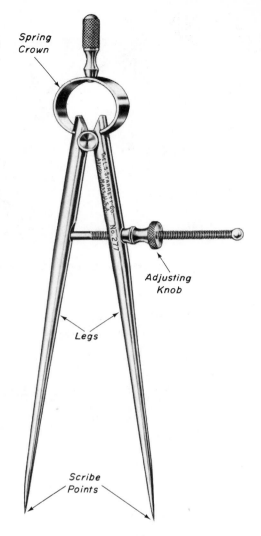

FIGURE 5.17. Dividers. Courtesy of The L. S. Starrett Company, Athol, Massachusetts.

Spring Crown

THE L.S. STARRETT CO. ATHOL, MASS. U.S.A. No. 277

Adjusting Knob

Legs

Scribe Points

taining the first divider tip firmly in place, the other is swung around the pivot point inscribing an outline of the circumference of the hole in the dye. The divider should be positioned so that the spring crown is perpendicular to the work surface as the divider is rotated (Fig. 5.18a). If a high degree of drilling accuracy is required, two proof hole outlines are scribed about the center point. The inner circle has the diameter of the pilot drill and the outer circle has the diameter of the undercut drill. (See Chapter 6 for a description and application of these drills.) Indentations with a prick punch are made at every 45 degrees around the circumference of the undercut outline. If the center punching and drilling are accurately done, one-half of these indentations should be perceptible around the circumference of the hole after the undercut drilling operation is completed. Figure 5.18b illustrates this technique.

(a)

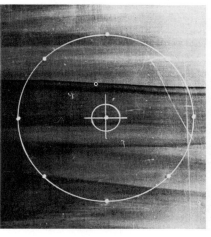

(b)

FIGURE 5.18. Layout for drilling and punching. (a) Divider position for scribing circle. (b) Proof circle.

To divide an uneven dimension into several equal parts, the divider is used with the technique illustrated in the solution to the following problem:

It is necessary to position five holes along the front panel of the amplifier, the total length of which is 14 inches. The spacing between each hole and the distance from each end of the panel must be equal.

Solution: A sheet of quadruled paper is placed lengthwise and a point is marked on a grid intercept close to the bottom and halfway between either side (Fig. 5.19). (Although 17- by 22-inch paper with four divisions to the inch is used for this solution, the size and divisions per inch are not important as long as linear graph paper is used.)

At a point approximately 3 inches directly above the first reference point, another point is marked. Several points are then marked at 1-inch intervals both to the left and to the right of the second point on the same horizontal grid line. For this problem, seven points, including the center point, were made to account for the six spaces necessary. Straight lines are now drawn through each of the seven points and intersecting at the first point at the lower edge of the paper. It can now be seen that each horizontal grid line above the first reference point represents a division of six equal parts, progressively increasing with each vertical increment. To determine the correct distance between hole centers for this problem, the edge of a scale is placed over the construction line on the extreme left. With the scale aligned with the horizontal grid lines and maintaining the reference point on the scale edge, the scale is moved up until the 14-inch mark on the scale intercepts the construction line on the extreme right. (Refer to Fig. 5.19.) Dividers can now

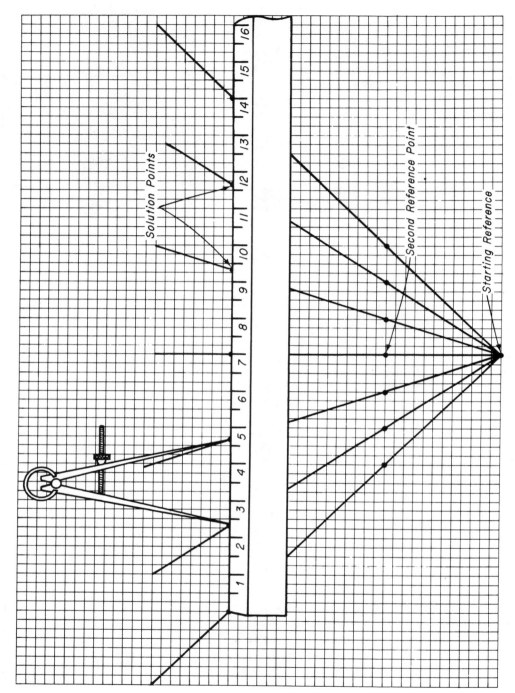

FIGURE 5.19. Construction technique for equal spacing of location marks.

be set to one of the equal horizontal dimensions on the edge of the scale and this dimension can be transferred to the metal. A scribed center line on the work is necessary to locate the height of the position of the holes. The dimension set on the divider can now be transferred along this center line. One scribe tip is brought gently to the edge of the panel where the center line runs off the surface. Care must be exercised that this tip does not drop below the face of the panel; otherwise, an error in layout will result. With the first scribe tip held carefully in this position, the other scribe tip is brought down on the center line and a small arc is drawn on the dye. The same scribe tip is maintained at the intersection of that arc and the center line and the divider is gently rotated until the other scribe tip is brought along the center line to cut another arc. By such successive 180-degree rotations of the divider, the required number of hole centers will be accurately located (Fig. 5.20).

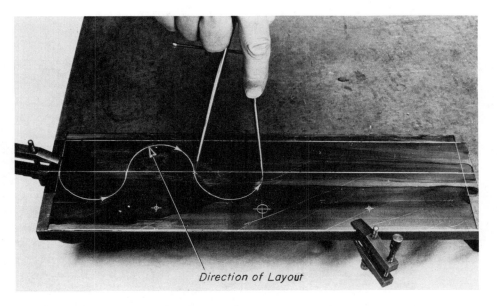

Direction of Layout

FIGURE 5.20. Successive 180-degree rotations of the divider to achieve accurate hole-spacing layout.

The three amplifier chassis elements, shown in Fig. 5.21, have been scribed with all layout lines in accordance with the procedures outlined in this chapter. The following chapter provides detailed information concerning the proper selection and use of drills, bits, and chassis punches.

FIGURE 5.21. Three amplifier chassis elements with layout completed.

PROBLEM SET

5.1 Using the layout techniques discussed in this chapter and quadruled graph paper, locate six evenly spaced hole locations across 11 inches.

5.2 Scribe two bend lines on the piece of stock cut for a soldering iron stand in Problem 4.4 at 2 inches from either end and perpendicular to the length of the stand. Bending is accomplished in Problem 8.2.

5.3 Lay out the bend lines on the stock cut for the heat sink in Problem 4.5. The bend lines are positioned from A to B, B to C, C to D, and D to A. This type of heat sink is suitable for device case styles such as DO-4, DO-5, TO-3, TO-61, and TO-68. From Appendix III, determine the hole spacing required for the TO-3 style and locate and center punch this hole location pattern on the center section of the heat sink. Drilling and deburring is performed in Problem 6.2 and bending in Problem 8.3.

5.4 Shear a $3\frac{1}{2}$- by 8-inch piece of 16-gauge aluminum to lay out on oil pressure, battery, and temperature gauge panel shown in Fig. 5.22. The gauge hole locations are to be evenly spaced across the panel. Drilling and deburring is completed in Problem 6.1, hole cutting and bandsaw operations are performed in Problem 7.4, bending is done in Problem 8.1, and finishing in Problem 9.3.

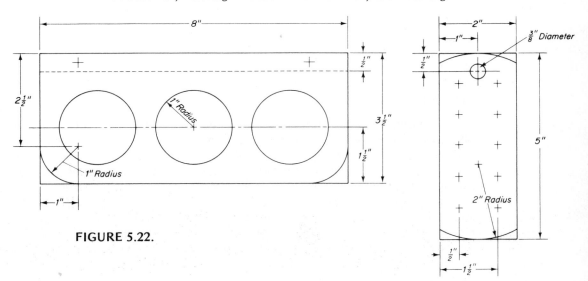

FIGURE 5.22.

FIGURE 5.23.

5.5 Lay out a 2- by 5-inch piece of 14-gauge steel as shown in Fig. 5.23 for a drill gauge. Using the layout techniques discussed in this chapter, center-punch five pairs of evenly spaced hole locations along the length of the gauge in addition to the $\frac{3}{8}$-inch hole position. Drilling and deburring of the gauge is done in Problem 6.3; radius cutting, punching, and filing is performed in Problem 7.3; and number stamping is completed in Problem 9.1.

Drilling, Reaming, and Punching

6.0 INTRODUCTION

In preceding chapters we have discussed the detailed techniques for preparing sheet metal for chassis construction. All the layout work, including the accurate positioning of all required holes, has been located on the metal blanks. Cutting operations can now begin. This chapter presents the basic cutting operation techniques of *drilling*, *reaming*, and *punching*.

In this chapter drill bit design is discussed, including *spiral flutes* for twist drills, *straight* and *taper* shanks, *standard drill sizes*, and *drill gauges*. Drill point configurations and sharpening techniques are also presented. To implement the drill bit, the *drill press*, together with its associated hardware and general operating instructions, is considered with respect to obtaining accurate hole size for common sheet metal applications. In addition, the proper lubrication for drilling specific metals is tabulated.

Deburring tools also are discussed, in addition to the techniques of reaming, including the use, advantages, and disadvantages of the *hand-expansion reamer*.

Practical operational information on punches, including the *hand*, *draw*, and *turret* punches is presented. Techniques for using *square*, *rectangular*, and extremely large *circular hole* punches are discussed and punching and drilling are compared.

6.1 BASIC TWIST-DRILL CHARACTERISTICS

Drill bits are manufactured of *carbon* or *tungsten–molybdenum* steel rods. The carbon steel bits are made for drilling into soft metals such as aluminum, copper, brass, or mild-tempered steel. These bits become dull after a relatively short period of use. Tungsten–molybdenum bits, commonly referred to as *high-speed* (HS) bits, are used for drilling either soft metals or high-tempered steel, stainless steel, and other hard metals. High-speed bits will produce more satisfactory results than carbon bits when used on fiberglass, Bakelite, or other plastic materials. Some plastics, especially glass-cloth laminates, tend to dull carbon bits very quickly.

A drill bit is manufactured by milling two *spiral grooves* called *flutes* into the rod stock. The flutes are positioned directly opposite each other, beginning at the *point* and extending along the entire *body length* of the bit. The area between the flutes is called the *web*. Figure 6.1 identifies the basic parts of the twist drill.

Along the outer surface of the *land* generated when the flutes are milled, a *margin* is ground to provide clearance on the cutting edge and also to ream and maintain drill alignment. When the point is ground to the correct angle, as described in Sec. 6.2, a *chisel edge* is formed at dead center in addition to two *cutting lips* at the beginning of each flute.

Twist drills, as shown in Fig. 6.2a, are available in *short lengths* for center-hole or starting purposes, *jobber lengths* for common drilling applications, and *extended lengths* for deep-hole drilling. Twist drills are also available in a variety of shank configurations, the most common of which are the *straight* and *tapered* shanks. These shanks, shown in Fig. 6.2b, are extensively used in sheet metal and panel stock drilling applications in electronic packaging. Smaller-diameter drills are made with a straight shank since they are accommodated by the conventional drill press or powered hand-drill chucks. Although more expensive for the smaller sizes (less than ½ inch), the taper shank has several advantages. It is more convenient to insert the bit by pressing it into a *tapered collet* on a drill press or lathe. In addition, bits can be removed and replaced much more quickly from a collet than can straight-shank bits used with a key-type chuck. The prices of comparable size drills of each type in the larger sizes are equivalent.

FIGURE 6.1. Twist-drill nomenclature.

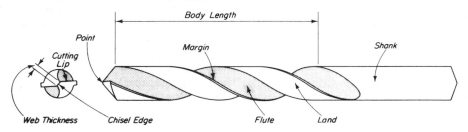

FIGURE 6.2. Twist-drill styles.

(a) Standard length comparison for twist drill bits.
   (1) Short, (2) Jobber, and (3) Extended lengths.

(b) Straight and tapered shank twist drills.

Three size systems are used for drill bits: *fractional*, *numerical*, and *letter*. The three lengths previously described with straight or tapered shanks are available in each system.

Fractional bits range from $\frac{1}{64}$ (0.0156) inch to $\frac{1}{2}$ (0.500) inch in diameter in increments of $\frac{1}{64}$ inch. A complete set of drills in this system would therefore include 32 bits.

The *numerical* system consists of 80 bits ranging from number *80*, whose decimal equivalent size is 0.0135 inch in diameter, to a number *1* bit, 0.228 inch in diameter.

Letter designation of bits starts with *A* size (0.234 inch in diameter) through *Z* size (0.413 inch in diameter). For a composite tabulation of the standard drill size systems, showing the fractional, number, and letter designations along with their decimal equivalents, refer to Appendix IV.

The larger drill bits have their size stamped on the shank in accordance with the particular system used. For the smaller bits, *drill gauges*, such as the one shown in Fig. 6.3a, must be used to determine the size because the stamping would be too small and therefore difficult to read. Each hole in the drill gauge is designated by the size bit that will fit snugly into it. Drill sizes are checked by passing the flute portion of the bit into the gauge hole. The shank end is never used for measuring since any burr or wear incurred in its use will result in an incorrect size determination. Drill gauges are also available for the larger bit sizes to determine diameters when the stamping on the shank has worn from extensive use. Larger-fractional-size drill gauges are shown in Fig. 6.3b and c. Drill bit diameters may also be measured with a micrometer. It is important, however, that this measurement be taken across the margins. This procedure is shown in Fig. 6.3d.

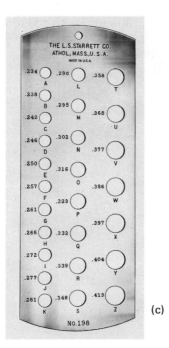

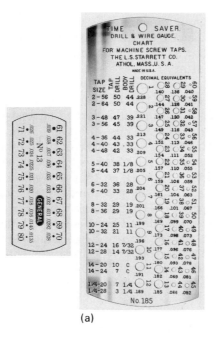

(a)

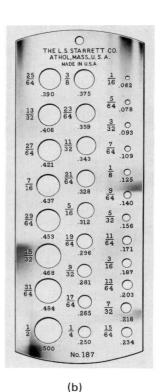

(b)

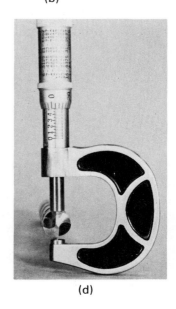

(c)

(d)

FIGURE 6.3. Tools for drill size measurement. Courtesy of General Tool Company and The L. S. Starrett Company, Athol, Massachusetts.

6.2 DRILL BIT POINTS

The points of twist drills have several features that must be examined in order to completely understand the cutting phenomenon. Figure 6.4a identifies the drill point nomenclature.

When drilling, the bit is introduced into a center-punched hole, and the chisel edge at dead center of the point is the first cutting edge to come in contact with the work. As the rotating bit is forced against the work, the cutting lips begin to chip metal away. These chips are channeled from the work through the flutes. The flutes also allow cutting lubricants to be applied directly to the work area at the chisel edge. Once the drill bit has penetrated deeply enough into the work so that the entire length of the cutting lips are making contact, the sharp leading edge of the margin will begin to ream the inside of the hole to the final dimension. Notice in Fig. 6.4b that the body clearance is achieved by the margin being *slightly raised* above the surface of the land. Lip and body clearance are essential so that the bit will not jam or bind in the work during the drilling operation. Once the point has passed through the work, the bit should be slightly raised and lowered before removing it so that the hole formed will be completely reamed.

The degree of drill point angle depends on the type of material being worked. In general, the harder the material, the flatter the point. For general-purpose applications, such as drilling aluminum and soft steels, a drill point angle of 118 degrees, a lip clearance of 12 to 15 degrees, and a standard chisel edge angle of 120 to 135 degrees is recommended. When drilling materials such as Bakelite or epoxy-glass (used extensively in printed circuit work), a steep drill point angle of 90 degrees and a lip clearance of 12 degrees is recommended.

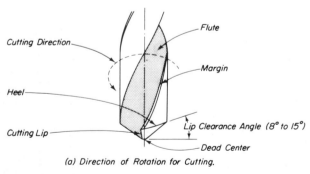

(a) Direction of Rotation for Cutting.

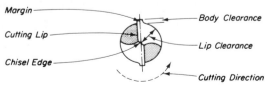

(b) Cutting Edges and Clearances.

FIGURE 6.4. Drill point nomenclature.

The point configurations described above result in reasonably round holes, especially in the smaller sizes. Many times, however, chassis construction requires the drilling of larger holes (greater than $\frac{3}{8}$ inch). Because aluminum is a soft metal, conventional drills tend to cut a hole that is out of round, especially in the larger sizes. The *sheet metal bit*, as shown in Fig. 6.5a, reduces this problem with its *double-spurred* cutting lips. The point or chisel edge at the center of the web is used to align the bit in the center-punched hole. As the bit is forced against the work, the sharpened spurs on each side of the lands cut into the material in a circular fashion. Continued pressure on the bit causes the spurs to completely cut through the metal, leaving a slug the diameter of the hole. The margin then reams the hole to the desired dimension.

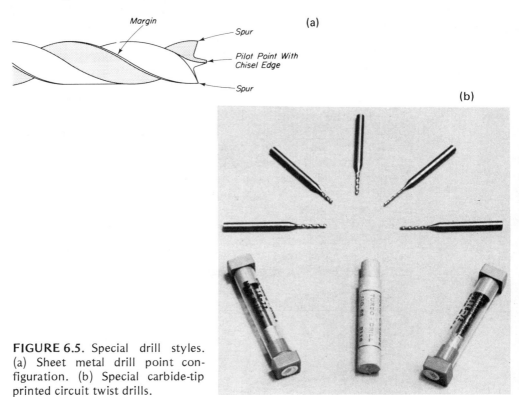

FIGURE 6.5. Special drill styles. (a) Sheet metal drill point configuration. (b) Special carbide-tip printed circuit twist drills.

For drilling printed circuit boards, a specially constructed drill bit is available. This bit, shown in Fig. 6.5b, consists of a short length of twist bit with an enlarged shank of approximately $\frac{1}{8}$ inch. Since common sizes of bits used in printed circuit work are of the order of No. 55 to No. 67, difficulty may be encountered when attempting to rigidly secure and align these small shanks into a conventional chuck. The enlarged shank on the printed circuit drill bit overcomes this problem. However, the shorter length bit increases the possibility of breaking. This problem is overcome by using straight-shank bits. The printed circuit bit, used for drilling through epoxy-resin boards, is usually made of carbide. Although carbide is a brittle material, it does not dull so quickly as HS drills. Extreme care must be exercised when using small-diameter drills because of their delicate structure.

6.3 SHARPENING TWIST DRILLS

Carbon steel bits frequently need sharpening, especially when the work is hurried and the point overheats. Heating removes the temper and hardness from the bit. When this happens, the metal takes on a *blue color*, and continued use of the bit, without sharpening, will cause an irregularly shaped hole.

Experience plays an important part in developing the proper sharpening technique. An experienced machinist can sharpen a drill on a grinding wheel by "eye" but for the beginning technician, guides will be established to aid in this operation.

When sharpening twist drills, four equally important drill point characteristics, as shown in Fig. 6.6, must be achieved. These are *equal-length cutting lips*, *equal-* and *proper-size drill angles*, *proper cutting lip clearance*, and *correct chisel edge angle*. It is important to first test the face of the grinder wheel to see that it is square with the sides, makes a true circle, and is free of contaminants. The tool rest, shown in Fig. 6.7, should be positioned to within $^1/_{16}$ inch of the wheel. This distance will ensure that the bit will not become wedged between the tool rest and the face of the wheel. *Safety glasses must be worn when using the grinder.* In this example exercise, shown in Fig. 6.7, a 118-degree drill point angle with a 12-degree lip clearance will be ground. The procedure outlined, of course, can be adapted to any required grinding angles.

FIGURE 6.6. Important drill point characteristics for sharpening.

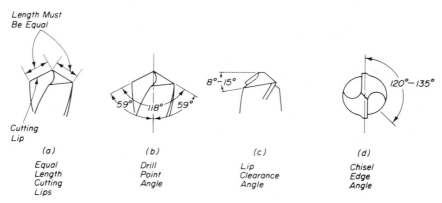

(a)	(b)	(c)	(d)
Equal Length Cutting Lips	Drill Point Angle	Lip Clearance Angle	Chisel Edge Angle

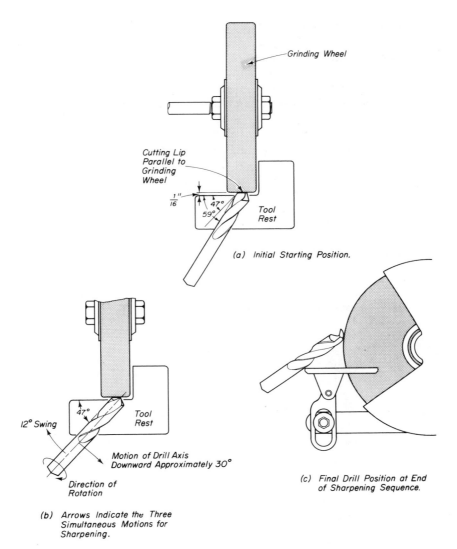

FIGURE 6.7. Drill point sharpening sequence.

Initially, two marks are made on the tool rest to the face of the wheel, one at 47 degrees and one at 59 degrees. (These marks should be made in pencil since other angles may be desired and scribe lines could cause confusion.) The 59 degree mark represents one-half of the required drill point angle of 118 degrees. Before starting the grinding operation, the grinder motor should be allowed to come up to full speed. In the starting position the cutting lip is held approximately parallel to the face of the wheel and the axis of the drill is aligned with the 59-degree mark on the tool rest (Fig. 6.7). Bringing the entire cutting lip lightly against the wheel entails *three simultaneous motions*. The shank end must be smoothly moved to the left by 12 degrees (to the 47-degree mark) and downward approximately 30 degrees (raising the point, resulting in the desired lip clearance angle) while being rotated clockwise until the trailing edge of the cutting lip is reached. This procedure is repeated for the second cutting lip. The result is checked with a *drill-grinding gauge*, as shown in Fig. 6.8. If the proper motions have been executed, the four essential characteristics, as shown in Fig. 6.6, will be achieved. If the gauge indicates otherwise, the grinding procedure is repeated.

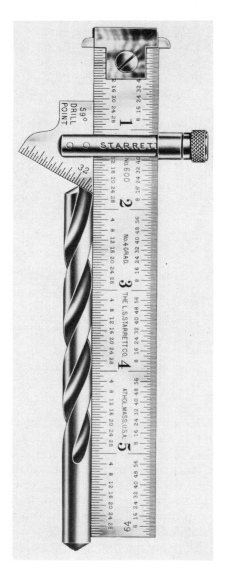

FIGURE 6.8. Drill point angle grinding gauge. Courtesy of The L. S. Starrett Company, Athol, Massachusetts.

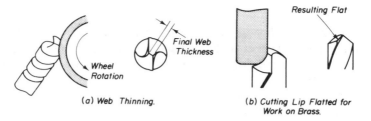

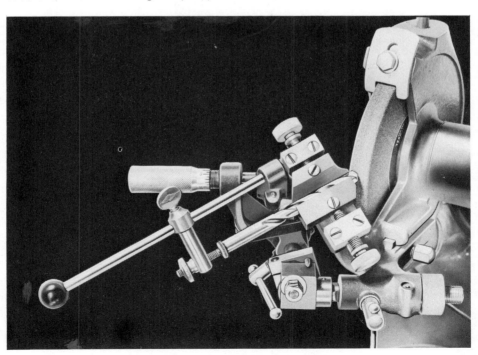

FIGURE 6.9. Specialized point grinding.

The web of the drill must not be allowed to become too thick or the chisel edge will become too large, creating an inefficient cut. To thin the web, the drill is held gently against a round-faced wheel. This technique is shown in Fig. 6.9a. If thinned properly, the wheel will not reduce the angle of the cutting lip enough to affect drill performance.

When drilling brass, more of a *scraping action* is required for a smooth cut. A specially ground drill must be prepared for this application. By holding the cutting lips against the right-hand side of the wheel, a small *flat* in line with the axis of the drill bit is ground (Fig. 6.9b). This greatly reduces the cutting lip angle and results in the desired scraping action. Drill bits that have not been prepared in this way will tend to chatter and bind in brass, especially when drilling through pilot holes.

If difficulty is encountered in sharpening drill points by the freehand method just described, *tool grinders*, as shown in Fig. 6.10, are available.

FIGURE 6.10. Tool grinder attachment for sharpening drills. Courtesy of Rockwell Manufacturing Company, Power Tool Division.

The tool grinder is secured in front of the wheel to the lower portion of the wheel guard and allows the bit to be clamped into the correct position as well as providing adjustments for precision settings for various angles and clearance.

Care must be taken during all grinding operations not to overheat the drill point. Once the temper has been drawn out, the bit will be unable to maintain its sharp edge for any reasonable period of time. To avoid overheating, *very light* pressure for short periods of time is applied to the grinding wheel. Dipping the point in water during the grinding operation should be avoided because water adversely affects the metal.

6.4 DRILL PRESS

One of the most useful power tools employed in electronic fabrication is the *drill press*. A bench-type drill press is shown in Fig. 6.11. Mounted on the vertical *column* is a horizontal *base table* to support and secure the press to the workbench. The base table may also be used as a supporting

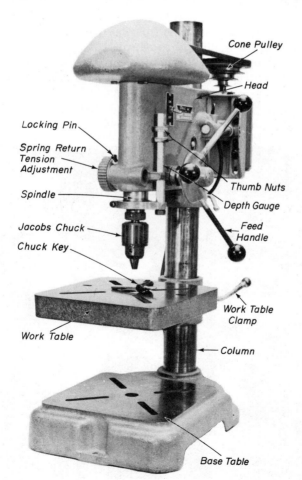

FIGURE 6.11. Bench-type drill press.

surface when unusually long pieces are to be drilled. Uppermost on the column is the *head*. Between the base table and the head, a movable *work-table* is positioned on the column and held in place by a *lever-operated clamp*. The work to be drilled is first positioned on the worktable, which is moved vertically along the column by first loosening the lever clamp. When the table is conveniently positioned and the hole in the center of the table is aligned with the *spindle* axis, the clamp is tightened, rigidly securing the table.

Most standard drill press heads are equipped with a three-jaw *Jacobs chuck*, accommodating straight shank drill bits up to ½ inch. The chuck jaws are opened by a counterclockwise rotation of the *chuck key* inserted into the chuck. When the shank of the drill bit is inserted between the jaws of the chuck, clockwise rotation of the key will securely clamp the jaws onto the drill (Fig. 6.12a).

FIGURE 6.12. Drill press chuck and tapered collet. (a) Securing straight-shank drill in Jacobs chuck. (b) Removing tapered-shank drill with a drift key.

(a)   (b)

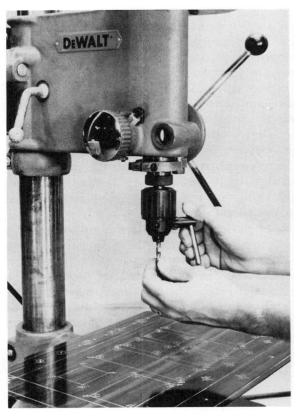 

Drill presses may also come equipped with *tapered collets* to accommodate the tapered shank drill bit. A *drift key* is used to remove the tapered shank drill from the collet (Fig. 6.12b).

Chucks for both straight and tapered shanks are secured to the lower threaded end of the spindle. The drilling pressure is applied to the spindle by pulling the *feed handle* toward the operator. This action provides the *downward* motion of the normally stationary spindle and revolving chuck toward the work. The feed handle is coupled to the spindle with a *spindle return spring* that causes the spindle to return to its normal position after the downward force is removed. This upward motion should be guided with the feed handle to prevent needless jarring of the spindle and head assemblies. When the spindle does not automatically return to its normal position, *spring return tension adjustment* is necessary. The *locking screw* or *pin* on the left side of the head is loosened and the spring return tension knob is turned counterclockwise to increase tension. Since the knob is under constant spring tension, this adjustment should never be made when the drill press is on or when a drill is in the chuck. Should the knob be inadvertently released with the locking screw loose, the spindle will plunge downward (if the feed handle is not held), resulting in damage to the work, the drill, and possibly to the operator.

When a specified depth of hole is to be drilled into the work rather than a complete hole, a *depth gauge* is provided along the right side of the head. Figure 6.13 shows how the depth gauge limits penetration of the drill below the work. The depth gauge has graduations in $\frac{1}{16}$-inch divisions and is threaded to provide for two *locking thumb nuts*. As the feed handle is turned for downward motion, the depth gauge moves through the stop, which is a machined hole in the head casting. The gauge moves with the spindle. With the thumb nuts positioned on the gauge, downward motion stops when they contact the *depth stop*. This prevents any further downward motion beyond the desired position. This position is set by feeding the spindle downward with the drill in the chuck to the desired drill point depth and then tightening the locking nuts until they contact the depth stop. Two locking nuts ensure that the vibration during drilling operations will not change the adjustment.

The spindle is rotated by a belt-driven pulley assembly powered by a motor mounted to the rear of the head. The motor base is secured to two steel rods that pass through machined holes to the rear of the head casting. The position of the motor base can be adjusted by loosening the *clamping bolts* that hold the rods in place. The belt tension between the *cone pulleys* on the motor and upper end of the spindle should not be excessive in order to avoid needless spindle and motor bearing wear. An acceptable means of testing belt tension is to depress either side of the belt at its midpoint. Proper tension exists if approximately $\frac{1}{2}$ inch of inward deflection is attained. More or less than this optimum deflection

FIGURE 6.13. Depth gauge set for drilling sheet metal to eliminate unnecessary spindle travel.

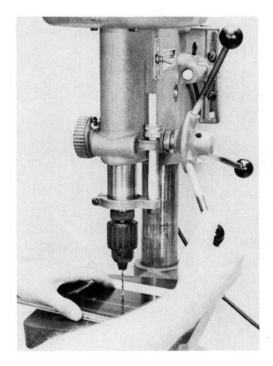

indicates a need for adjustment. Belt tension is increased when the motor is moved away from the head and is decreased when brought toward the head.

The spindle speed can be changed by moving the belt directly opposite to the motor and spindle pulleys. The belt must always operate in a horizontal position. When the belt position on the cone pulleys is to be changed, the motor should be brought toward the head to provide sufficient slack to avoid stretching or damaging the belt. When the largest motor pulley and smallest spindle pulley are used, the fastest spindle speed is obtained, which is approximately 3600 revolutions per minute (rpm). The slowest spindle speed of approximately 400 rpm is obtained using the smallest motor pulley and largest spindle pulley. These combinations are achieved by the motor pulleys and the spindle pulleys being inverted with respect to each other.

6.5 DRILLING OPERATION

All drilling operations begin by aligning and securing the stock to the worktable. *C clamps* provide good stability for drilling metal used in chassis fabrication. This arrangement is shown in Fig. 6.14a. The center hole in the worktable should always be aligned with the spindle axis. In

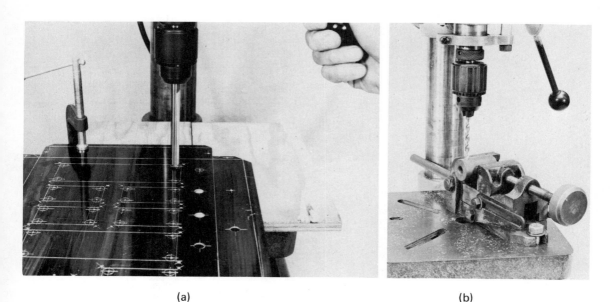

(a)   (b)

FIGURE 6.14. Clamping techniques for drilling.

addition, when drilling sheet metal, a piece of wood of uniform thickness is used between the worktable and the stock. This allows the drill bit to completely pass through the work without contacting the worktable and provides complete support so that burring is minimized as the drill point breaks through the work (see Sec. 6.6). When drilling small parts or components, a *drill press vise* provides the necessary support. This vise is shown in Fig. 6.14b. For holes less than $\frac{1}{4}$ inch, holding the vise manually against the table is usually sufficient. However, for larger holes, the vise should be mechanically secured. Drill press vises are available with a slotted side or end sections that can be bolted to the table through these slots. For a further measure of security against work rotation, a *safety bar*, as shown in Fig. 6.14b, is employed. This bar threads into the front of the vise and must rest against the *left side* of the column.

The correct drill bit size, drill point angle, and lip clearance must be compatible with the material to be drilled. Table 6.1 shows the type of drills with cutting speeds, drill point angles, lip clearance angles, and proper lubricants to use for various materials. The proper drill is secured in the drill press chuck with the drill point serving as reference to align the work. Without securing the work to the table, the spindle is lowered so that the drill point nearly touches the work surface. With the drill point held stationary in this position, the work is adjusted so that the center-punched hole is aligned with the dead center of the drill bit. The power is then turned on and the drill point is allowed to gently touch the center hole and remove a small amount of material. The spindle is then returned to its normal position and the power is turned off. Without moving

TABLE 6.1

Drilling Recommendations for Various Materials

Materials	*Cutting Speed in Feet/Minute for High-speed Drills	Drill Point Angle in Degrees	Lip Clearance Angle in Degrees	Lubricants
Stainless steel	40	136-150	8-12	Sulphur and lard oil
Carbon tool steel	60	136-150	8-12	Sulphur-based oil Mineral lard oil
†Epoxy glass	80	90-110	12-15	Dry
Cold roll steel	90	120-130	12-15	Sulphur-based oil Mineral lard oil
Bakelite	100	60-90	12-15	Dry
Cast iron	120	90-118	12-15	Dry or air jet
Copper	200	90-110	12-15	Kerosene or dry
Hard brass	200	110-125	12-15	Dry, kerosene, or lard oil
Bronze	200	110-125	12-15	Dry or mineral oil
Aluminum	250	90-118	12-15	Kerosene, lard oil, or turpentine

*Drilling speeds for carbon steel drills are approximately one half of those for high speed drills.
†Use carbide tip drill for long tool life.

the work, the test cut is inspected for uniformity about the center lines of the hole. If the alignment is proper, the work is carefully secured to the table. When working with soft metal and twist drills, it is advisable to use a small bit (No. 28 to No. 32) to drill a *pilot hole* through the material to aid in alignment. Also, in order to ensure as round a hole as possible, a drill bit several hundredths of an inch smaller than the final dimension should be used before the final hole is drilled. This progressive *undercutting* will allow the lips to cut rather than the chisel edge and thus avoid the possibility of the drill's "walking." Of course, the use of a sheet metal drill will eliminate the need for undercutting. However, the use of a pilot hole is recommended for either type of drill.

From Table 6.1 the average cutting speed is selected for the given material. (Notice that the maximum cutting speeds for HS drills are approximately twice those for carbon-steel drills.) This information is given in units of *feet per minute* (fpm). These speeds should be considered *maximum* and should not be exceeded in order to minimize wear and not overheat the point. It is therefore important to adjust the spindle speed to the closest available speed without exceeding the maximum as prescribed in Table 6.1. This is done by the proper choice of cone pulleys on the motor and spindle. To convert average cutting speed from feet per minute to rpm, the following relationship is used:

$$\text{rpm} = \frac{12 \times \text{cutting speed (fpm)}}{\pi \times \text{diameter of the drill (inches)}} \approx \frac{4 \times \text{cutting speed}}{\text{diameter}}$$

See Appendix V for a table of drill cutting speeds in feet per minute with the corresponding rate in revolutions per minute.

Continuous lubrication is necessary during operation to minimize friction between the point and the work, thus reducing wear. Table 6.1 lists the recommended cutting compounds and lubricants for several common materials. Lubrication should *not* be used sparingly if the life of the drill is to be prolonged.

The following example illustrates the use of the information in Table 6.1.

Example: It is necessary to drill a $\frac{3}{8}$-inch hole using twist drills through a piece of 16-gauge aluminum. The selected drills are as follows:

Pilot size	No. 31 (0.120 inch)
Undercut size	P (0.323 inch)
Final	$\frac{3}{8}$ (0.375 inch)

From Table 6.1, the necessary information is obtained as follows:

Drill Bit Size	Cutting Speed (HS)	Drill Point Angles	Lip Clearance	Lubrication	RPM	Closest RPM on Drill Press
#31	250 ft/min	118 degrees	12-15 degrees	Kerosene or turpentine	*7959	3600
P	250 ft/min	118 degrees	12-15 degrees	Kerosene or turpentine	2956	2800
3/8	250 ft/min	118 degrees	12-15 degrees	Kerosene or turpentine	2546	2400

*Notice that although this calculated rpm is 7959 and that the highest available speed on the conventional drill press is 3600 rpm, this is acceptable because the maximum recommended cutting speed will not be exceeded.

6.6 DEBURRING

The fabrication of a chassis requires numerous *through* holes in which the drill body passes completely through the stock. In soft metals, such as aluminum, copper, and brass, the result is a rough edge generated on the reverse side of the stock. This *burr* is the result of the flex in the remaining thin metal as the trailing edges or *heels* of the drill push this material aside along the circumference of the hole instead of cutting it away as the chisel edge breaks through the work. The amount of burr can be minimized by using a wooden block as a support. In addition, by lowering the spindle more slowly, the remaining metal will tend to flex *less* and allow the drill to chip more of the remaining material away. This reduces the size of the resulting burr.

Since burrs can prevent flush mounting of components, in addition to marring rubber or plastic hardware, they must be removed. Two common hand deburring tools are the *countersink tool*, shown in Fig. 6.15a, and the *deburring tool*, shown in Fig. 6.15b. The standard ½ -inch countersink is attached to a handle by a press fit for positive gripping action and control, and is constructed with three to five flutes. These flutes do not spiral as on a twist drill. The cutting lips run parallel to the flutes and converge at dead center of the point. The cutting lips have a clearance angle of approximately 15 degrees.

The deburring tool, shown in Fig. 6.15b, has three interchangeable *cutting heads* that pivot when fully seated in the handle. These heads are designed to deburr and form *chamferred edges* on holes in sheet metal or tubing and edges on channel stock.

The countersink is commonly employed for holes less than ¼ inch in diameter. This tool requires only moderate hand pressure to impart a circular rotation of the cutting lips against the burr. The result is a slight chamfer about the circumference of the hole. It is important to hold the countersink as perpendicular to the work as possible to ensure uniform cutting.

When harder materials need deburring, hand pressure may not be sufficient. In this application, a countersink can be used in a drill press chuck to obtain maximum pressure. The countersink should be rotated at the slowest possible spindle speed to avoid chatter, which will roughen the surface of the chamferred portion. Before the deburring operation begins on the drill press, the hole center must be aligned with the countersink point. The procedure is similar to aligning a drill point to a center-punched hole. With the drill press power turned off, the spindle is lowered until the countersink uniformly seats into the hole. The spindle is then returned to its normal position and the work is clamped securely into place. With the power on, the countersink is brought into contact with the edge of the hole to begin the deburring operation. Continual inspection is required so that only the burr is removed. Excessive spindle pressure will cause too much material to be removed from around the edge of the hole. This will result in a sharp edge around the hole on the reverse side of the work and will weaken the material around the edge of the hole because of the excessively large chamfer. Also, if the countersink is allowed to pass too far into the hole, the inside diameter will be altered.

For holes larger than ¼ inch, the deburring tool shown in Fig. 6.15b may be used. The proper cutting head is inserted into the work from the *reverse side* of the burr. The cutting head has a *hook-shaped blade* that is brought down against the burr. With the handle firmly gripped and held perpendicular to the work, it is rotated in a circular

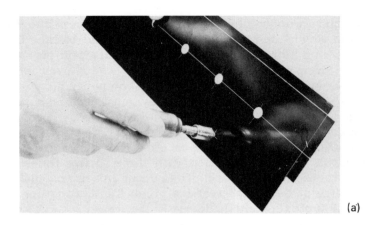

(a)

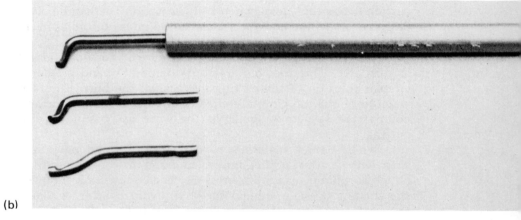

(b)

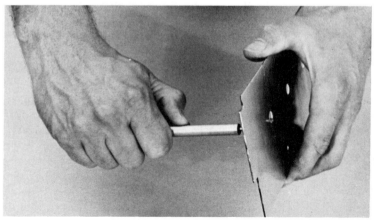

(c)

FIGURE 6.15. Sheet metal deburring. (a) Deburring small holes with a countersink. (b) Deburring tool with interchangeable cutting heads. (c) Deburring tool held at proper cutting angle.

motion about the circumference of the hole (Fig. 6.15c). The cutting head will pivot in the handle, allowing the blade to remove the burr. It may take several turns to completely remove the burr and produce the desired chamfer. Caution must be exercised to avoid gouging or removing excessive material about the edge of the hole.

6.7 REAMING

It occasionally becomes necessary to expand the diameter of a previously drilled hole to a slightly larger size to correct an error in the selected size of a drill. Since this error is not usually detected until the assembly phase of construction, redrilling becomes cumbersome and, in some cases, impossible. For this reason the *hand expansion* reamer, shown in Fig. 6.16 is an extremely convenient tool for enlarging holes.

Enlarging drilled holes introduces an important point in prototype chassis construction, especially if the unit being fabricated is the first of many to be produced. The undersized holes could be the result of an error on the original construction drawings. If this is the case, the drawings must be revised to show the required change in dimension of these holes so that later reference to these drawings will not result in the same error. An erroneous estimate of a hole dimension is not uncommon, but it should be corrected immediately once detected.

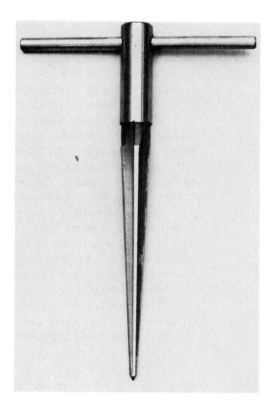

FIGURE 6.16. Hand expansion reamer.

The hand expansion reamer consists of a *T handle* and *tapered fluted blades* running lengthwise along the body of the reamer. There are *six cutting blades* that converge at the taper point. At least six cutting blades are necessary to minimize chatter along the edge of the hole as the tool is used. Cutting is accomplished by a scraping action, and each cutting blade has a *body clearance* for relief similar to that of a countersink.

To expand a previously drilled hole, the reamer is inserted into the hole and the handle is rotated clockwise while a moderate downward pressure is applied. The result is a slightly enlarged hole tapered at the same angle as the reamer blades. When hole tolerance is critical, the tapered reamer should not be used to remove more than 0.030 inch because the resultant taper will weaken the surface of the metal about the hole.

The use of the tapered hand expansion reamer is a crude but effective method of enlarging holes. It also generates a burr, because of the downward pressure, which must be removed.

Straight-sided reamers with either straight or spiral flutes are available for precision work. These are commonly used in conjunction with a drill press or milling machine.

6.8 PUNCHING

Of all the methods available to the technician for fabricating holes in chassis, the *sheet-metal punch* will produce by far the neatest and most accurate results. Holes larger than $\frac{1}{2}$ to 3 inches in diameter fabricated by punching are superior to those formed by drills, hole saws, or fly cutters. (Hole saws and fly cutters are discussed in Chapter 7.) The punch forms a hole that is not only more circular but produces a negligible burr. (A slight chamfer around the edges of punched holes may still be desired to break the sharp edge in preparation for finishing.)

The common *round hand-operated punch* is shown in Fig. 6.17a, consisting of a *punch*, having a threaded hole; a *die* with an unthreaded hole; and a *screw*. The punch is constructed of tempered steel with the cutting edges having two sharpened *spurs* aligned 180 degrees apart. As the screw is tightened, drawing the punch and die together, the spurs and cutting edges *shear* the metal.

Punching with a hand punch begins by drilling a guide hole to accept the screw. This guide hole is drilled in the center of the hole to be formed. The screw size is either $\frac{1}{4}$ inch, $\frac{3}{8}$ inch, or $\frac{3}{4}$ inch in diameter, depending on the size of the punch. The guide hole thus drilled should provide a $\frac{1}{16}$-inch clearance for the screw. Table 6.2 shows the screw size and guide hole size for punching holes from $\frac{1}{2}$ inch to 3 inches.

The punch has two parallel flat sides at its base to secure it in a vise so that it will not turn. Once the correct guide hole has been drilled and the punch secured, the screw is passed through the closed end of the die.

(a)

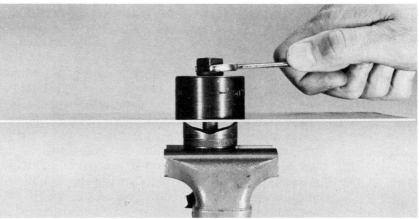

(b)

FIGURE 6.17. Hand-operated chassis punch. (a) Round chassis punch.
(b) Chassis punch properly assembled for cutting.

TABLE 6.2

Screw and Guide Hole Sizes for Chassis Punches

Chassis Punch Size (in.)	Screw Size (in.)	Guide Hole Size (in.)
1/2 9/16 5/8	1/4	5/16
11/16 3/4 13/16 7/8 15/16 1 1-1/16 1-1/8 1-5/32 1-3/16	3/8	7/16
1-7/32	3/4	13/16
1-1/4 1-5/16 1-3/8 1-1/2	3/8	7/16
1-5/8 1-3/4 1-7/8 2 2-1/8 2-1/4 2-1/2 2-25/32 2-3/4 3	3/4	13/16

As shown in Fig. 6.17b, the sheet metal is positioned over the punch. The screw is then passed through the guide hole and threaded into the punch. The screw is tightened to clamp the punch and die against the stock, making sure that the die is centered. (If the die is not centered, the resulting hole will only be off-center by a maximum of $\frac{1}{32}$ inch, which, for most applications, is tolerable.) The head of the screw is square and should be tightened with a properly fitted open-end wrench. As the screw is further tightened, the wide edge of the die applies uniform pressure on the sheet metal about the circumference of the hole being punched. As the metal is forced onto the punch, the spurs start the initial cut. When the punch clears through the stock, a definite "snap" will be heard and the stock will drop down about the punch. When this occurs, the screw should not be tightened further. This would cause the cutting edges to be forced against the inside of the die resulting in unnecessary damage.

In addition to the round configuration, hand punches are also commonly available in *square*, *key*, and *D* styles in various common sizes. These special-application punches, as shown in Fig. 6.18, consist of four parts: the *punch*, *die*, *screw*, and *nut*. *Keys*, *keyways*, or *flats* are located on portions of the dies, punches, and screws for automatic alignment and to hold the parts in the correct cutting position. These punches operate on the same principle as the round hand punches.

Although the use of hand punches requires drilling a guide hole and assembling the various parts, the results obtained completely justify the time and effort involved.

FIGURE 6.18. Special-application chassis punches.

(a) Square.　　　　　(b) Key.　　　　　(c) "D" Style.

Draw punches and *turret punches* are preferable when many round holes of various sizes are to be punched. The draw punch assembly is secured by bolts to a work bench. This lever-operated tool has a punch and die, but the screw common to hand punches is replaced with a *draw bolt*. The draw bolt is fastened to an *off-center moment arm* of the lever. The alignment of the draw punch is identical to that just described for the hand punch in terms of positioning the metal between the punch and the die. Some draw punches have a punch with a threaded center hole, whereas others use an additional knurled thumb nut to secure the punch in position on the draw bolt. To punch the hole, the lever is drawn toward the operator, which forces the punch through the metal and into the die. Die and punch sets are interchangeable on the draw punch, providing a wide range of hole sizes.

The most versatile of all the punches is the *turret* punch. No guide hole is necessary and a large number of die and punches are mounted on rotating drums and are readily available to the operator. A typical lever-operated turret punch is shown in Fig. 6.19. Both manual and hydraulically powered turret punches are available. The action of each is identical, with the exception of the application of power. Before any punching can be performed, a matched die and punch set of the desired size must be properly aligned. To accomplish this, the *pin locks* of both the die and punch turrets are disengaged. The two turrets are then rotated so that the selected die and its associated punch size are aligned and positioned

FIGURE 6.19. Lever-operated turret punch.

directly below the lever arm. The locking pins are then engaged and the punch is ready for use. To operate, the metal is inserted between the punch and die. To position the metal for proper hole location, the operating lever is brought downward until the center point on the punch fits into the center-punched hole in the metal. Further downward motion of the lever engages the *cam* and drives the punch through the sheet metal into the die. The sheet metal then drops and clears the die for immediate duplication of the operation. In addition to providing for rapid setup or size changes, no guide holes are required with the turret punch and no wrench or vise is necessary.

FIGURE 6.20. Amplifier's metal elements after drilling, punching, and deburring.

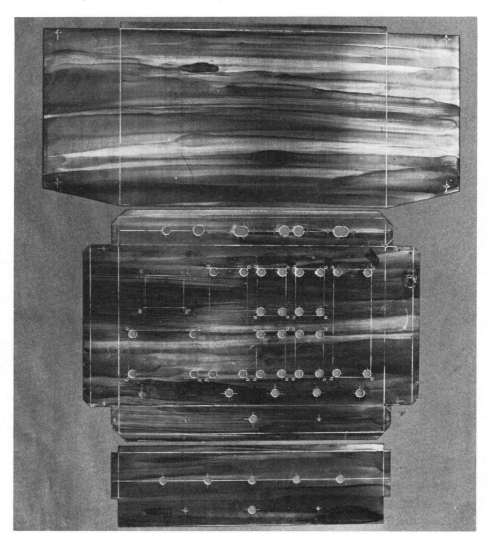

The capacities of all the punches discussed in this chapter are from approximately 12-gauge aluminum to 16-gauge tempered steel.

The amplifier's metal elements have been drilled and punched in accordance with the methods prescribed in this chapter. The results are shown in Fig. 6.20. The next phase in the fabrication of the amplifier's sheet metal components will be to cut all the irregular shapes for mounting items, such as printed circuit boards, transformers, connectors, and associated hardware. The following chapter describes the tools and procedures for completing the chassis cutting operations.

PROBLEM SET

6.1 Find the clearance hole size for a number 8 sheet metal screw from Table 18.2 and drill and deburr the mounting holes in the apron of the gauge panel of Problem 5.4. Gauge holes and radius corners will be cut in Problem 7.4.

6.2 Determine the drill sizes necessary to mount the TO-3-style case, whose outline appears in Appendix III, onto the heat sink of Problem 5.3. (Device lead holes should be drilled approximately 0.040 inch oversize.) All holes must be carefully deburred because flush mounting is critical. The ends of the heat sink are bent in Problem 8.3.

6.3 Tabulate the clearance and tap drill sizes for machine screw sizes 2, 4, 6, 8, and 10. Drill pairs of clearance and tap drill size holes along the length of the gauge in Problem 5.5 for the five machine screw sizes listed. Also, punch the indicated $\frac{3}{8}$-inch hole at the end of the gauge. Deburr all holes formed. Bandsaw and filing operations are done in Problem 7.3.

6.4 Grind a drill having a drill point angle of 118 degrees to a drill point angle of 136 degrees.

6.5 Reshape a standard drill point into a sheet metal drill point configuration.

7

Metal Cutting

7.0 INTRODUCTION

With the drilling and punching operations completed, the next step involves cutting irregularly shaped holes and slots and large round holes in the chassis elements in preparation for bending the metal into the desired forms. Information about the proper selection and use of each tool is provided.

In this chapter various metal cutting blades used in conjunction with a vertical bandsaw are examined together with the proper blade tension, guide settings, and cutting speeds. Practical cutting operations are illustrated on the bandsaw. The hand nibbler used in cutting irregularly shaped holes is described and operational information for proper use of this versatile cutting tool is supplied. Fly cutters used with a drill press to cut large circular holes are covered and the various types of files used to provide finish to the edges of cut metal are described.

7.1 METAL CUTTING BLADES

The technician often encounters the problem of cutting round, extruded, or angle stock, as well as interior holes or irregular shapes in sheet metal.

The shear is not suitable for these tasks. The vertically operated bandsaw with the proper metal cutting blade is ideally suited.

The shape and thickness, as well as the type of material to be cut, determine the proper blade selection. Cutting blades have three specific characteristics: *set*, *tooth pitch*, and *profile*. *Set* is the manner in which the teeth are aligned or bent away from the blade edge to provide adequate clearance between the blade and the work while cutting. Sufficient set is essential to generate adequate *kerf* (slot) so that the blade will not bind in the work during cutting. *Tooth pitch* indicates the number of teeth per inch, and *profile* is the style or shape of each tooth.

The three most common types of set are *alternate*, *raker*, and *wavy*. In blades with *alternate* set, every other tooth is tilted to the same side. The alternate teeth are bent in the opposite direction. This type of set is shown in Fig. 7.1a and is used primarily for cutting the softer, nonferrous metals. In the *raker* set blade, shown in Fig. 7.1b, one tooth is tilted left, the next is tilted right, and the third tooth is straight. This arrangement is continued throughout the length of the blade. The raker set blade will remove chips more easily during cutting operations than the alternate set blade and is used for cutting hard metals such as iron and steel. *Wavy* set, as shown in Fig. 7.1c, is characterized by groups of teeth tilted first left and right and then right and left along the entire blade length. Beginning with one tooth that is not offset, the first half of the group tilts increasingly until the maximum offset is reached in one direction. The second half of the group tilts decreasingly until a tooth with no offset is reached.

FIGURE 7.1. Saw blade set syles. Courtesy of The L. S. Starrett Company, Athol, Massachusetts.

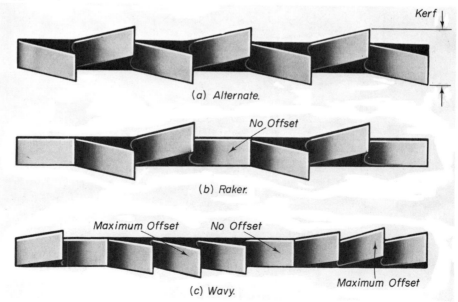

(a) Alternate.

(b) Raker.

(c) Wavy.

This arrangement of outward and inward tilt is then repeated on the opposite side of the blade edge, in sequence, throughout the length of the blade. Wavy set is applicable to cutting thin sheet metal stock and tubing.

The tooth pitch must be suited to the work to result in a clean and efficient cut and to prevent damage both to the work and to the blade. For cutting thin sheet metal or tubing, blades with 32 teeth per inch are recommended. At least two teeth must contact or engage the work surface at all times. This prevents any tooth from becoming hooked on an edge or corner of the work and being stripped from the blade. The wavy set readily provides adequate kerf in thin stock. When soft or mild materials such as brass or copper or low carbon steel are to be cut, 14 or 24 teeth per inch are recommended. When cutting large cross sections of these materials, a coarse 14-teeth-per-inch blade is best and ensures proper chip removal. When these materials are in the form of angle iron, conduit, or pipe, a finer 24-teeth-per-inch blade should be used because of the thinner cross sections. High-carbon and high-speed steels, as well as tool steel, should be cut with an 18-teeth-per-inch blade to provide for adequate chip clearance.

The profile of the blade also affects the cutting operation. Figure 7.2 shows the three most common styles of tooth configuration: *standard straight face tooth*, *hooked tooth*, and *skipped tooth*.

The *standard straight face* design moves down over the work and scrapes the metal, removing the chips. This style of tooth is used for general-purpose work. Blades with *hooked teeth* are used almost exclusively for cutting large cross sections of steel, soft metals, and plastics.

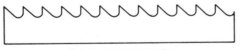

(a) Standard Straight Face.

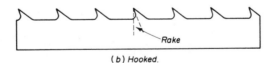

Rake

(b) Hooked.

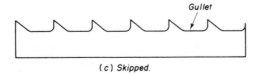

Gullet

(c) Skipped.

FIGURE 7.2. Saw blade profiles.

The slight *rake* on the teeth digs into the work instead of scraping it, thus removing metal more quickly than the standard straight-faced blade. For work on thin sheet metal, especially aluminum, the *skipped-tooth* blade is preferable. Fashioned with a large *gullet* (space) between each tooth, this style allows chips to form within the gullet and clears easier than the standard or hooked blade.

Both carbon-steel and high-speed steel blades are available. Carbon-steel blades do not maintain their sharpness as well as high-speed blades and should be used only for relatively soft materials. In addition, carbon-steel blades must be operated at slower speeds than high-speed blades to avoid impairing normal blade life. In general, hard materials should be cut slower than softer materials to prevent overheating and loss of temper. Most bandsaws are equipped with belt and pulley speed shift or gear reduction arrangements to provide a variation of blade surface speeds of approximately 50 to 4500 feet per minute. Table 7.1 lists the recommended speed for various materials.

Since the blade is very often utilized to cut arcs and circles, blade width determines the minimum radius of cut. The ¼-inch and ½-inch blades generally used in electronic fabrication will cut minimum radii of ⅝ inch and 2½ inches respectively. The ¼-inch blade width is preferred for general-purpose work (Table 7.2).

TABLE 7.1

Recommended Speeds for Carbon-Steel Bandsaw Blades

Materials	Surface Speed (ft/min)
Stainless steel	40−80
Carbon tool steel	100−150
Epoxy glass	90−200
Cold roll steel	125−200
Bakelite	800−1000
Cast iron	75−150
Copper	500−1000
Hard brass	200−400
Bronze	150−300
Aluminum	3000−4500
Plastics	3000−4000

TABLE 7.2

Bandsaw Blade Cutting Radius Guide

Bandsaw Blade Width (inch)	Minimum Radius of Cut (inch)
1	7-1/4
3/4	5-7/16
5/8	3-3/4
1/2	2-1/2
3/8	1-7/16
1/4	5/8
3/16	5/16
1/8	1/8
3/32	3/32
1/16	Square

7.2 VERTICAL BANDSAW

A vertical bandsaw, as shown in Fig. 7.3, is a power tool found in most electronic shops. Although larger and more sophisticated models are used in production, the principles of operation and use are similar. The bandsaw consists of a *drive wheel* mounted below the *worktable* and an *idler wheel* above the table, both *V-pulley*-driven. The motor is positioned within the pedestal that supports the saw. The blade is positioned around both the idler and drive wheels. The teeth are protected by the hard-rubber outer surfaces on the wheel rims that also provide friction to prevent blade slippage. To permit proper operation, the idler wheel is equipped with a *tension adjustment* and a *tracking* or *blade centering adjustment*.

To replace a worn or broken blade, the protective wheel covers are removed and the blade tension is reduced by rotating the spring-loaded tension adjustment knob counterclockwise. This will lower the idler wheel (Fig. 7.4). The used blade is removed and the new blade is passed into the *table slot*. (On some bandsaw tables, a large set screw is threaded into the table at the leading edge of the slot. This screw must be removed before a blade can be replaced.) The blade is then passed around the drive wheel and carefully over the idler wheel so that it is positioned between the *upper* and *lower blade guides*. These guides are located immediately above and below the worktable (Fig. 7.5a). With the blade centered on both wheels, the tension-adjustment knob is rotated clockwise, thus increasing the tension. Optimum tension is achieved when

FIGURE 7.3. Vertical bandsaw nomenclature.

Idler Wheel

Blade Guard

Throat Clearance

Work Table

Drive Wheel

Safety Cover

DeWALT

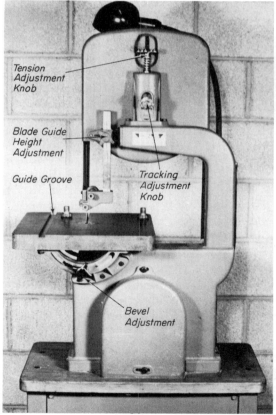

Tension Adjustment Knob

Blade Guide Height Adjustment

Guide Groove

Tracking Adjustment Knob

Bevel Adjustment

FIGURE 7.4. Blade tracking and tension-adjustment controls.

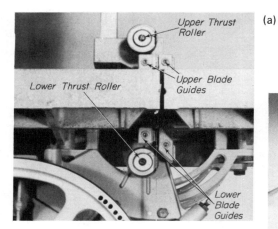

(a)

Upper Thrust Roller

Lower Thrust Roller

Upper Blade Guides

Lower Blade Guides

$\frac{1}{4}$" Displacement

(b)

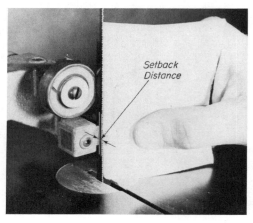

Setback Distance

(c)

FIGURE 7.5. Bandsaw blade adjustments. (a) Bandsaw blade guides and thrust rollers. (b) Bandsaw blade displacement test for proper tension. (c) Testing setback distance and side clearance.

approximately ¼-inch displacement is obtained when finger pressure is applied to the unsupported portion of the blade (Fig. 7.5b). This displacement is typical for blades whose width is from ¼ to ⅜ inch. For blades ½ to ¾ inch wide, a displacement of ⅜ inch is recommended.

It is important that work be fed into the blade from the front edge of the table when the throat of the bandsaw is to the left of the operator. Therefore, the teeth of the blade must be facing the operator when viewed from the front edge of the bandsaw table. Also, the blade is designed so that the teeth will cut when they move in a clockwise direction. The distance in inches from the side of the cutting blade to the inside face of the column supporting the idler wheel is called the *throat clearance*. This clearance is used to designate the bandsaw sizes, which range from 16 to 60 inches. Bandsaw capacities are also given in terms of height from the table to the underside of the top blade guides when this assembly is in the uppermost position. These capacities range from 6¼ to 12 inches.

With the proper tension applied to the blade, the tracking is next inspected by manually rotating the idler wheel and observing if the blade deviates from its original center position on the wheels. Any deviation can be corrected by changing the tilt on the idler wheel by rotating the *tracking adjustment knob* clockwise to correct for a forward deviation and counterclockwise to correct for a blade that has moved back toward the rear of the wheel and table slot. Tracking is properly adjusted if the blade remains centered when manually rotated. Correct blade tracking is extremely important. Blades installed without regard for tracking could ride out from between the guides or completely off the wheels, thus damaging the work, the blade, and possibly the operator. *Note: Never test the tracking by switching on the power.*

Finally, the *blade guides* and *blade thrust rollers* are inspected and adjusted if necessary. The blade guides are bearings or short lengths of hardened steel with square ends. Two pairs of guides are positioned along both sides of the blade above and below the worktable. They retard any twisting motion that is incurred during the cutting operation and yet provide clearance for the blade to pass freely. The setback distance, which is a small clearance maintained between the bottom of the gullet and the front edges of the guides, must be sufficient to allow the teeth to clear the guides (Fig. 7.5c). If sufficient clearance is not provided, the set of the teeth will be damaged as the blade passes through the guides. The blade guides are spaced properly if a single sheet of paper can be inserted as a feeler gauge between each roller or guide bar and the blade. To check the setback distance, the blade must be pressed back against the thrust rollers, as would be the case while cutting. The thrust rollers are bearings located behind the blade guides and, by their position, govern the depth of blade entry into the guides. Pressure can be applied to the blade by forcing a piece of wood against the teeth until the back edge of the blade touches the rollers. In this position, $1/32$ to $1/16$ inch is an adequate setback distance. The thrust roller associated with each pair of blade guides is also properly positioned if the blade does not touch either roller while free-running. The space between the blade and either roller should be uniform. Normal thrust of work against the blade should cause a *light* simultaneous contact of blade and rollers. All the aforementioned adjustments can be made by loosening the associated set screws that clamp these elements into place (refer to Fig. 7.5a and c). Both guide and roller adjustments are critical if normal blade life and set are to be maintained and if quality work is to result.

To cut sheet metal with the bandsaw, the work is placed flat against the worktable and about $1/4$ inch from the blade. The *blade guard clamping screw* is loosened and the guard is lowered to within $1/8$ inch of the stock. The guard is secured and the power switch is turned on. After allowing sufficient time for the blade to attain its operating speed, the stock is fed into the blade with the right hand, as shown in Fig. 7.6. Notice that

FIGURE 7.6. Proper positioning of hands when cutting sheet metal.

both hands are kept clear of the blade to prevent injury. The work should *never* be forced. If cutting requires excessive pressure, the blade should be inspected for dullness and replaced if necessary. At the end of a *through cut* (a cut that traverses the entire length of the stock), care must be taken not to close the kerf generated by the blade set. This will cause the back edge of the blade to bind and could result in damage to both the blade and the work. A through cut is made whenever possible because backing the work from the blade when the power is on could cause blade damage.

To saw a curve or an irregular shape, the layout of the cut is first scribed onto the work surface (see Sec. 5.2). Figure 7.7a is an example of a series of separate cuts necessary to remove a rectangular opening. One continuous cut is first made to remove the bulk of the rectangle. This cut is made perpendicular to the edge and close to one finished scribed line.

FIGURE 7.7. Cutting sequence to form a rectangular opening with a bandsaw.

As the first corner is approached, the smallest possible radius is cut in such a manner so that the blade is positioned close to the *inside* finish line. This process is continued to the other corner with a second radius, allowing the blade to exit along the remaining finish line. This will remove the largest possible portion of the metal from within the scribed area. To square the corners, the edge of the blade is placed along the edge of the first straight cut and the remaining curved area is fed into the blade until the 90-degree corner is encountered. The work is rotated and this procedure is repeated until all the metal is removed from the scribed area in that corner. These cuts are duplicated for the other corner (Fig. 7.7b).

Inside openings in sheet metal cannot be cut with a bandsaw unless it is equipped with a *welding unit*, which is available on the larger industrial machines. These units include a *power grinder*, *blade shear*, and *thickness gauge*. Figure 7.8 shows a bandsaw with these attachments.

To cut an inside opening, it is necessary to drill a hole within the finish lines of the area to be removed. This hole should be large enough to accommodate the blade width. (See Chapter 6 for sheet metal drilling techniques.) The blade is then broken with the shear attachment and

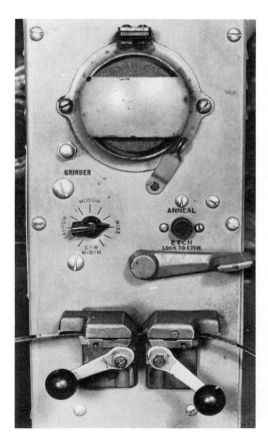

FIGURE 7.8. Welding attachment.

passed through this drilled hole in the work. With the welding attachment the blade is reformed and replaced into the bandsaw and the cut is made. When the desired opening has been cut, the blade is again broken to retrieve the work. After welding and replacing the blade, the saw is again ready for use. *Note* that since the welded joint will be thicker than the blade, it must be ground with the power grinder attachment in order to pass freely through the blade guides. The thickness is tested on the thickness gauge attachment.

When cutting inside openings, the first cut made must be continuous with a full revolution of the work piece about the blade. This will permit a large section to be removed to provide working room for the blade for all finish cuts. Of course, only one cut would be necessary for a round hole.

For cutting and contouring materials such as fiberglass and epoxy resins, special blades are available. The *spiral tooth blade*, shown in Fig. 7.9, will cut when work is fed into it from any direction. This feature is especially desirable for internal cutting. The *line-grind band* consists of a steel band that has an abrasive bonded to one edge and extending onto each side. This blade is extremely convenient when contouring, for the abrasive will act like a fine belt sander for finish work. *Silicon carbide* blades are best suited for cutting fiberglass and epoxy resins, whereas *aluminum oxide* blades are preferable for cutting metals.

Another special-purpose blade used for shaping and contouring is the *band file*. This blade consists of a series of small file segments mounted on a continuous steel backing strip. These band files are available in several different widths as well as surface shapes and cuts. Coarse file segments that range from 14 to 20 teeth per inch are used for cutting soft or mild materials. For fine finishes, file segments with 100 or more teeth per inch are employed. The band file possesses a definite advantage over other types of saws for internal cutting in that it can be easily taken apart with a mechanical joint, thereby eliminating the necessity for breaking and welding.

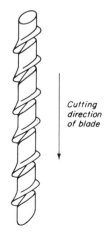

Cutting
direction
of blade

FIGURE 7.9. Spiral tooth blade for contour sawing.

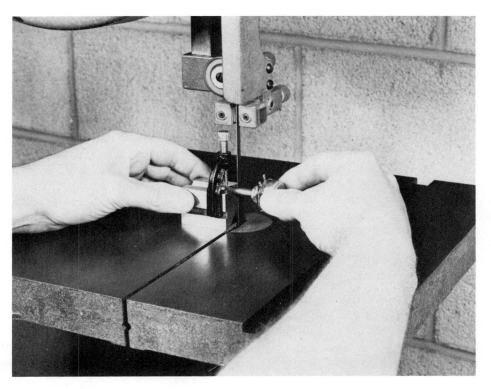

FIGURE 7.10. Use of V-block to aid in cutting round stock.

When irregularly shaped objects are to be cut on a bandsaw, special holding and clamping devices or jigs are necessary. For example, to cut the shaft of a potentiometer or switch, a *V-block* is needed for support and as a base to guide the work. In all cases, a solid support should be provided as close as possible to the cut to minimize saw and component damage owing to unnecessary vibration. Figure 7.10 illustrates the correct technique for cutting a round shaft using a V-block.

Not only does the profile, pitch, and set of a blade affect the cutting action, but the speed at which the work is fed into the blade also influences the characteristics of the resulting cut. In general, a higher rate of penetration will result when a coarse saw and a heavy feed rate are applied. For fine finish work within close tolerances, the finest pitch allowable for the particular material should be used and the feed rate should be reduced. The appropriate feed rate is best determined not only by visual inspection of the work but also by audible strain of the motor speed. Too fast a feed rate will reduce the motor speed. At all times, attention should be directed to (1) avoiding excessive feed rates that crowd or bind the blade needlessly, and (2) being constantly alert to the inherent dangers associated with the use of this power tool.

7.3 HAND NIBBLER

Cutting internal openings in sheet metal is a frequent requirement in electronic fabrication. Bandsaws not equipped with welding attachments cannot be used in this application. Those with welding attachments require, moreover, a considerable amount of setup time. For this reason, the *hand-operated nibbler* is ideally suited for cutting internal holes because of its operating simplicity.

The hand nibbler, shown in Fig. 7.11, is basically a miniature hand shear consisting of two blades. The lower blade engages the underside of the work when the cutting head is passed through and pressed against the edge of the access opening. The upper blade is actuated through a spring located in the handle. As the handle is squeezed, the upper blade is pulled down to contact the upper surface of the work. Further pressure on the handle causes the shearing action between the two blades. Each shearing

FIGURE 7.11. Hand-operated nibbler. (a) Sheet metal nibbler. (b) Interior stock removal simplified through the use of a nibbler.

(a)   (b)

stroke removes a small rectangular piece of metal approximately $1/16$ inch by $3/8$ inch. Thus, this tool "nibbles" a $3/8$-inch-wide strip through sheet metal. Removing the applied pressure from the handles automatically raises the upper blade to its original position and the metal chip freely falls away. The nibbler requires that a hole, approximately $7/16$ inch in diameter, be initially drilled into the area of metal to be removed to accommodate the cutting head through the stock.

When the opening has been completed, the edges must be filed to remove the irregularities that result from the use of this tool. (The various types of files and their uses will be discussed in Sec. 7.5)

7.4 FLY-CUTTERS AND HOLE SAWS

The use of a fly-cutter in conjunction with a drill press will produce a neater and more precise hole with less finish filing required than either a bandsaw or a nibbler. In addition, the fly-cutter requires considerably less set-up time than the bandsaw.

The fly-cutter, shown in Fig. 7.12a, consists of a shaft with a reduced shank that will fit all standard drill press chucks. The shaft accommodates a small-diameter twist drill that is secured in place with a clamping screw. A horizontal crossbar passes through the shaft above the vertically positioned drill. One end of the crossbar holds a tool bit parallel to the drill. Both the crossbar and tool bit positions are adjustable to obtain holes ranging from 1 inch to $5\frac{1}{2}$ inches in diameter. Fly-cutters are generally limited to stock thicknesses up to $1/8$ inch because the heel of the tool bit will begin to contact the work surface in thicker materials, owing to the small clearance angles of these tools.

The radius of the cut can be varied by loosening the clamping screw that secures the crossbar to the shaft. The tool bit height is adjusted to the work by loosening the set screw that secures it to the crossbar. The point of the tool bit must be positioned so that there is at least a distance equal to twice the thickness of the stock above the drill point. This ensures that the drill will pass completely through the work before the tool bit begins to cut. This setting is essential because the drill acts as a center pilot, keeping the tool bit in constant alignment during the cutting operation. When all adjustments have been made and the clamping screws tightened, the fly-cutter is ready for use.

Work to be cut with the fly-cutter must be placed on a block of wood and both secured to the worktable. Simple hand pressure is not sufficient, especially if the tool bit should bind in the work. C-clamps or similar securing devices are absolutely essential.

The center hole is first drilled with the crossbar removed. The drill press spindle is then raised and the power turned off. The crossbar is adjusted so that the tool bit point contacts the scribed circumference line of the hole. After the clamping screw has been secured, the spindle is rotated immediately over the work one complete revolution *manually* to

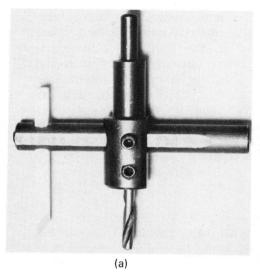

(a)

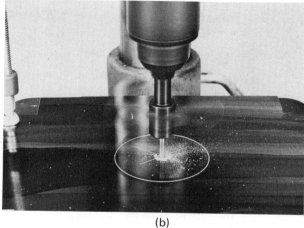

(b)

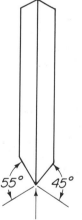

55° *45°*

Cutting Edge (30° angle hidden)

(c)

(d)

FIGURE 7.12. Large-diameter hole-forming tools. (a) Fly-cutter. (b) Fly-cutter used to form large-diameter holes. (c) Fly-cutter tool bit angles. (d) Hole saw. Courtesy of The L. S. Starrett Company, Athol, Massachusetts.

see that the bit has been properly positioned and there is no interference between any clamps and the assembly. The power may now be applied and the hole cut (Fig. 7.12b).

The cutting speed of the fly-cutter must be as slow as possible, owing to the unbalanced characteristics of this tool, which create a hazardous vibration in addition to *chatter*. Too slow a speed, however, will cause the tool bit to bind into the metal irrespective of the amount of feed pressure imparted to the spindle. For safety and optimum cutting operations, fly-cutters are best suited for use with a variable-speed drill press in which the cutter can be brought up from zero rotation to the most suitable speed for the type of material, thickness, and hole diameter that is being worked.

An angle of up to 30 degrees is ground back from the cutting edge of the tool bit (Fig 7.12c). It is important to provide sufficient clearance between the tool and the work, but the angles should be kept as small as possible to avoid excessive thinning and structural weakening of the cutting edge of the tool bit.

Other tools used for cutting large internal holes are the *hole saw*, shown in Fig. 7.12d, and rotary boring tools. The hole saw, similar to the fly-cutter, also has a centering drill to serve as a guide, and a circular cutting edge with generally 6 teeth per inch. It will cut a clean, round hole in any machinable material to a depth of approximately 1 inch. Hole saws are available from $\frac{1}{2}$ inch to 6 inches in diameter.

The *two-lip rotary boring tool* is available from $\frac{3}{16}$ inch to $\frac{3}{4}$ inch in diameter and the *four-lip type* from $\frac{13}{16}$ inch to 2 inches in diameter in increments of $\frac{1}{16}$ inch. These tools are made of molybdenum-alloyed high-speed steels and perform well on all machinable materials, including abrasive phenolics such as hard fiber printed circuit board.

The disadvantage of these tools over the fly-cutter is that different cutters are required for each hole diameter. However, they can be used with any drill press because speed is not critical in terms of tool balance, as is the case with the fly-cutter. Although multiple-blade hole saws are available, they are prone to chatter in certain applications.

When any of the hole-cutting tools are employed, it is essential, as previously described, that a piece of wood be clamped between the work and the drill press table. This prevents the tool from damaging the machined surface of the drill press table when it breaks through the underside of the work.

Safety glasses *must be worn* when using these metal cutting tools because of the danger from flying chips.

7.5 FILES

Of all the mechanical skills necessary in electronic fabrication, perhaps the one that requires the most practice for proficiency is *filing*. In order to gain this proficiency, a familiarity with file characteristics is essential so that the most suitable file for the particular application and material can be selected and used properly.

Filing is primarily a finish operation and is employed to improve the appearance of a previously made cut in a workpiece. It is also used to remove irregular and rough edges or burrs present after certain metal-cutting operations. Also critical for a quality appearance and to minimize damage to components, wiring, and possible injury to the operator is the technique of *breaking* or *rounding* all sharp corners and edges. This is especially true if a workpiece is to be finished by painting. Breaking ensures a better bonding of the finish over the edges of the work and prevents chipping (see Chapter 9).

Files are generally constructed of hardened high-carbon steel and consist of six basic parts: *handle, heel, face, edge, point*, and *tang* (Fig. 7.13). The sharp tang must be inserted into an appropriate size handle prior to its use to prevent injury to the operator. The wooden handles used for files have a metal *ferrule* around the end at which the tang is inserted. This prevents the wood from splitting when the tang is forced into the handle. The file is properly seated by tapping the end of the handle on a wooden block or bench after the tang has been firmly inserted by hand. This process is reversed to remove the file from the handle. By striking the edge of the ferrule next to the tang against the edge of a work bench, the file will loosen sufficiently to be withdrawn. Of course, this procedure can be avoided if a handle is provided for each file.

Files are classified with respect to *length, cut of teeth*, and *cross-section*. The length of a file is measured from the heel to the point. For electronic fabrication uses, files vary in length from 3 inches to 12 inches.

Files have four basic cuts: These are *single cut, double cut, curved tooth*, and *rasp cut*. The curved tooth and rasp cuts are not normally employed in electronics work. The single-cut and double-cut files are shown in Fig. 7.14. The single-cut file is characterized by parallel rows or courses of teeth that are set between 65 and 85 degrees with respect to the axis of the file. This type of file is preferable when a smooth surface finish is desired. A double-cut file has *two* courses of teeth on the same

FIGURE 7.13. File nomenclature.

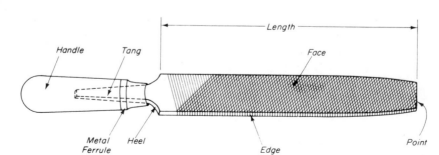

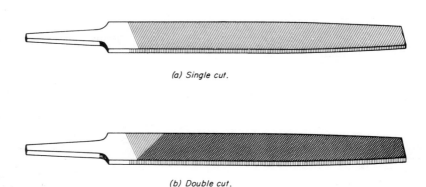

(a) Single cut.

(b) Double cut.

FIGURE 7.14. Basic cuts of files used in electronics work. (a) Single cut. (b) Double cut.

face that are set at different angles. One course is set between 40 and 45 degrees and the second course traverses the face at between 70 and 80 degrees to the axis of the file. These angles are characteristic of files used for general-purpose applications. Double-cut files are also available with first and second courses at 30 degrees and 80 to 90 degrees, respectively, for finer work. Double-cut files are used primarily in *second-cut* work (i.e., in removing larger amounts of material prior to finishing with a single-cut file). The double-cut configuration produces a slicing effect that requires less effort than the single-cut file.

Teeth spacing determines the relative *coarseness* of a file. Files are categorized relative to coarseness as *rough*, *coarse*, *bastard*, *second cut*, *smooth*, and *dead smooth*. The tooth spacing is the largest with *rough cut* and becomes progressively smaller with the *dead smooth cut*. These spacings for the six categories are 20, 25, 30, 40, 50, and 100 or more teeth per inch, respectively. It is important to mention that each of the six grades are coarser for long files (10 to 12 inches) than for the equivalent grade of short files (6 inches). Consequently, length as well as grade of tooth coarseness determines the number of teeth per inch. Figure 7.15 shows the six grades of coarseness for a 6-inch file. When selecting a particular grade of file coarseness, the technician must consider the type of material to be worked. Steel and other hard metals are worked well with the less coarse grades, such as *second cut* and *smooth*. When soft metals such as aluminum, brass, or copper are to be worked, coarser grades, such as *coarse* or *bastard*, are more effective. These choices are dictated by a condition called *loading*, which is the accumulation of waste metal that clogs the file teeth. This condition builds up rapidly when using too fine a file on soft material. These bits of metal impede normal file operation and should be periodically removed with a *file cleaner*. This cleaner consists of a wooden handle to which are attached short, stiff wire bristles called the *card* and a regular bristle brush on the reverse side

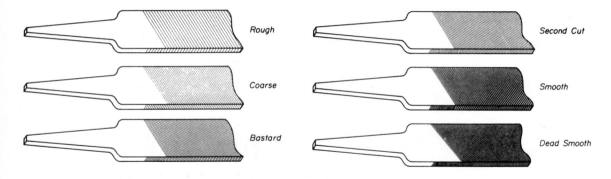

FIGURE 7.15. Comparison of single-cut tooth spacing grades.

of the card (Fig. 7.16). To use the cleaner, the regular bristle brush is first employed to remove the loose waste material on the file face. If any material remains after briskly brushing parallel to the teeth, the cleaner handle is inverted and the card is rubbed over the file teeth. (To clean double-cut files, it is necessary to alternately brush the two coarses of teeth.) To remove any remaining bits of material (called *file points*) lodged between the teeth after these two operations, a pick must be used. The pick is made of a softer metal than the file so that it will not dull the file teeth. To minimize the rate of loading, *chalk* may be rubbed into the face of a file prior to its use. The chalk will not impede the filing operation.

Many types of file shapes and cross sections are available to accommodate the numerous applications of this tool. Figure 7.17 shows most of the common types frequently used. Following is a brief description of each.

Mill File. A single-cut rectangular file used primarily for lathe work and draw filing. It is available in round or smooth edges and tapers in thickness and width for one-third of its length. Other applications are tool sharpening and removing burrs and rough edges. This file is not recommended for filing flat work because of loading and the resulting marring of the work surface.

FIGURE 7.16. File cleaner.

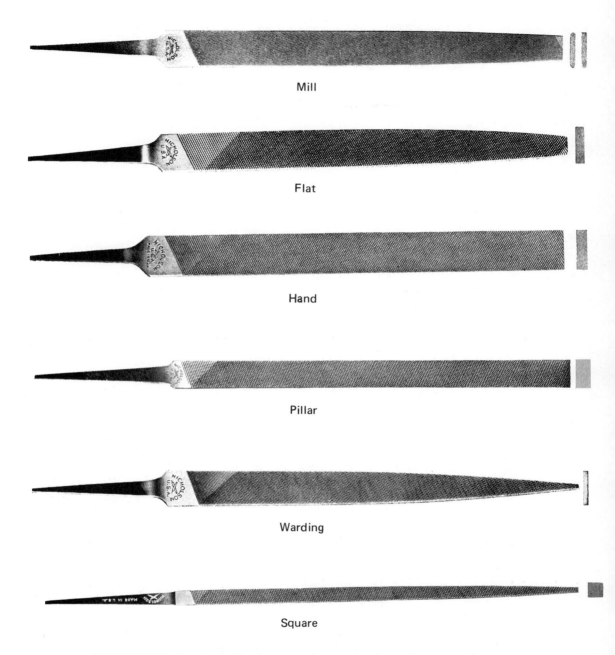

Mill

Flat

Hand

Pillar

Warding

Square

FIGURE 7.17. Standard file shapes and cross sections. Courtesy of Nicholson File Company.

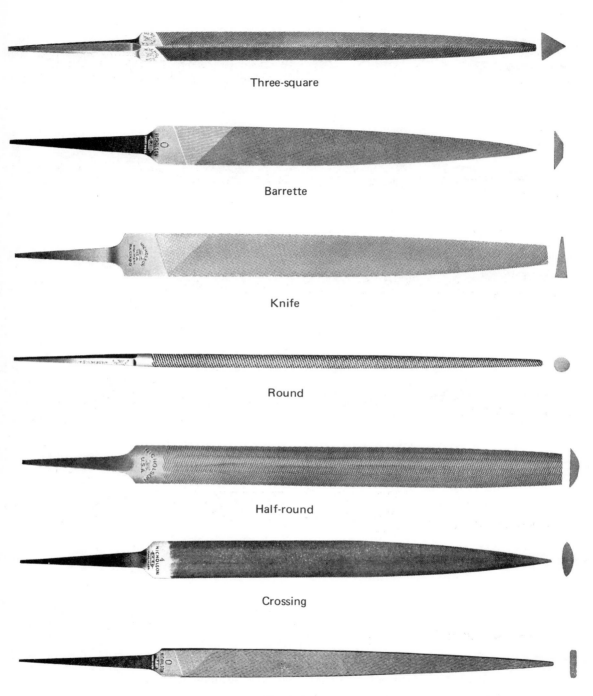

Three-square

Barrette

Knife

Round

Half-round

Crossing

Crochet

FIGURE 7.17. (cont.)

Flat File. A double-cut rectangular file used for general-purpose work in finishing flat surfaces. It has single-cut edges and tapers in thickness and width toward the heel and point from the center of the face.

Hand File. A double-cut rectangular file also used to finish flat surfaces. It is sometimes preferred over the flat file because of its parallel edges. One edge is *safe* (no teeth) and the other is single cut. This file is extremely useful for filing into 90-degree corners. The faces taper slightly toward the heel and point.

Pillar File. A double-cut rectangular file similar to the hand file and used for general-purpose work. It is slightly thicker but not so wide as the hand file.

Warding File. A double-cut rectangular file much thinner than those previously described and tapers to a point. It finds wide application in filing slots and notches.

Square File. A four-sided double-cut file that tapers toward the point and is used primarily for filing or enlarging square or rectangular holes.

Three-Square File. A three-sided double-cut file that tapers toward the point. It is generally used for smoothing internal angular surfaces or corners, especially those having angles less than 90 degrees.

Barrette File. A four-sided double-cut file, tapering to a point, having sharper edges than the three-square file. Only two faces cut; the other two faces are safe. This file is used for finishing in grooves, slots, and sharp angles or corners, especially where cutting must be done while an upper face is contacting a surface from which no further material is to be removed.

Knife File. A file with a safe back edge, a single-cut thin edge, and two double-cut faces. It is often used in place of the Barrette file.

Round File. A round double-cut file tapering toward the point and used for finishing or enlarging round holes or small radii.

Half-round File. A semicircular file usually double cut on the flat face and single cut on the curved face. This unique shape makes it one of the most versatile files for general-purpose work such as finishing curved edges or large-diameter holes as well as flat surfaces.

Crossing File. An oval-shaped file with one face having a radius similiar to the half-round file and the other face having a larger radius.

It tapers to a point in both width and thickness and is double cut on both surfaces. This file also finds application in finishing curved edges, but affords a selection of the most appropriate face for the work radius.

Crochet File. A flat file with rounded edges, double cut on all faces and edges. It tapers in both width and thickness and is used primarily for finishing slots, rounded corners, and filleted shoulders.

Most of the file cross sections just described are available for extremely delicate work. *Swiss pattern files*, shown in Fig. 7.18, find wide application in electronic packaging. The file selected is generally determined by the contour to be worked.

When filing, the metal to be worked should be firmly secured in a vise and protected from damage with *jaw protectors*. This is especially true when working with sheet metal. These protectors are often fashioned from scraps of soft sheet metal, such as aluminum or copper. The metal is bent at 90 degrees and inserted on each jaw to either side of the work. The 90-degree bend keeps the protectors in position when the jaws are opened to release the work. To minimize vibration of the work, which causes the file to "skip" across the surface resulting in *chatter* and an irregular surface appearance, the metal should be gripped as close to the working edge as possible.

FIGURE 7.18. Swiss pattern file sets. Courtesy of Nicholson File Company.

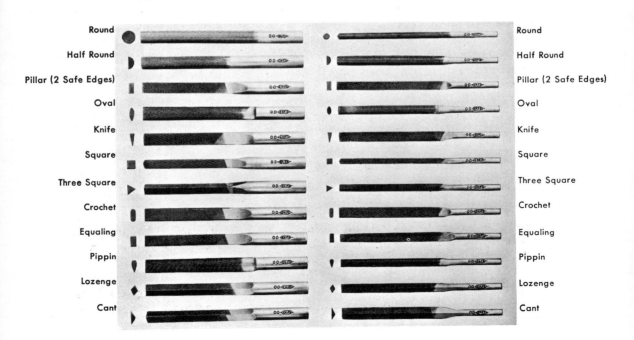

Cross-filing is the technique of pushing the file, under pressure, over and along the edge of the work. The file handle is held in the right hand with the thumb on the ferrule and the file point in the left hand with the base of the thumb resting on the face. The right hand provides the pushing action while the thumb controls the direction and pressure. For best results, a smooth, even *cutting stroke* must be maintained as the file is forced *away* from the operator. Improved cutting results if the file is swung at between 30 and 45 degrees to the work and level to the filing surface (Fig. 7.19). A common mistake of the beginner is to *rock* the file on the cutting stroke, thus leading to a curvature of the surface. The file must be kept level along the entire length of cut. Cutting pressure is applied on the forward stroke *only*. No pressure is applied on the return stroke.

To obtain a fine finish such as on the edges of an exposed portion of a chassis, the method of *draw-filing* is used. The handle is gripped with the right hand and the point with the left with both hands perpendicular to the file axis and both thumbs pressing against the closest file edge. This technique is shown in Fig. 7.20. With the file axis at right angles to the work surface, the file is passed *back* and *forth* with uniform pressure. Cutting is accomplished on both strokes. Single-cut files are considered superior to double-cut files for draw-filing.

Single-handed operation of a file is advisable only when using Swiss pattern files. Since these are so much smaller than the standard type, the file can be completely controlled with one hand. The knurled file handle should be held firmly with the index finger along the uppermost face or

FIGURE 7.19. Proper file position for cross-filing.

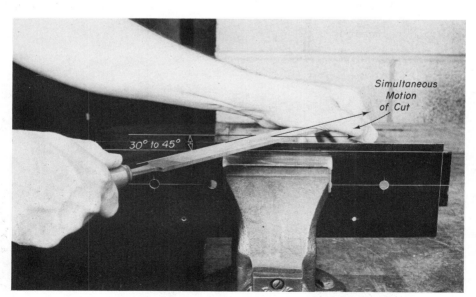

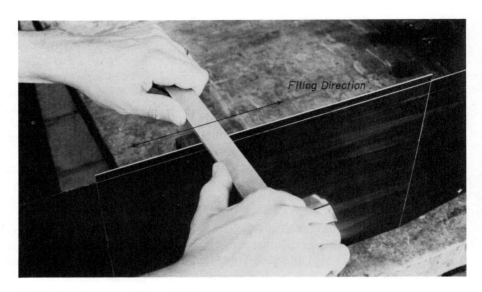

FIGURE 7.20. Draw-filing technique.

edge. The procedures for cutting are similar to those described for the standard-size files. Figure 7.21 shows a knife file being used to remove burrs along the tab edges of the amplifier chassis that resulted from the bandsaw cutting operations.

The proper care of files is just as important as their correct use. Files must not be allowed to come in contact with other files or tools in storage. This will cause tooth chipping and dulling. Also, the face or edge must never be struck on a surface to attempt to remove accumulated

FIGURE 7.21. Swiss pattern files are used where access space is limited.

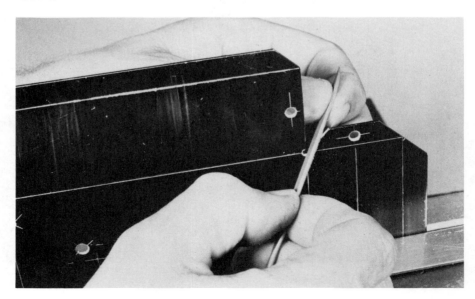

waste material that is loading the file. Although the proper care and maintenance of tools is time consuming, it is more than worthwhile in terms of longer tool life and quality workmanship.

The amplifier chassis elements cutting operations have been completed. These are shown in Fig. 7.22. The final process necessary for the completion of the chassis prior to applying a finish requires forming these elements into the desired configurations. Metal-forming tools and techniques are discussed in the following chapter.

FIGURE 7.22. Amplifier chassis elements ready for bending.

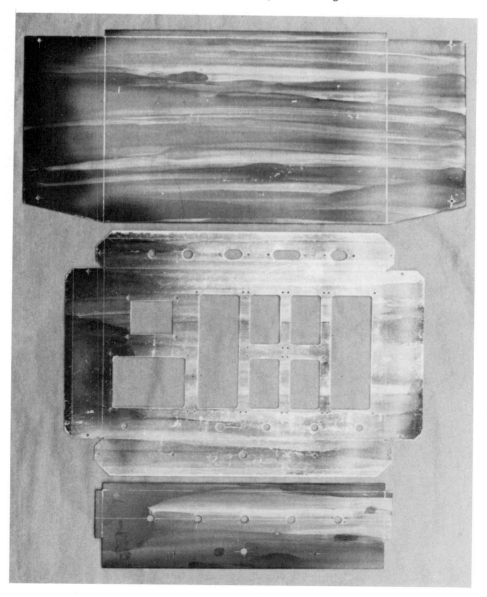

PROBLEM SET

7.1 Determine the maximum bandsaw blade width and speed for cutting a 2-inch radius in a piece of 16-gauge aluminum.

7.2 Fabricate a clamp-sink soldering aid discussed in Sec. 14.6 for hand soldering flatpacks to printed circuit boards. Using a bandsaw, cut a 1-inch length of $3/32$-inch-thick aluminum channel. Channel dimensions should be approximately $3/8$ inch by $3/8$ inch by $3/8$ inch. Break all sharp edges and points with the appropriate file.

7.3 Using a bandsaw, cut the 2-inch radii of the drill gauge laid out and drilled in Problems 5.5 and 6.3. Break all sharp edges with a file. Number stamping and finishing is accomplished in Problem 9.1.

7.4 Using a hole saw, nibbler, or fly-cutter, cut the three holes into the gauge panel laid out and drilled in Problems 5.4 and 6.1. Cut the 1-inch radii with a bandsaw, deburr the holes, and break all sharp edges with a file. Bending is performed in Problem 8.1.

7.5 List the most appropriate type, coarseness, and cut of file for each of the filing operations required in Problems 7.2 through 7.4.

Bending

8.0 INTRODUCTION

After the cutting operations are completed, the next step in the fabrication of a chassis is to form the metal into the desired configuration. Industrial mass production techniques utilize hydraulic press brakes, which are very similar in mechanical action to the power shear (see Chapter 4). These presses can be set to automatically produce *all* the required bends in one step. The hydraulic presses are not considered in this chapter because the cost of tooling makes such automation prohibitive in prototype and small-volume work.

This chapter deals exclusively with the *hand-operated finger brake* (also called the *box* and *pan* brake), which is common to most sheet metal shops and allows the technician to produce a good-quality chassis. Information is provided on layout bend allowance, types of chassis configurations, bending sequence and brake nomenclature, adjustment, and operation. In addition, the critical limitations of most standard brakes will be discussed.

8.1 BEND ALLOWANCE

When sheet metal is bent, there is an inherent distortion of the metal that results in a dimensional change. To maintain a degree of precision when bending, several characteristics of the metal and the equipment must be examined. The result of these examinations will be an adjustment in layout measurements termed *bend compensation* or *bend allowance* to overcome the distortions generated by bending. Since aluminum and mild steel are the most common metals used for chassis fabrication, the factors that affect the bending of these metals with a hand-operated finger brake will be considered.

Bending sheet metal *stretches* the outer section of the bend and *compresses* the inner section. The effect of these modifications on the bend area is to alter the position of the metal edges with respect to the bend lines established in the layout operations (see Chapter 5). Figure 8.1 shows the deviation of an edge from the original established line after bending. Notice that in Fig. 8.1a an inside dimension of distance x measured from edge C is required after bending. This is impossible because the metal possesses a thickness t. After bending, the compression of the metal leaves an inside dimension of $x - B$ (Fig. 8.1b). Moreover, measured from edge C is a new outside dimension, of $x + A$, the result of stretching. It therefore becomes necessary to determine the desired total inside and outside finished dimensions *after bending* by computing the values A and B and compensating for the change in measurements due to the metal-bending characteristics. If an accurate outside dimension is required, the bend line is positioned at a distance $x - A$ from the reference edge C. This procedure is shown in Fig. 8.2a. When bent, the resulting outside dimension will be the desired dimension x (Fig. 8.2b). The *stretching* of the metal induced by the bending has been compensated for. If an *inside* dimension is critical, the bend line is positioned at a distance of $x + B$ from the reference edge C. This compensates for the *reduction* of the inside dimension stemming from the compression of the metal when bent.

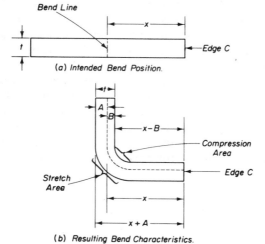

(a) Intended Bend Position.

(b) Resulting Bend Characteristics.

FIGURE 8.1. Dimensional changes due to metal thickness.

FIGURE 8.2. Bend allowance methods to obtain accurate outside and inside dimensions. (a) Bend allowance for accurate outside dimension. (b) Resulting bend. (c) Bend allowance for accurate inside dimension. (d) Resulting bend.

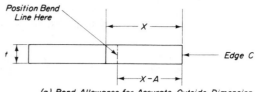

(a) Bend Allowance for Accurate Outside Dimension.

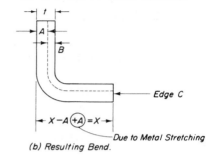

(b) Resulting Bend.

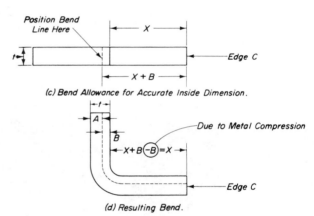

(c) Bend Allowance for Accurate Inside Dimension.

(d) Resulting Bend.

The desired dimension x, as shown in Fig. 8.2c and d, is accurately obtained because of this bend allowance.

The dimensions A and B depend on (1) the type of metal, (2) the radius of the bend, and (3) the angle of the bend. In chassis fabrication, 90-degree bends and bend radii equal to the thickness of the metal are constructed almost exclusively. Therefore, for our purposes, the values of A and B will be a function of metal type only.

For 16- and 18-gauge aluminum alloys of the 1100 and 3000 series, commonly used in chassis work, a value of A equal to two-thirds the thickness (t) of the metal ($A = \frac{2}{3} t$) is a good approximation for most applications. It follows, then, that B is equal to $\frac{1}{3} t$ ($B = \frac{1}{3} t$).

Steel presents a high opposition to compression variations. As a result, A is considered to be the *total thickness* of the metal ($A = t$), whereas B can be approximated to be zero ($B = 0$). An inside bend measurement on steel will be approximately correct if laid off directly without regard to bend allowance. To compensate for an outside measurement caused by stretching, the bend line dimensions must include the thickness of the metal.

Typical sheet metal gauges, together with their associated thicknesses, are shown in Appendix I to aid in evaluating bend allowance.

8.2 BENDING BRAKE

Sheet metal forming by bending is accomplished primarily in prototype work with the aid of a *hand-operated finger brake*, as shown in Fig. 8.3. The basic parts of the brake are the *table, wing, fingers, crosshead, clamping handle, angle stop, operating handles, clamping-tension adjustment screws*, and *finger-position adjustment screws*. The metal to be bent is positioned on the brake table under the fingers and secured in place by pulling the clamping handle *forward*. This engages the crosshead, to which the fingers are attached, and applies clamping pressure to the metal. Once positioned, the metal is bent by raising the operating handles attached to the wing. This causes the wing to move to the desired angle with respect

FIGURE 8.3. Finger brake. Courtesy of Di-Acro.

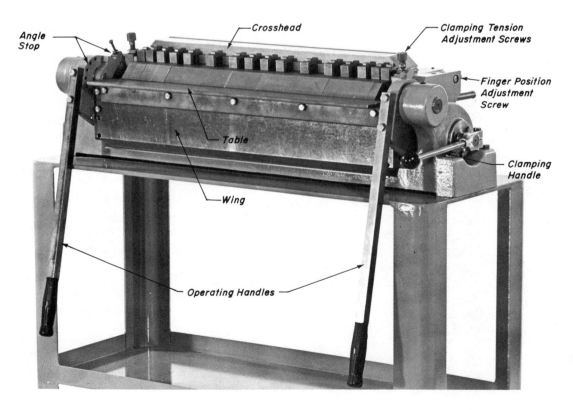

to the table. The adjustable angle stops are located on either end of the brake frame. These contact the wing when it is swung up to the desired angle and impede any further motion. When the metal is bent, the wing is returned to the downward position and the work is released by again operating the clamping handle in an *upward* motion to raise the fingers and release the clamping pressure.

Before any bending operations are made on the brake, *four* adjustments must be checked and corrected, if necessary, to protect the brake and the work from damage. These adjustments, which are described in the following paragraphs, are (1) *finger seating*, (2) *clamping pressure*, (3) *finger clearance (setback distance)*, and (4) *wing stop angle*.

The fingers must first be checked to ensure that they are all properly seated and securely clamped to the crosshead. The seating is correct if all the fingers are parallel to one another with their leading edges in line and parallel to the edge of the wing. A *box-end wrench* should be the only type used when adjusting any of the hexagonal clamping or adjusting screws on the brake. Open-end adjustable wrenches should be avoided because they tend to slip and can damage the screw head.

After the fingers are properly positioned, the clamping pressure is then tested. This is done by first shearing a piece of stock (of the type used in forming the chassis) to a size equal to the brake's full lateral capacity and at least 2 to 3 inches wide. This sample piece is placed between the fingers and the table with at least 1 inch of the stock extending over the upper edge of the wing. The clamping handle is engaged and the pressure is checked by attempting to pull the metal from the brake. If the metal can be moved easily at either end, the clamping pressure must be increased. This is done by releasing the locknut securing the appropriate pressure adjustment screw. These screws are located on the frame at either end of the crosshead. The adjustment screw is rotated *clockwise* to lower the crosshead, thereby increasing the pressure. This test procedure of pulling the metal and adjusting the crosshead is continued until sufficient pressure is applied to the metal to just secure it to the table without damaging its surface. If the pressure appears to be excessive, the adjusting screws are turned *counterclockwise* to raise the crosshead. When the optimum pressure is achieved for the type of metal being worked, the locknuts are again secured. It is important that equal finger pressure be applied across the entire length of the metal to maintain positive gripping.

The finger clearance distance is tested with the same sample stock used for testing the clamping pressure. The edge of this stock is placed perpendicular to the table and touching the leading edges of the fingers, as shown in Fig. 8.4. The wing is then rotated to the 90-degree bend position. The setback clearance of the fingers should be just equal to the thickness of the metal to be worked. Any distance less than this thickness will cause the fingers to dig into the metal along the bend line as the wing is engaged. This will severely weaken the work and possibly damage the fingers. A setback distance larger than the thickness of the metal will

FIGURE 8.4. Finger setback clearance test.

result in a "sweeping" 90-degree bend. The radius of this type of bend may be greater than normally acceptable. If adjustment is necessary, the two locknuts securing the *finger-position adjustment screws* are released. These are located at either end and to the rear of the brake frame. Clockwise rotation of the screws will drive the crosshead forward, causing the fingers to move closer to the wing. *Counterclockwise* rotation will cause the fingers to be drawn away from the wing. With the fingers uniformly positioned at the desired distance across the entire length of the table, the locknuts are tightened to secure the adjusting screws.

The adjustments for setting the bending angle stops are made via the assemblies located immediately above and to either side of the wing. A rough adjustment is first made by inserting a pin into one of four holes provided on the inside face of each assembly. These rough angles are typically 30, 45, 60, and 90 degrees. Adjusting screws secured with locknuts are also associated with these assemblies. These screws allow for fine angle adjustment among the four rough adjustments obtained with the pins. The pins extend outward from the assemblies when seated in the desired hole position and prevent any further upward movement of the wing beyond the pin's angular position with respect to the brake table. It is important to remember that because of the flexing characteristics of most sheet metal, called *"spring back,"* the metal must be bent slightly *beyond* the desired angle to compensate for this effect. The angle adjustment settings, therefore, must be *slightly greater* than the intended angle. After initial adjustments are made, a sample piece can be bent and the resulting angle tested with a protractor head and blade (refer to Chapter 5). Final angle adjustments can then be made if necessary.

The four brake adjustments outlined can be made independently and do not influence one another. However, each of the adjustments affect any resulting bend.

The work must be carefully aligned and inserted into the brake for an accurate bend. To obtain a sharp bend, the scribed bend line is positioned parallel to the leading edges of the fingers and aligned so that it *bisects* the finger clearance distance. This positioning is critical to achieve the proper dimensions resulting from the bend allowance considerations previously discussed. When bending harder or thicker metals than 18-gauge mild steel, a larger radius is necessary to prevent fracture along the axis of the bend as the metal is formed. For this application a *mandrel* may be used. This is a piece of metal that is inserted into the brake over the leading edges of the fingers. The front edge of the mandrel has a radius equal to the radius of the desired bend (Fig. 8.5). When mandrels are used, clamping tension must be readjusted to account for their thickness. The finger clearance distance must also be changed to provide for the additional thickness of the mandrel. When sweeping bends are not critical, they can be formed without a mandrel. By adjusting the finger clearance distance to *twice* the thickness of the stock to be bent and bisecting this distance with the scribed bend line, a uniform sweeping bend will result.

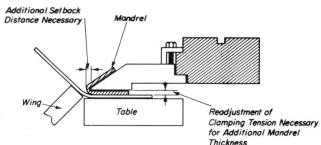

FIGURE 8.5. Mandrels can be used to obtain sweeping bends in thick stock.

The advantage of the finger brake, having individual fingers of varied widths, over the larger type brakes (*cornice* or *leaf type*), which operate with only the wing and one solid bending bar or leaf, will become apparent in the following section.

8.3 BENDING SEQUENCE

Any bending operation, irrespective of its simplicity, requires preplanning concerning the sequence and positioning of bends because of the limitations and restrictions of the brake. Random bends, without regard to sequence, may result in an impossible bending situation and waste of a time-consuming layout. Even with the simple U-shaped chassis shown in Fig. 8.6a, one critical limitation could result as a consequence of the dimensions of the work and the size of the brake. This configuration is to be bent from the scribed blank of Fig. 8.6b. The first bend can be either line *A* or line *B* and involves no problem as long as the width of the metal is less than the lateral capacity of the brake. In either case, however, the dimension *x* is critical. If the distance *x* is less than the distance measured from the table to the top of the fingers, a second 90-degree bend is impossible. Attempting to form the second bend by placing the unbent side section under the fingers will result in the previously bent section contacting the top surface of the fingers before the desired bend can be completed. A bend less than 90 degrees will result, as shown in Fig. 8.6c.

FIGURE 8.6. Bending sequence for U-shaped configuration.

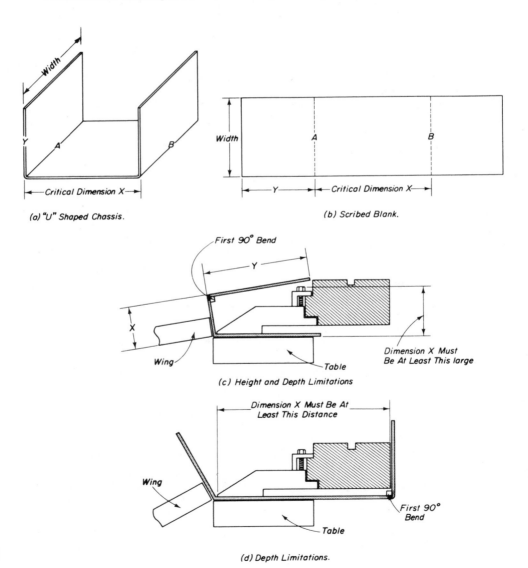

(a) "U" Shaped Chassis.

(b) Scribed Blank.

(c) Height and Depth Limitations

(d) Depth Limitations.

To bend this type of chassis, the distance x must be at least equal to the height of the finger-securing bolts. Again referring to Fig. 8.6c, note that dimension y is also critical if finger clamp assembly or crosshead interference is to be avoided.

An alternative method of making this second 90-degree bend is to position the chassis under the fingers with the first bent section to the *rear* of the crosshead and the second bend line aligned with the finger edges (Fig. 8.6d). This technique introduces still another limitation of the brake. That is, dimension x must be greater than the distance measured from the leading edges of the fingers to the rear face of the crosshead.

Even with the simple U-type configuration, the aforementioned restrictions of the brake limit the minimum size of the chassis. These restrictions should be considered during the planning stages in order to easily produce a quality chassis. When a bend cannot be made on a brake or completed to the desired angle through an oversight in the bending sequence, a great deal of effort, skill, and ingenuity with wooden blocks, clamps, and soft-faced mallets will be required to obtain an acceptable bend.

To introduce further brake limitations and bending problems, a cover, as shown in Fig. 8.7a, for the U-shaped chassis will be formed. The layout of this cover, shown in Fig. 8.7b, will require four 90-degree bends. (Two of these bends, B and D, will produce tabs.) A bending sequence that would appear to be reasonable is to begin with bend A and to continue bending clockwise or counterclockwise around the work. This sequence, however, will introduce another bending limitation. If bend A is formed first, the tabs associated with side E *cannot* be formed on the brake. Owing to this limitation, the following sequence must be followed to form this chassis. Bends B and D are formed first, followed by bends A and C. Figure 8.7c shows the last bend, C, being formed in accordance with this sequence. Since this cover has the same critical dimensions x and y as the U-shaped chassis, similar brake limitations apply. Notice also in Fig. 8.7c that just enough fingers are used to cover the bend line A or C between the previously bent tabs along lines B and D. If it is necessary to leave any length of bend line uncovered by the fingers, this distance should be limited to $\frac{1}{16}$ to $\frac{1}{8}$ inch between fingers or from the ends. This spacing will not cause any noticeable distortion along the bend.

It is always good practice to provide *relief holes* where stresses are incurred during bending, especially at corners where more than one bend is to be made. These relief holes, shown in Fig. 8.7b, prevent bulging or fracturing of the metal at these points of high stress. Notice that four relief holes were drilled at the intersections of the bend lines. The diameter of these holes should be at least equal to the stock thickness.

The bending sequence for a box-type chassis, as shown in Fig. 8.8a, will be formed from the layout shown in Fig. 8.8b. This chassis consists of a *top*, four outside *aprons*, and four *subaprons*. The bends are numerically labeled to show the proper bending sequence to produce the desired

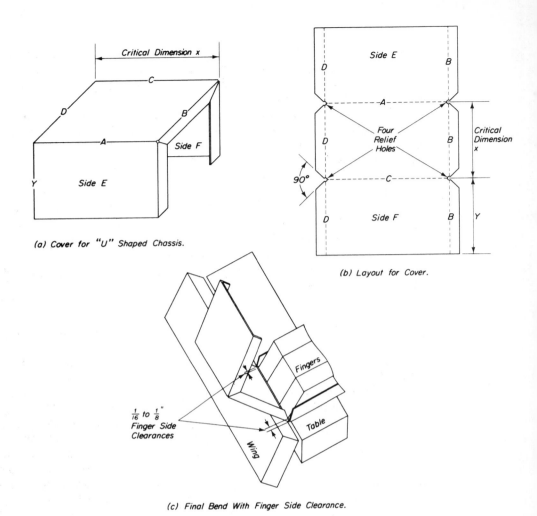

(a) Cover for "U" Shaped Chassis.

(b) Layout for Cover.

(c) Final Bend With Finger Side Clearance.

FIGURE 8.7. Bending sequence for U-shaped cover.

results. All four subaprons (bends 1, 2, 3, and 4) are bent first. No special order or sequence of bending is necessary for these subaprons because their relative location and short width cause no restrictions on the brake. The four inside bends (5, 6, 7, and 8) are then made to form the aprons. Notice that the first dimensional restriction of the chassis is that the subapron height must be limited to the *heel clearance* height of the fingers because the subaprons must be placed *under* the fingers when bending the aprons (Fig. 8.9). A further restriction is the dimension y, also shown in Fig. 8.9. This is the inside height of the finished apron as measured from the inside face of the subapron to the underside of the top. For the four aprons to be formed, this distance y must be greater than the distance measured from the leading edge to the heel of the fingers. Figure 8.9 shows the final bend being made with all the aforementioned dimensional restrictions satisfied.

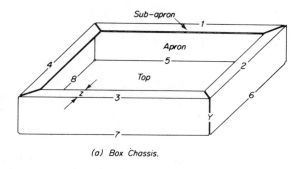

(a) Box Chassis.

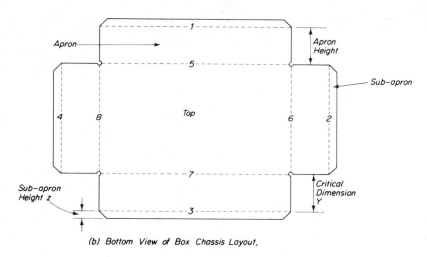

(b) Bottom View of Box Chassis Layout.

FIGURE 8.8. Bending sequence for box chassis.

FIGURE 8.9. Finger and heel clearance restrictions.

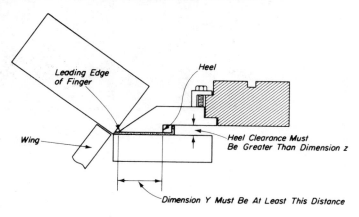

147

8.4 CORNER TAB BOX CHASSIS

The need for a box chassis with secured corners has already been considered in the planning stages of construction (see Chapter 1). The U-type chassis and the unsecured box chassis are too weak to support heavy components. For this reason, the box chassis with secured corners is the most widely used.

Figure 8.10 shows the unformed blank of the amplifier's main chassis element. This is a *corner tab box chassis*. The bending operations begin by forming the four outside subaprons (bends 1, 2, 3, and 4). As in the unsecured box chassis, these subaprons may be bent in any order. It is next necessary to make the inside bends for the tabs and aprons, beginning with bend 5. This will include bending *both* tabs to either side of the apron associated with bend 5 as well as the apron itself. Since the wing of most brakes is a single, straight unit, the tabs cannot be bent separately. As a result, all bends along bend line 5 must be formed at the same time. If this bend was formed in the usual manner, bends 7 or 8 would make the edges of the corner tabs contact the adjacent end apron edges, making it impossible to complete the 90-degree bend. For this reason, some means must be devised to provide the corner tabs with sufficient clearance so that they may be bent along a line that is offset behind the apron by a distance equal to the thickness of the metal. By providing this clearance, bends 7 and 8 can now be made, and the tabs will pass *behind* the end aprons where they may be mechanically secured. To properly form line

FIGURE 8.10. Bending sequence for a corner tab box chassis with subaprons.

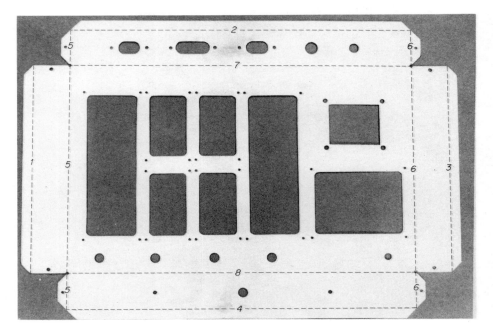

5 to allow for this clearance, it is first necessary to reposition the fingers with the *finger-position adjustment screws* and to use *shims* to provide the proper finger offset and bend. (A shim is nothing more than a strip of sample stock used to determine the necessary setback distances.) The finger clearance is first set to a distance of *twice* the thickness of the metal. The work is next fed into the brake from *behind* the crosshead. With bend *5* aligned with the fingers, there will be enough fingers to cover *just* the length of the apron without extending beyond either relief hole onto a tab. *These* fingers are then moved *forward* from the crosshead by an amount equal to the thickness of the metal by placing a shim between the crosshead and the finger supports (Fig. 8.11). The shims used in this procedure have a width of approximately ¾ inch. The shim length depends entirely on the width of the fingers that it is used with but should be no longer than the length of the section of the bend line with which it is associated. (In this example, this length is bend line 5 *within the relief holes*.) It would be informative to demonstrate what would happen if the bend were performed at this point. The apron section would be bent normally whereas the tabs would have a bend radius *twice that of the apron* and *with no setback distance*. To achieve the desired setback distance, additional shims are placed between the tabs and the upper edge of the wing. These shims should *not* extend onto the brake table. Also, they should not extend laterally beyond a relief hole on either apron end. This technique is shown in Fig. 8.12. The bend resulting from this arrangement is shown in Fig. 8.13. Notice that the tabs are offset the exact required

FIGURE 8.11. Finger and shim arrangement to bend the short side of a corner tab box chassis.

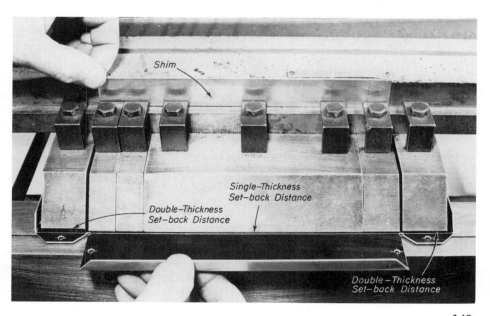

FIGURE 8.12. Tab shims are used to obtain the offset single thickness setback distance for tabs.

amount behind the apron. Bend *6* can now be formed in the same manner as bend *5*. The chassis is placed on the brake table with bend *5* (just completed) positioned to the rear of the crosshead. After bend *6* is completed, it will be necessary to remove all shims and fingers from the crosshead to retrieve the chassis.

To complete the bending of the chassis after bend *6* has been formed, the fingers are realigned and reset to a setback distance equal to the thickness of the metal. By using a total finger length that will accommodate the lengths of bends *7* and *8*, these bends can be formed. When bend *7* or *8* is undertaken, the subaprons previously formed by bends *1* and *3* will prevent the leading edges of the outermost fingers being placed at the extreme ends of these remaining bend lines. If bend *7* is placed under the fingers from the front of the brake, a set of fingers covering the entire length of the bend can be employed to partially set the radius. A 90-degree bend from this position, however, cannot be formed. When the apron is bent to approximately 30 degrees, the tabs will contact the faces of the fingers. This partial bend, however, is sufficient to allow the re-

FIGURE 8.13. Bend results in tab offset.

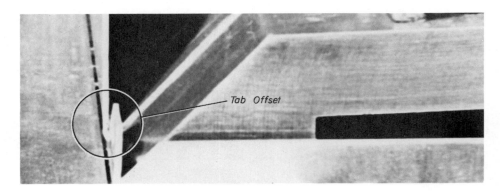

mainder of the bend to be completed by reducing the total finger width to allow for the necessary side clearance for the subaprons. Although the subaprons now prevent the fingers from covering approximately $\frac{3}{8}$ inch at either end of the bend line, no noticeable distortion will result when the bend is completed because the radius was basically established by the first partial bend. Bend *8* is formed in the same manner to complete the chassis configuration.

The bending techniques discussed in this chapter are applied to form the amplifier chassis, front panel, and enclosure elements. These are shown in Fig. 8.14 in their completed form.

The final stage in the construction of a chassis is to apply a finish that will not only protect it but will also enhance its appearance. The various finishes and techniques of finishing are discussed in the following chapter.

FIGURE 8.14. Amplifier's chassis elements bent into desired forms.

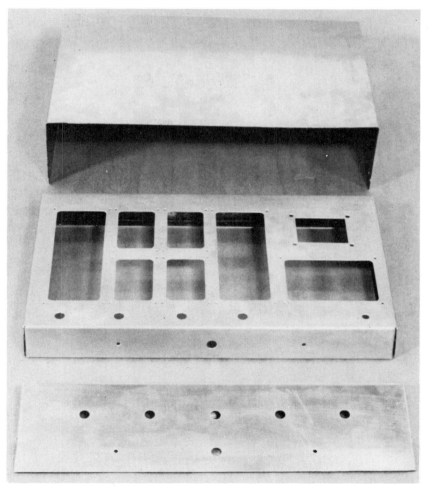

PROBLEM SET

8.1 Bend the apron of the gauge panel laid out, drilled, and cut in Problems 5.4, 6.1, and 7.4 to 90 degrees from the front section. The gauge panel is finished in Problem 9.3.

8.2 Bend the two ends of the soldering iron stand to 90 degrees from the base section along the bend lines scribed in Problem 5.2.

8.3 Bend the four ends of the heat sink laid out and drilled in Problems 5.3 and 6.2 to 90 degrees from the center section and all in the same direction.

8.4 Considering bend allowance design, layout and construct a utility chassis similar to the one shown in Fig. 1.6c, with final outside dimensions of 3 inches by 5 inches by 7 inches. The chassis surface is finished in Problem 9.2.

8.5 Considering bend allowance, design, layout and construct a box chassis similar to the one shown in Fig. 8.10, with $\frac{3}{8}$-inch tabs and subaprons and final outside dimensions of 2 inches by 6 inches by 9 inches. The chassis surface is finished in Problem 9.4.

9

Finishing and Labeling

9.0 INTRODUCTION

Most manufactured parts used in electronic packaging require a *finish* to enhance and protect their surfaces. This is especially true of metal elements whose corrosive characteristics require protection. A prerequisite for obtaining a sound finish is the proper preparation of the surfaces. Several preparation methods for removing common types of surface contaminants and for improved bonding of the finish are discussed in this chapter.

The type of finish used depends on the material and the desired properties of the completed product. To help select and use the most appropriate finish for common applications, we will discuss in this chapter the characteristics of several finishes. Part of the chapter will include information about applying literal and numerical reference markings on chassis and panels. The type and procedure of several methods of labeling also will be discussed. Cleaning, finishing, and labeling will be limited in their application to *aluminum* and *steel*, the most common materials used in fabricating chassis and panels.

9.1 SURFACE PREPARATION FOR FINISHING

Metal surfaces that are to be finished must be free of contaminants. The most common *soils* absorbed by aluminum and mild steel are (1) *water vapor*, (2) *fingerprint deposits*, (3) *oils*, (4) *lubricants*, (5) *paints*, and (6) *mill scale* (this term is used to describe the contaminants present on the sheet metal as a result of the manufacturing processes). If these soils are not removed, an inferior bond between the surface and the finish will result. There are two general classifications of surface cleaning methods. These are *chemical* and *mechanical*, which are described here.

Chemical cleaning makes use of *alkalines*, *solvents*, and *acids* applied by such methods as *hand wiping*, *dipping*, *spraying*, *vapor degreasing*, and *electrocleaning*. The solvents and acids used depend on the material to be cleaned and the process selected. The most common acids used are *muriatic*, *nitric*, *phosphoric*, and *sulfuric;* typical solvents are *alcohols*, *carbon tetrachloride*, *acetone*, *ketone*, *cyclohexanone*, *trichlorethylene*, and *kerosene*. Alkalines generally used for cleaning include *sodium hydroxide*, *sodium carbonate*, and *trisodium phosphate*.

These cleaning chemicals must always be used very carefully. Some release toxic fumes and should be used only in well-ventilated areas. Carbon tetrachloride is especially toxic and should be used only under a laboratory hood that is equipped with an exhaust fan. In addition to toxicity, some solvents are highly flammable, and acetone and ketone are explosive. For this reason, open flames and the possibility of sparks must be completely eliminated in areas where these solvents are to be used. In addition, acids and solvents should not be allowed to come into contact with the skin because of their highly caustic or irritating characteristics. Rubber or plastic gloves should be worn when handling the chemicals used for surface cleaning purposes.

The chemical used for cleaning depends on the material's resistance to corrosion. See Table 9.1 for commonly used cleaning chemicals and their applications. Although solvents are used for general-purpose cleaning, they are not so effective as the other chemicals in removing some of the more resistant contaminants, such as mill scale and paints. Normally, the alkaline cleaners are used to clean aluminum and the acids for cleaning steel. Phosphoric acid is used with a concentration of *demineralized water* at 15 to 30% by weight, and hydrochloric acid is generally diluted with a ratio of three parts of water to one part of acid. Other acids used must also be diluted to reduce their strength to a workable level. Although acids are inherently the most dangerous chemicals to work with, they should be used to clean steel whenever possible because they are more effective in removing contaminants, especially mill scale and paint.

Chemicals not only clean but also improve bonding properties of surfaces. For example, if paint is to be applied as the finish, the application of a *conversion coating* improves the adhesion qualities of the

TABLE 9.1

Chemicals Commonly Used for Cleaning Purposes

Chemical	Type	Application
Alcohol, ethyl	Petroleum solvent	Cleaning solder connections Thinner for shellac and rosin
Acetone	Petroleum solvent	Removal of oily films, paints, and lacquer Lucite cement
Bright dip	Acid mixture containing sulfuric, hydrochloric, and nitric acids	Cleaning metal surfaces after etching or soldering
Butyl cellulose	Petroleum solvent	Thinner and wash-up for epoxy resin inks
Cyclohexanone	Petroleum solvent	Vinyl solvent and cement thinner
Hydrochloric acid	Diluted acid	Remove mill scale from steel Bright dip ingredient
Isophorone	Petroleum solvent	Wash-up for vinyl inks
Kerosene	Petroleum solvent	Machine cutting fluid
Ketone, methyl ethyl	Petroleum solvent	Lacquer thinner and paint remover
Lacquer thinner	Petroleum solvent	Thinner and wash-up for lacquer and lacquer ink
Mineral spirits	Petroleum solvent	Wash-up and thinner for rubber, oil, ethyl cellulose inks, and alkyd enamels
Perchlorethylene	Chlorinated solvent	General-purpose cleaner and vapor degreaser
Phosphoric acid	Diluted acid	Remove mill scale from steel
Sodium hydroxide	Alkaline solvent	Cleaning and etching aluminum
Toluene	Petroleum solvent	Wash-up and thinner for rubber, oil, ethyl cellulose inks, and alkyd enamels
Trichloroethane	Chlorinated solvent	Wash-up layout dye and screen inks Ultrasonic cleaning
Trichlorethylene	Chlorinated solvent	General-purpose cleaner and vapor degreaser
Turpentine	Petroleum solvent	Machine cutting fluid
Xylene	Petroleum solvent	Thinner for acrylic printing inks Wash-up for synthetic enamels and photo resist ink

surface. This coating chemically generates a film from the original metallic surface that promotes bonding. This coating is obtained through the application of acid solutions such as *phosphate, chromatic,* and *anodic* coatings. (See Sec. 9.2 for a further discussion on anodizing.) Anodic and chromate coatings and chromate-phosphate coatings are the most widely used of all conversion coatings because they can be applied before metal-forming processes are undertaken (i.e., they will not crack, chip, or flake if applied prior to bending operations on sheet metal).

The alkaline solutions used to clean aluminum may also provide a *caustic etch* on the surface of the metal. (Etching provides improved finish bonding by generating minute scratches in the surface.) In order to bring about this etch, however, the chemicals must be prepared and applied under specific conditions. The etchant often used for aluminum is a 20% *sodium hydroxide* or household lye solution at a temperature of 110 to 115°F (43 to 46°C). The stronger the solution, the more rapid the etching time. One-quarter to ½ cup of lye to 1 gallon of water at a temperature of 140 to 180°F (60 to 82°C) will produce an etch in ½ hour to 2 hours. Thirteen ounces of lye to 1 gallon of water at these same temperatures will reduce the etching time to 1 to 3 minutes. The solution should be prepared in a plastic or enameled container with a wooden stirrer used to dissolve the crystals. Because of substantial bubbling and fuming action, this process should be performed in a well-ventilated area with rubber gloves worn for protection. Wooden sticks or plastic tongs are used to handle the work to avoid undesirable marks on the metal surface. The work is placed into the solution with the surfaces completely submerged. After the prescribed time (depending on the strength and temperature of the solution), the work is removed from the bath. The surface will be covered with a black residue that can be easily removed by first rinsing it in water and then dipping it in a *bright dip solution.* (This solution will be discussed later in this section.) After removal from this solution, the work is rinsed in hot water and then blotted with lint-free paper towels to prevent water stains. Any remaining stains can be removed with acetic acid or vinegar. The surface now has a velvety silver-sheen appearance and is highly susceptible to stains. The finish can now be applied. If the silver appearance is to be preserved, a clear plastic vinyl or acrylic spray coating may be applied. For other colors, the etched surface will readily accept a dye with excellent results. The amplifier's main chassis element, as shown in Fig. 9.19, was processed by this etching method.

If surfaces are to be cleaned and not etched, chemicals may be wiped by hand using a cloth or cotton swab. Hand wiping, however, should be limited to alkalines only. Acids are extremely corrosive and chlorinated solvents are highly toxic and create a hazardous condition if used by hand. In addition, hand wiping is only suitable if an extremely clean surface is not required. An example of this is the application of the lay-out dye (see Chapter 5) wherein this coating is only temporary. In hand wiping there is always a strong possibility of spreading a thin film of impurities over the entire surface as the solvent is applied.

Dipping parts into a cleaning solution does not produce contaminant-free results. While submerged, the surfaces will be extremely clean. However, as the work is removed from the bath, a thin film of the contaminants will be deposited onto the work surface. In addition, dissimilar metals must not be used for handles or clamps when working with dip cleaning. The *galvanic action* generated by the contact of these metals will discolor the work surface.

A more elaborate and effective system of cleaning is *acid spraying*. This system requires more complex apparatus in order to contain the acid and fumes. The work is clamped in a vertical position and a jet spray of acid removes all foreign material from the surface. The acid may be heated or used at room temperature. Sufficient time must be provided in this process for the lower portion of the surface to be completely cleaned of its own contaminants in addition to those that flow over it from the upper work surface.

Another effective method of chemical cleaning is by a process termed *vapor degreasing*. The parts to be cleaned are suspended in an enclosed tank in which solvents such as *trichlorethylene* or *perchlorethylene* are heated until they vaporize. As the vapors rise and pass over the cooler surface of the work, they condense into droplets. As these droplets increase in size they stream over the surface, carrying the contaminants with them, and drip back to the bottom of the tank. Approximately 30 seconds is the normal length of time necessary for this process to completely remove the impurities from the work. If several parts are to be cleaned together, they must be positioned in such a manner that the contaminants washing off one surface do not drip on another. Vapor degreasing is commonly used on parts to be electroplated, which is discussed in Sec. 9.2.

Electrocleaning is still another method of surface cleaning. This process uses alkaline solutions almost exclusively. Depending on the type of material, the work acts as either a cathode (negative) or anode (positive) suspended in the alkaline solution and the tank acts as the opposite electrode. Metals that tend to form protective oxide films are cleaned cathodically. As current is passed through the system, hydrogen is released at the cathode and oxygen at the anode as the water in the solution is electrolyzed. This formation of gas agitates the cleaning solution sufficiently to remove all surface contaminants from the work. When some forms of alloyed aluminum are cleaned in this manner, a black residue forms on the surface of the work. This is easily removed by a final acid dip termed *bright dip*. This bright dip solution contains sulfuric, nitric, and hydrochloric acids in demineralized water. *Caution:* When any acids are to be mixed with water, extreme care must be taken that the *acid* is added *very slowly* to the *water*. Water added to an acid causes a violent splattering reaction and an extremely rapid generation of steam. A common mixture for the bright dip solution contains 1 part of sulfuric acid, 4 parts of nitric acid, and several drops of hydrochloric acid. This acid mixture is then added to 12 to 16 parts of water. The black residue on the surface of the aluminum can usually be removed in this acid dip in approximately 15 to 20 seconds. Another mixture for removing this residue is chromic acid in a solution diluted 50% by weight. Chromic acid leaves a surface less susceptible to fingerprints.

The final procedure in the use of chemical cleaners is the removal of these chemicals from the work, by rinsing thoroughly with demineralized water. One final precaution should be mentioned here. If the part being cleaned is not a completely flat surface, cleaning chemicals can become entrapped and eventually corrode the surface. Areas of concern are spot welds, rivets, swage-mounted nuts, and recesses formed under mounted parts. If these areas are inaccessible to washing with water, the contaminants should be removed by mechanical means.

Mechanical cleaning methods include *abrasion*, *flame*, *steam*, and *ultrasonic* cleaning. Abrasion cleaning may be accomplished by *wire brushing*, *sand* or *grit blasting*, or *tumbling*. Wire brushing may be performed by either manual or powered processes. Various brush stiffnesses are available in both hand and rotary motor-mounted brushes. In sand or grit blasting, abrasives are sprayed over the work surface through high-pressure nozzles. Tumbling performs several procedures simultaneously. In addition to cleaning, it will dull sharp edges and provide smooth surfaces. The metal parts are placed in rotating drums containing sawdust and stone chips. The drums are rotated and the abrasives scour the metal surfaces. This process may take up to several hours to achieve the desired results, however.

Gas flames are sometimes used when mill scale or paint is to be removed from a metal surface. This method, however, must not be used if exposure to high temperatures will damage the metal.

Paints, oils, and lubricants may also be removed through steam cleaning. The steam is forced under high pressure through a nozzle onto the work. For some applications, the steam may contain a detergent.

The most commonly employed mechanical method of cleaning electronic parts and hardware is ultrasonic cleaning. This is accomplished through the use of ultrasonic energy provided by an ac generator having power capabilities of from 100 to 5000 watts. (The power output depends on the size of the cleaning unit.) The frequencies used range from 20 to 40 kilohertz (kHz). The generator is coupled to a rigid tank, which contains a detergent solution, by means of ultrasonic transducers. The acoustic energy produced by the generator and transformed through the transducers causes *voids* to be set up in the cleaning solution. These voids collapse, causing pressures of over 1000 pounds per square inch. When a part to be cleaned is immersed into the bath, these high pressures cause the solution to penetrate into joints and cavities formed by spot welds, rivets, or swage nuts. Soils are literally torn away from all surfaces exposed to the solution. Any type of glassware detergent may be used by an ultrasonic rinse in demineralized water to remove any detergent film. Figure 9.1 illustrates the use of an ultrasonic cleaner.

FIGURE 9.1. Ultrasonic cleaning system.

9.2 FINISHING

The choice of finish selected for a part depends on many factors, the most important of which are (1) resistance to corrosion and moisture, (2) ability to withstand wear, and (3) appropriate appearance.

Anodizing is the most common process of generating a conversion-type coating on aluminum and results in a surface that is resistant to abrasion, corrosion, and stain and has high electrical resistance properties. Although anodizing is a finish process in itself, it also provides an excellent base for applying paints and dyes. The work to be anodized acts as the positive electrode, or *anode*, in a solution having a high oxygen content. This solution is called the *electrolyte*. A common electrolyte is a 15 to 25% sulfuric acid solution. The negative electrode, or *cathode*, is generally a lead-lined stainless steel tank. The system is energized by a dc power supply. One lead from the positive terminal of the supply is connected to the work and the lead from the negative terminal is connected to the tank lining. This system is shown in Fig. 9.2. The voltage of the power supply is generally between 12 and 24 volts, with a current density of between 12 and 18 amperes per square foot of work surface. It is estimated that the total surface area of the amplifier's main chassis ele-

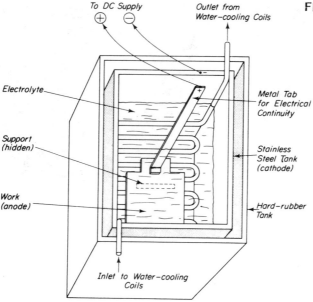

FIGURE 9.2. Anodizing system.

To DC Supply

Outlet from
Water-cooling Coils

Electrolyte

Metal Tab
for Electrical
Continuity

Support
(hidden)

Stainless
Steel Tank
(cathode)

Work
(anode)

Hard-rubber
Tank

Inlet to Water-cooling
Coils

ment is 382 square inches. If the larger cutouts are considered, this figure reduces to 275 square inches, or 1.9 square feet. The power supply for anodizing will have to provide between 23 and 35 amperes. When energized, large amounts of oxygen are liberated on the work surface and form a dense oxide film. Within 20 to 30 minutes, a film thickness of from 0.5 to 2 mils can be formed. Since the electrolyte dissolves some of the outermost film as new film is being formed at the surface of the metal, there is a practical maximum thickness obtainable. The point at which the rate of dissolution is the same as the rate of oxide formation occurs within 1 to 2 hours under the conditions described. Temperature is a critical factor in the anodizing process. The solution must be maintained at 70 ± 2°F, the lower temperature producing a thicker and denser film. Because of the electrochemical action, the temperature of the solution will tend to rise. An increase that causes the temperature to rise excessively above 70°F (21°C) will dissipate the energy of the electric current in electrolysis and little of the oxide film will form. In addition, the film will become porous. To avoid these excessive temperature rises, cooling coils are included in the tank. Cold water passing through these coils is normally sufficient to maintain an even temperature within the specified limits. For effective anodizing, the work must be completely submerged in solution during the entire process. Dissimilar metals must not contact the work to avoid discoloring the surface. The work normally rests inside the tank on ceramic supports. Superior anodizing will result if the work has been previously caustic etched (see Sec. 9.1). Because of the fumes generated, the anodizing process should be carried out in a well-ventilated area.

After the work has been anodized, it is rinsed with water and then dipped in a weak solution of *sodium* or *ammonium carbonate* to neutralize any acid that may remain on the surfaces. A mixture of 2 ounces per gallon of water is sufficient. The anodizing process results in a finished silver color similar to a caustic etch surface. To produce a colored surface, the work is first rinsed in water and then dipped in a 1 gram per gallon solution of *analine dye* and water at a temperature of approximately 140°F (60°C) for 5 to 15 minutes. The depth of the resulting color will depend on (1) the thickness of the oxide film, (2) the concentration of dye in solution, and (3) the length of time the work remains in the dye. When the desired color is achieved, the dye must be *sealed*. To seal the dye, the work is removed from the dye, dipped in water, and then placed in an 8-ounce-per-gallon solution of *nickel acetate* and water at a temperature of 200°F (93°C)for 3 to 5 minutes. The work is removed from this solution and placed in boiling water for 30 minutes to an hour to seal the pores of the oxide film. For all solutions and rinsing purposes, demineralized water should be used exclusively.

Electroplating is a process for depositing less corrosive metal coatings on a work surface through electrochemical means. The resultant finish has improved corrosion and wear resistance and a more decorative appearance. Electroplating is similar to the electrolytic process for anodizing except that a metal is deposited on the surface of the work rather than an oxide film. In addition, the work is connected to the power supply as a cathode and the type of metal to be deposited, in the form of a rod or plate, is the anode. The work and the metal are immersed in an electrolyte. The electrolyte used will depend on the type of metal to be plated. This electrolyte is a *salt* of the metal to be deposited. For example, *silver nitrate* is used for plating silver and *zinc nitrate* for plating zinc. Steel, which is commonly electroplated, needs only to be degreased (cleaned) prior to processing. Steel chassis are often *cadmium-* or *zinc*-plated to improve corrosion resistance and soldering characteristics while maintaining good electrical conductivity. Zinc plating results in a more corrosive-resistant finish than cadmium. However, both deposits are vulnerable to deterioration through oxidation. In addition, neither of these plated surfaces furnish good adhesive qualities as a paint base. Consequently, if the resultant finish is to be paint, the work must be dipped into a *chromate* solution after zinc or cadmium plating to produce the necessary adhesive qualities. For decorative purposes, steel may be plated with *chromium*. Unlike cadmium or zinc, electroplated chromium is quite porous. Therefore, it is first necessary to plate the steel surface with a very thin film of copper followed by a thin film of nickel before the final chromium plating. To obtain a polished chromium finish, the nickel plating should be highly polished before applying the chromium plating. (This highly polished nickel surface is very decorative but it tarnishes quickly.) The result of this total plating combination is an extremely durable and esthetic finish.

Aluminum also lends itself very well to electroplating for protection and decorative purposes. However, to provide adequate adhesive properties for the plating, the aluminum must first be cleaned and then given an immersion-type zinc coating. The solution for this zinc coating is a mixture of *caustic soda* and *zinc oxide*. For ease in soldering to the surface, aluminum is often copper- or brass-plated.

Electroplating is used very widely in electronics to protect printed circuit conductor patterns prior to the etching process.

Of all the materials available for finishing applications, the most economical and easiest to apply are *paints*, *enamels*, *varnishes*, and *lacquers*. They also offer the widest selection of colors and textures among all surface finishes. These finishes are made from *binders* or *vehicles* (which are oils or synthetic resins), *pigments*, and *solvents*. The composition of these finishes is as follows:

Paint — vehicle and a pigment

Enamel — paint and varnish

Varnish — vehicle and solvent

Lacquer — vehicle, solvent, and nonvolatile coloring additive

Paints and enamels are used extensively on metal surfaces common to electronic packaging. They provide a heavier and more durable coating than either lacquer or varnish and have the advantage of hiding minor surface imperfections such as scratches or nicks in metal. Paint and enamel finishes range from *lusterless* (flat) to *high gloss*. The degree of gloss is determined by the proportion of pigment to vehicle. Semigloss to flat finishes are most commonly used in electronic packaging applications. Some of the more common paint textures available are *wrinkle*, *crackle*, and *hammertone*. These finishes are shown in Fig. 9.3. They not only hide surface imperfections but also enhance the overall appearance of the chassis. The *wrinkle*-type enamel tends to *expand* as it cures. This causes small, randomly oriented wrinkles of paint to form over the entire surface. The *crackle*-type enamel is applied over a color base. As this finish cures, it *contracts* and forms very fine random lines. The base color contrasts with the finish color producing an extremely attractive package. The *hammertone* enamel leaves a very smooth surface when it cures, yet provides a hand-hammered appearance. This finish is available in a variety of colors.

Varnishes are used when a surface is to be protected from moisture and mildew.

Lacquers are often used clear or unpigmented to protect labeling from abrasion or to help protect etched or plated surfaces from staining or tarnishing. Lacquers are water resistant but will not stand up to saltwater or prolonged exposure to sunlight. These critical environments will cause the lacquer to discolor and become brittle. Plastic sprays of the acrylic or

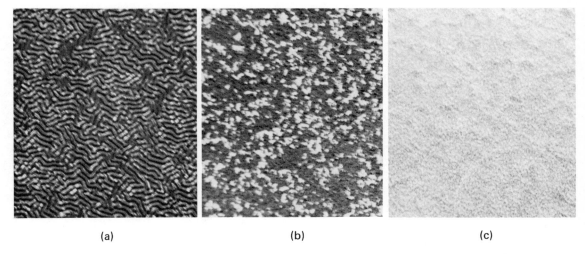

FIGURE 9.3. Textured paint finishes. (a) Wrinkle. (b) Crackle. (c) Hammertone.

vinyl type are readily available in aerosol cans. These sprays do not possess many of the undesirable characteristics of lacquer and varnish. Plastic sprays are available either clear or in a choice of colors.

Various means of applying paints, enamels, varnishes, and lacquers are by *brushing, dipping, flow coating, aerosol-pressurized spray cans, compressed-air spraying*, and *roller coating. Brush painting* leaves brush marks. Special low-viscosity paints are available, however, that will minimize the appearance of brush marks. The work should always be held *horizontal* to prevent the paint from *sagging.*

Dip painting is extremely effective for coating parts with irregular surfaces. This method, however, can cause the paint to build up at the edges closest to the bath as the excess paint drips off the work.

Flow coating is a production line method that pumps paint under pressure in streams over the work from various angles. This process also distributes paint unevenly over the surface.

Aerosol-pressure spray cans are limited to painting small surface areas because of their cost. These cans are pressurized to approximately 40 pounds per square inch with a refrigerant, usually *Freon*, which acts as the propellant for atomizing the paint. Applying paint with aerosol cans is similar to *compressed-air spraying*, which meets the requirements of a wide range of surface configurations. Figure 9.4 shows a suction-type spray gun. Spray painting must be done in a well-ventilated spray booth that will exhaust both the solvent and excess paint particles. In addition, atomized paints are highly combustible and should be used with caution. In the spray gun shown in Fig. 9.4, the paint is drawn up from a canister into a high-velocity stream of air that passes through the gun and nozzle assembly. The nozzle causes the atomized paint to fan out as it leaves the

FIGURE 9.4. Spray gun.

spray gun. Compressed air is the propellant that atomizes the paint. A high-pressure rubber hose connects the spray gun to the air compressor. Compressed air pressures for suction guns usually range from 30 to 65 pounds per square inch (psi). The optimum pressure depends on the viscosity of the paint. The higher the viscosity, the more air pressure required for effective spraying. The pressure rating of the gun must not be exceeded. Suction feed is not suitable for high-viscosity paints. For this application, pressure-can-type spray guns are used. These operate at lower pressures than suction-type guns, in the order of 20 to 35 psi with a maximum allowable pressure of 50 psi. Paints used in either type of spray gun must be thinned with highly volatile solvents to allow the paint to be properly atomized and prevent the paint from sagging while curing. Each type of paint, lacquer, or enamel has its recommended type of solvent for reducing the viscosity for spray painting. The specific type is generally designated by the manufacturer. Common solvents are *lacquer thinner*, *enamel reducers*, *mineral spirits*, *toluol*, and *toluene*. The solvent evaporates between the time that the paint leaves the nozzle and contacts the work surface. This increases the viscosity of the paint and prevents it from running or sagging. To aid in obtaining a uniform coating, especially over irregularly shaped surfaces, the work may be placed on a turntable. This eliminates having to handle the work to paint the reverse side. For best results, the spray gun (or aerosol can) is held perpendicular to the work surface, which is held vertically. Spraying is begun at the top of the work with the nozzle held 6 to 10 inches from the surface. The spray is swept across the work in a straight-line motion continuing beyond the edge of the work. To ensure proper coverage, each sweep of spray should slightly overlap the last pass when moving the spray gun to either the left or right. Once the surface has been completely covered, the paint must be allowed to cure. If a single coating does not provide sufficient coverage, a second coat may be applied *only after* the first coat is completely cured. For superior results, it is always good practice to apply two coats rather than one thick coat. This reduces the possibility of the paint sagging or running. For uniform paint thickness, the more difficult and unexposed areas are sprayed first and then the open surfaces, if the work is irregularly shaped. Some finishes cure at normal room temperatures but others require a drying oven. Most paints can be dried in about 20 minutes at 225 to 250°F (107 to 121°C). Manufacturers' curing specifications are available in their literature. Figure 9.5 shows the amplifier's front panel being spray-painted using an aerosol can.

Roller coating is the method of applying by means of a roller. This method is performed manually or by machine and is limited to flat surfaces.

A textured finish on a chassis can be obtained by purely mechanical means. Figure 9.6a shows the arrangement for applying a uniform textured finish onto an aluminum surface by means of an abrasive rubber tip, such as a pencil eraser secured in a drill press chuck. The resulting finish is shown in Fig. 9.6b. The rubber tip is rotated and pressed against the work surface, which is positioned on the drill press table against a clamped guide bar. The result of a single contact is a perfectly round polished area. By moving the work to the left or right (the chosen direction must be consistent) successive polished areas are made to slightly overlap the preceding one. Once a row has been completed, the guide bar

FIGURE 9.5. Use of aerosol-pressure spray cans simplifies finishing small parts.

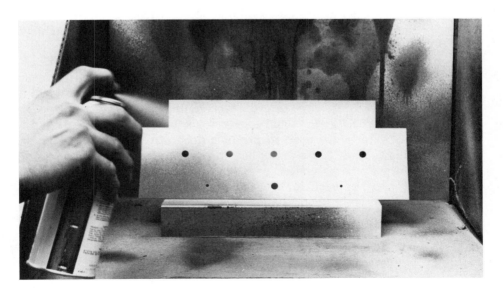

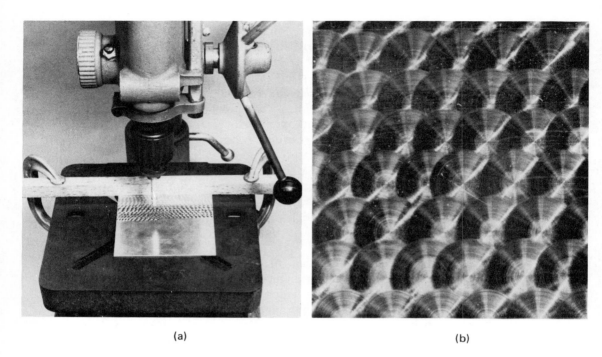

(a)   (b)

FIGURE 9.6. Mechanical technique for finishing. (a) Drill press arrangement for textured finish. (b) Resulting finish.

is moved back a sufficient distance so that the overlap is consistent. One-third to one-half the diameter of a single polished circle will provide an optimum overlap for circles *within* a row. Each row should overlap the previous one by one-fourth of the diameter of a single circle. Exact measurement with a scale is unnecessary. After several practice trials on a sample piece of stock, uniform spacing can easily be achieved through visual comparison.

Other mechanical-type surface textures such as striations or etched-like finishes can be obtained by hand-sanding or with the use of steel wool. The mechanical textured finishes are excellent in hiding surface imperfections. They do, however, require a final clear protective coating.

Adhesive films and *vinyls* in a wide variety of colors and textures are also extensively used to produce decorative effects on chassis and panels. This type of finish is applied in basically the same way as *contact paper*. The vinyl is precoated with a pressure-sensitive adhesive that is protected by a paper *backing* to avoid contact with contaminants while in storage and in preliminary handling. This backing is peeled off at the time of application. To ensure proper adhesion, the surface to be covered must be smooth and thoroughly cleaned. Dust particles, grease, or wax films will prevent positive bonding. The only tools required for applying these vinyls are a felt squeegee or a 2-inch rubber roller, standard household scissors, and a sharp cutting tool such as a razor blade or knife. Optimum adhesion and contour conformability is obtained when the temperature of the surface is between 70 and 90°F (21 and 32°C). Surface temperatures below 70°F require the application of an adhesive activator to the pressure-sensitive adhesive coating. This is done by first removing the paper backing and applying a light, even coat of the activator with a 3-inch or 6-inch squeegee.

To remove the paper backing, a corner of the vinyl is first lifted and separated. The vinyl is placed face down on a clean flat surface and held down with the flat side of a knife blade while the backing is smoothly pulled away with a 180-degree motion. Under hot or humid conditions, a slight jerking motion helps remove the backing. It is important that the vinyl not become stretched while the backing is being removed. If stretched, the vinyl will wrinkle and not conform smoothly to the work surface. To avoid stretching, remove the backing while the vinyl is held firmly flat. Fingers should not come in contact with the adhesive coating. In this respect, tweezers are extremely helpful for applying small pieces.

To achieve the best results, the vinyl should be prefitted over the work surface, marked, and cut to the desired measurements prior to removing the backing. To apply relatively small sections of vinyl, the entire backing is removed from the adhesive. One edge of the vinyl is then aligned and pressed against an edge of the work surface. Using a squeegee or roller, as shown in Fig. 9.7, overlapping strokes are made over the vinyl,

FIGURE 9.7. Application of small vinyl pieces using overlapping strokes of roller.

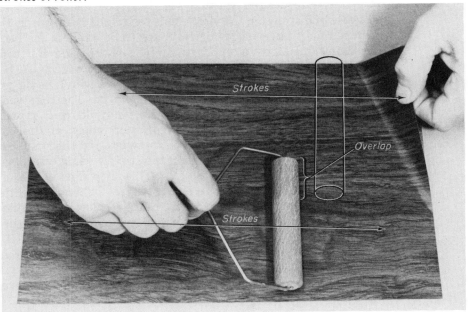

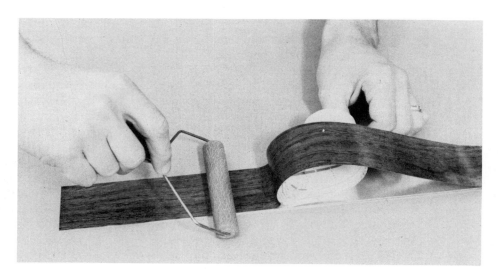

FIGURE 9.8. Technique of applying narrow vinyl strips.

gradually pressing down the remaining adhesive portion to the surface. It is also important that the vinyl not be stretched during this operation and that all pressure strokes overlap to avoid any trapped air bubbles. If a narrow strip of vinyl is to be applied, a short length of the backing is first peeled back to expose the adhesive. This edge of the vinyl is pressed firmly onto the desired starting position. The backing is then gradually removed while applying the vinyl. This technique is shown in Fig. 9.8.

When very large pieces of vinyl are to be applied, the backing is first lifted from one end to approximately one-third of the total distance across the sheet. This amount of backing is then sliced off, being careful not to cut the vinyl. Figure 9.9a shows the vinyl being prepared for application to the amplifier's enclosure element. The removed portion of the backing is then replaced onto the vinyl, leaving approximately 2 inches of exposed adhesive between the cut edges of the backing (Fig. 9.9b). The vinyl is next aligned with the work and the small exposed strip of adhesive is pressed firmly against the surface using overlapping strokes. This procedure is shown in Fig. 9.9c. This strip now acts as a hinge that maintains the alignment of the entire vinyl. The smaller section is raised and as the backing is removed, overlapping strokes with a roller are made, as shown in Fig. 9.9d. Once the smaller section of vinyl has been bonded, the same procedure is applied to the remaining section. If corners or embossed (raised) surfaces are involved, heat lamps or hot air guns may be used to heat the vinyl. When heated, the vinyl becomes very pliable and will adhere well to contours. Care should be taken, however, that the temperature of the vinyl does not exceed 200°F (93°C).

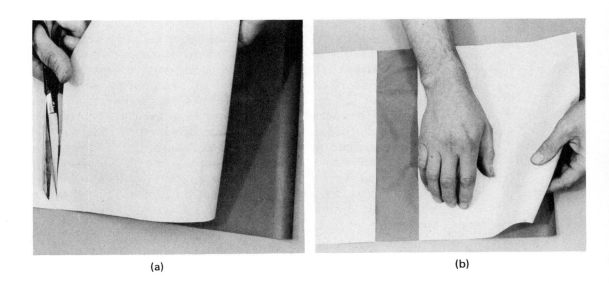

(a)

(b)

(c)

(d)

FIGURE 9.9. Recommended sequence for applying large sections of vinyl. (a) One-third of backing is first removed. (b) Replacement of backing exposing a strip of adhesive. (c) Vinyl alignment and initial application step. (d) Smaller section of backing is removed and vinyl pressed into position.

If vinyls are proving difficult to apply, first remove the entire backing and flood the adhesive surface of the vinyl and the work with a wetting solution of 2 teaspoons of any mild household detergent in 1 gallon of water. When the vinyl is applied to the work, it can then easily be slid into the desired position. (This method must not be used at temperatures below 70° F or with porous material.) Once properly aligned, the squeegee or roller is worked over the vinyl from the center toward the edges on horizontal surfaces and from top to bottom on vertical surfaces. After all air bubbles are removed and all edges are firmly adhered, the entire vinyl is dried with a clean cloth. Troublesome air bubbles can be removed by picking with a needle or cutting with a knife and working out with a squeegee. If a knife is used, the direction of cut should be selected to maintain optimum appearance. For example, if the vinyl has a grained appearance, the cut should be with the grain.

In all applications of vinyl, it is advisable to apply a size that is slightly larger than the working surface. This will ensure complete coverage of the surface, after which the vinyls can be trimmed to final size.

9.3 LABELING

The final step in completing a chassis and panel involves applying literal or numerical reference designations. The type and procedure of labeling will depend on several factors, including (1) the degree of clarity required, (2) cost, (3) time required for process, (4) durability, (5) materials and finishes involved, (6) availability of equipment and facilities, and (7) the complexity of the labeling layout pattern. The most common methods of labeling are by *stamping, engraving, metalized adhesive plates, decals, dry transfers,* and *silk screen.*

Stamping provides one of the fastest and most economical means of numbering or lettering. This process may be done with *rubber stamps* and *ink* or by *steel* number or letter stamps. Ink stamping is very effective when appropriate contrasting ink colors are used. Since ink is not durable, it requires a clear protective coating such as varnish or lacquer. Steel stamps provide extremely durable labeling since the characters are impressed into the work surface. For improved contrast, the indentations can be filled with a colored wax. When stamping cast iron or steel, a sample piece should be tested with a *mill file.* If the file will not cut the metal, steel stamps should not be used because they will be damaged. Nonferrous metals, even when heat-treated, may be stamped without pretesting because they are softer than the steel stamps. Since each character is struck individually, a guide, such as that shown in Fig. 9.10, should be used to aid in alignment and spacing. The work should be placed flat on a metal table and *only one* tap with the hammer should be applied. This will prevent characters from being "double struck." A sample piece of stock should be first tested to determine the correct amount of force necessary to obtain the desired depth of the character.

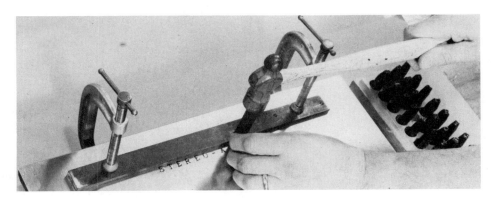

FIGURE 9.10. Guide bar aids in steel stamp alignment.

Both methods of stamping, especially for name plates or extensive labeling, are rather crude and limited.

Engraving is an extremely effective and durable method of labeling chassis. An engraving machine is used to cut characters into the work surface. These characters may then be filled with a contrasting colored wax-type material. Contrasting colored laminates are available for labeling purposes. When engraved, these laminates provide contrasting figures that do not require a wax fill.

The engraving machine, as shown in Fig. 9.11 consists of a *high-speed continuous-duty motor, metal router, pantograph lever system*, and *stylus*. The stylus is located on the lower end of the pantograph and the router is positioned on the upper end. The pantograph system causes the router to follow and duplicate any movement of the stylus. The stylus is used in conjunction with *metal templates*, which are flat plates in which have been indented a wide variety of individual letters, numbers, or symbols. These templates are clamped to the engraver table and the stylus is brought down to rest into a template groove. The router will begin to cut into the surface of the work, which is also clamped to the engraver table. As the stylus is slowly moved through the various grooves, the router cuts a duplicate character into the work.

The *six-ratio pantograph system* has six arm settings, designated *A*, *B*, *C*, *D*, *E*, and *F*. These literal designations represent the pantograph ratios of the finished engraved character to the template (master copy). These ratios are shown in Table 9.2. The finished engraved size is determined by simply dividing the template character size by the pantograph ratio. To find the ratio points on the pantograph arms, the height of the template characters is divided by the height of the desired engraved characters. The ratio thus obtained is the pantograph ratio used with Table 9.2. For example, if ¼-inch engraved letters are desired when using ½-inch template characters, a 2:1 pantograph ratio is required. This corresponds to setting *C* of Table 9.2. As shown in Fig. 9.12, the two

FIGURE 9.11. Engraving machine. Courtesy of Mico Instrument Company.

FIGURE 9.12. Pantograph ratio settings.

TABLE 9.2

Finished Sizes of Engraved Characters Obtained With Six Ratio Pantograph

		A	B	C	D	E	F
Pantograph Arm Setting / Pantograph Ratio		1.5:1	1-5/7:1	2:1	2.4:1	3:1	4:1
Ratio of Finished Engraving to Master Copy.		2/3 (.667)	7/12 (.583)	1/2 (.500)	5/12 (.417)	1/3 (.333)	1/4 (.250)
BLANK SIZE	TYPE SIZE						
3/4"	3/16"	1/8"	.109" (7/64")	3/32"	.078" (5/64")	1/16"	3/64"
	1/4"	1/6"	.146"	1/8" –	.104"	1/12"	1/16"
	5/16"	5/24" (1/5"+)	.182"	5/32"	.130"	5/48" (1/10"+)	5/64"
	3/8"	1/4"	.218" (7/32")	3/16"	.156 (5/32") –	1/8"	3/32"
	1/2"	1/3"	.291	1/4"	.208"	1/6"	1/8"
1½"	3/4"	1/2"	.437" (7/16")	3/8"	.312" (5/16")	1/4"	3/16"
	1"	2/3"	.583"	1/2"	.417	1/3"	1/4"
	1-1/4"	5/6"	.729"	5/8"	.521"	5/12"	5/16"
2¼"	1-1/2"	1"	.875" (7/8")	3/4"	.625" (5/8")	1/2"	3/8"
	2"	1-1/3"	1.167	1"	.834	2/3"	1/2"

The finished size is determined by dividing the master copy size by the pantograph ratio.

(Dimensioned center-to-center of engraved line, not over-all.)

Example: To obtain 1/4" engraved characters with 1/2" master copy the 2:1 pantograph ratio is used. This requires that ratio points "C" be matched on the pantograph arms.

The characters on Micro master type are self-spaced to eliminate the need of additional spacers.

When engraving, the width of line is controlled by the type of cutter used or the depth of cut made.

Cutters must be kept sharp to obtain satisfactory work.

pantograph positioning screws on opposite ratio arms are moved to the *C* positions on these arms and secured. Entire words may be set up on the engraver table by selecting the appropriate individual templates that are butted together and clamped into position. The templates are designed for character self-spacing, which eliminates the need for individual template spacers. Spacers are needed only between words. The width and depth of the engraved lines will depend on the type of router used. These routers must always be kept sharp in order to produce quality lettering.

Engraving is normally done *before* any surface treatment or finishing process. It is also a permanent process and care must be taken to avoid mistakes in the routing operation.

An extremely effective and simple method of labeling is through the use of *aluminum foil plates.* This process is shown in Fig. 9.13. These plates are available in thicknesses of either 0.003 inch or 0.012 inch, and can be imprinted with any standard typewriter. After typing, the plates can be attached to the work surface by soaking in water. Within approximately 1 minute, the protective cellophane liner is removed and the label is pressed into place with a squeegee or roller. Foils are also available that have a paper liner, similar to the adhesive vinyls, which is simply peeled away. The adhesive is of the dry-release or pressure-sensitive type. In all cases, the work surface must be free of contaminants for positive adhesion.

Pressure-sensitive decals and dry transfers, such as those shown in Fig. 9.14 provide a rapid means of labeling with crisp and sharp line characteristics. They are used extensively with the silk-screen and etching methods of labeling. Decals and transfers will adhere to practically any clean, smooth surface such as wood, paper, cardboard, plastic, glass, or metal. To improve their durability, they should be given a protective clear coating after they have been applied.

FIGURE 9.13. Application of aluminum foil label.

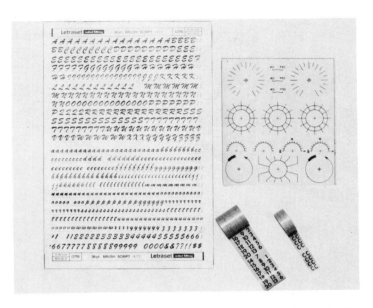

FIGURE 9.14. Typical selection of pressure-sensitive decals and dry transfers.

Pressure-sensitive decals have a protective backing that must be removed before they can be applied. Because literal and numerical decals are small, it is best to use tweezers or a knife edge to remove them from the backing, which is chemically treated for easy release. This technique reduces the possibility of touching the adhesive with the fingers and upsetting the tack. Once the backing has been removed, the decal is brought into the desired position, aligned, and pressed firmly into place, as shown in Fig. 9.15. Occasionally, it may be necessary to trim the clear support liner if overlapping occurs.

Dry transfers used in electronic packaging consist of drafting symbols, letters, words, numbers, and meter and dial markings printed on a transparent film with opaque black ink. This ink contains a pressure-sensitive heat-resistant adhesive. Before applying the transfer figure to a surface, the protective backing sheet is first removed. This sheet does not adhere to the film and therefore separates readily. The figure is brought into position with the inked side *toward* the work surface and is held firmly in place. The entire area around the figure is rubbed with a wooden burnisher, pencil, or ballpoint pen using overlapping strokes. Figure 9.16 shows dry transfers being applied to the amplifier's main chassis element. Since many figures or symbols are generally on the same sheet and closely spaced, care must be taken to avoid running over unwanted figures or portions of figures. By gently lifting one edge of the film, a check for complete transfer can be made. To assure perfectly transferred figures

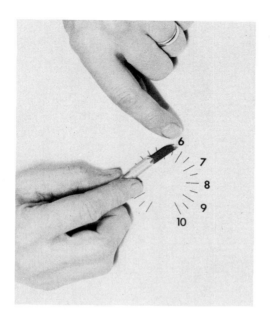

FIGURE 9.15. Positioning of dry transfer onto painted metal panel.

FIGURE 9.16. Application of dry transfers to metal surface.

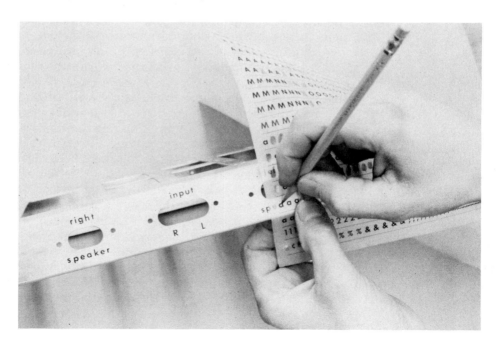

with sharp lines and no pinholes, the backing sheet should be placed back over the work surface and the figures re-burnished. A wax residue resulting from the transfer process is generally formed on surfaces such as glass or metal. This residue can easily be removed with rubber cement thinner and a soft cloth without damaging the transfers. When dry transfers are used for direct labeling on a surface, a *plastic*-type protective film is necessary. These acrylic or vinyl films should be selected carefully because they attack certain materials. Vinyls are often preferred over acrylics because they are more resistant to flaking and chipping. These films are available in aerosol cans in both gloss and matte finishes.

The silk-screen process for labeling requires the preparation of a *clear polyester positive master* when identical duplication of work is required. Unlike the previously discussed methods of labeling, the silk-screen process is intended for mass production. (The silk-screen operation for producing marking masks as well as printed circuit conductor patterns is described in detail in Chapter 13.) Decals or dry transfers allow the positive art work to be constructed quickly and eliminates time-consuming drafting procedures that would otherwise be necessary. A thin sheet of drafting *vellum*, which is slightly larger than the surface to be labeled, is first placed over the sheet metal layout drawing. The corners, mounting holes, and hole centers are carefully located by *pencil* on the vellum. Allowing for surface space to be taken up by mounting nuts, bezels, and control knobs, the required nomenclature and symbol locations together with the necessary spacing are *sketched* or *blocked* in. This technique is shown in Fig. 9.17a. Crowding should be avoided and the spacing should be standardized to avoid confusion. Whenever possible, nomenclature should be consistently above or below components or controls. For a balanced appearance, all lettering should be centered on a vertical center line of a mounting hole. After all labeling positions have been located, the vellum is placed over a comparable size sheet of graph paper and taped into place on a drafting board. (Graph paper with 10 divisions to the inch is recommended.) A *polyester sheet* is then taped over the vellum. With the aid of the grid lines and the blocked-in nomenclature and lettering positions, all decals or transfers can be accurately positioned and applied onto the polyester. This technique is shown in Fig. 9.17b.

For the silk-screen process, described in detail in Chapter 13, a *film negative* is contact-printed from the polyester master. The developed film, as shown in Fig. 9.18a, is allowed to dry on a fine mesh silk screen attached to a frame. Figure 9.18b shows the processed screen for the amplifier's front panel. The screen is positioned over the metal panel to be labeled and the appropriate paint is spread across the screen with a squeegee. The paint will penetrate the screen and be deposited onto the panel only through those areas seen as clear in Fig. 9.18b. This technique is shown in Fig. 9.18c. The screen is then removed and the characters allowed to dry. Figure 9.18d shows the amplifier's front panel labeled by the silk-screen method. (See Appendix VI for information on typical paints used for silk-screen labeling.)

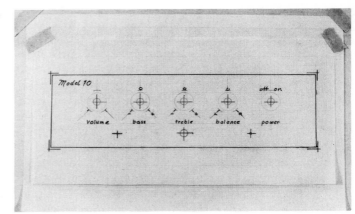

(b)

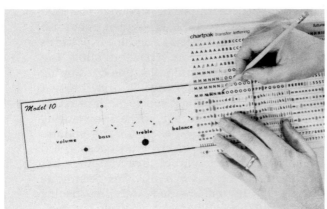

FIGURE 9.17. Construction of artwork master for the amplifier's front panel. (a) Nomenclature and hole position is first constructed on drafting vellum. (b) Dry transfers being applied to polyester sheet.

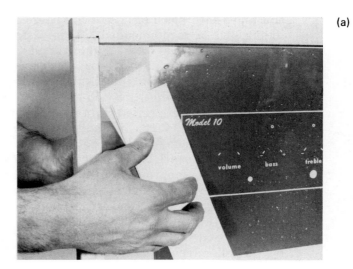

(a)

FIGURE 9.18. Silk-screen method for panel labeling. (a) Initial positioning of film onto silk screen.

(b)

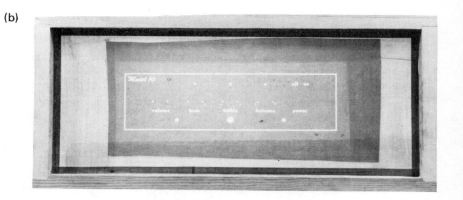

(c)

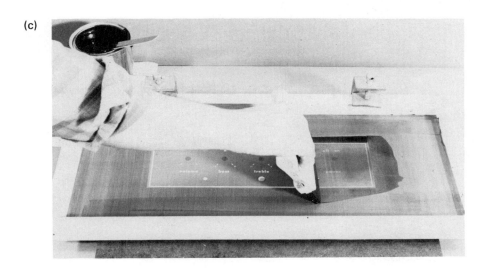

(d)

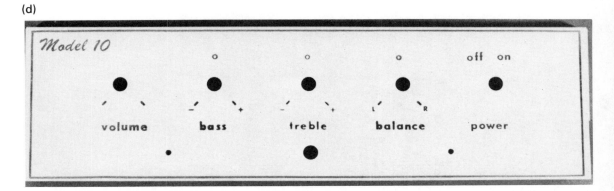

FIGURE 9.18. (cont.) (b) Processed screen. (c) Application of paint through the screen onto front panel. (d) Screened front panel of amplifier.

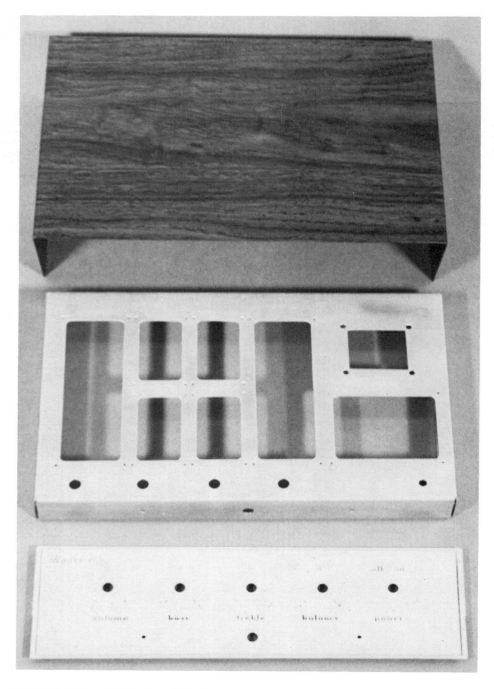

FIGURE 9.19. Finished amplifier's metal elements.

In this chapter only the most common types of labeling techniques have been discussed. Although many other processes are commercially available, they are not intended for prototype work. Of those labeling methods considered, decals and dry transfers are the most economical, least involved, and produce a highly acceptable appearance.

With so many variations and techniques available for finishing and labeling, the technician must consider the following factors prior to determining the most appropriate method to use: surface materials, number or complexity of markings, required sharpness of detail, equipment availability, cost, time required, and the technician's own skill. An evaluation of these factors will dictate the appearance of the finished product.

The amplifier's main chassis, front panel, and enclosure elements are shown in Fig. 9.19, processed by means of finishing and labeling procedures just discussed.

This chapter completes the discussion of chassis development. The following chapter begins a new phase in prototype construction with the introduction of printed circuit board material.

PROBLEM SET

9.1 Using steel number stamps, mark the size of each tap and clearance hole in the drill gauge laid out, drilled, punched, and cut in Problems 5.5, 6.3, and 7.3 for the machine screw sizes 2, 4, 6, 8, and 10. Buff surfaces after stamping with a piece of No. 00 steel wool.

9.2 Finish the outside surfaces of the utility box constructed in Problem 8.4 using the pencil-eraser technique discussed in Sec. 9.2.

9.3 Cover the apron and front surface of the gauge panel constructed in Problems 5.4, 6.1, 7.4, and 8.1 with adhesive wood grain vinyl.

9.4 Protect the outside surfaces of the box chassis constructed in Problem 8.5 with a spray paint hammertone finish.

9.5 Determine the size of engraved characters using a pantograph setting of B and a master copy size of $\frac{3}{8}$ inch.

10

Printed Circuit Board Materials

10.0 INTRODUCTION

With the advent of the transistor and the more recent heavy reliance on integrated circuits, electronic packaging and wiring problems have been compounded. No longer can the technician rely solely on hand-wiring methods; now designs must be oriented toward more extensive use of printed circuitry. Many modern electronic systems would be virtually impossible to package without incorporating printed circuits into their design.

Printed circuits are *metal foil conducting patterns*, usually copper, bonded to a *substrate* (insulating base material) for support. The metal pattern serves as the connecting medium for the electrical components that are assembled on the opposite side of the board. Component leads are fed into holes that are drilled or punched through the base material and foil. These leads are soldered to the conducting pattern to form the complete printed circuit.

The advantages of printed circuits over conventional wiring methods are that they (1) adapt easily to miniaturization and modular design, (2) provide for uniformity in production, (3) lower costs, (4) virtually eliminate wiring error, and (5) minimize assembly and inspection time.

Printed circuit techniques lend themselves readily to mass production, resulting in a highly reliable package. Although the format of this text is focused on prototype construction, it must be assumed that eventually the package will be released for production. Consequently, the technician should design the package with an eye toward this end by selecting materials and configurations that will reduce construction problems.

To completely understand the sequential process involved in printed circuit board design and fabrication, the next seven chapters will be devoted exclusively to these topics. This chapter discusses the initial step in printed circuit construction — the selection of the most appropriate type of printed circuit board material. Topics discussed are classification of printed circuit boards, insulating base materials, conductive foils, bonding, and tabulated information of board characteristics to aid the technician in selection.

10.1 CLASSIFICATION OF PRINTED CIRCUIT BOARDS

Printed circuits are designed as either *rigid* or *flexible* and are further classified into *single-sided*, *double-sided*, or *multilayer* conductors. Regardless of the type, all printed circuit boards are formed with an insulating base material on which a conducting foil is either chemically or mechanically bonded. When processed, the printed circuit board provides both an electrical wiring path and a mechanical support for the components.

Rigid printed circuit boards of the single-sided type are the most frequently used, finding extensive application in relatively uncomplicated circuitry. Figure 10.1a shows a rigid single-sided blank before processing.

FIGURE 10.1. Rigid single-sided printed circuit boards. (a) Copper-clad printed circuit blank. (b) Processed single-sided printed circuit conductor pattern.

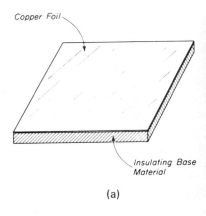

Copper Foil

Insulating Base Material

(a)

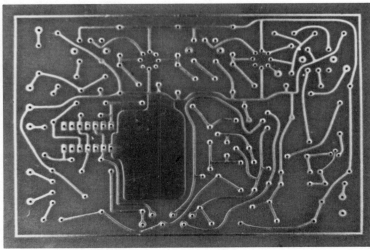

(b)

The completed conductor pattern ready for component mounting is shown in Fig. 10.1b.

Rigid printed circuit boards of the double-sided type consist of conducting foil bonded to *both* sides of the insulating base. This type of board is employed when circuit complexities make it difficult or impossible to develop practical wiring layouts on one side only. Double-sided boards also reduce the overall size of a comparable circuit constructed on a single-sided board, since components are mounted on both sides. Figure 10.2 shows an assembled double-sided board. Interconnections among the conductive patterns on both sides of the board are accomplished by *interfacial connections*, *eyelets*, or *tubelets*. (These and other printed circuit hardware will be discussed further in Chapter 14.) With the use of these interconnections, wires or component leads can be soldered between the foil patterns on opposite sides of the base materials.

Multilayer rigid printed circuit materials were developed to satisfy the needs of the more recent state of the art, which demands extremely complex wiring layouts in a relatively small space. High-density wiring, of the type shown in Fig. 10.3, can be achieved by stacking together preprocessed rigid boards and interconnecting them by flexible cable.

Flexible printed cables consist of conductors laminated between layers of insulation and exposed for soldering only at the terminal points,

FIGURE 10.2. Assembled double-sided rigid printed circuit board.

FIGURE 10.3. Rigid/flexible multilayer printed circuit design.
Courtesy of Fortin Laminating Corporation.

as shown in Fig. 10.4. Flexible printed cables are used in a manner similar to ordinary cable but are extremely flexible. In addition, the flat configuration of these circuits make them desirable for system interconnections where a round cable may prove to be too cumbersome. *Teflon, polymide, polyester, polyvinyls, polypropylene,* and *polyethylene* are typical insulating materials used in flexible printed cables. Table 10.1 shows a comparison of the various characteristics for common insulations. Since various methods and adhesives are used for the internal bond of these laminates, no one method of insulation removal will suffice for all types of these cables. The insulation is usually stripped by *friction wheel strippers, knife-edge strippers, chemical dipping,* or *melting* by a concentration of hot air. Figure 10.5 shows a friction wheel stripper. Table 10.2 shows the preferred stripping method for various insulating materials. Because flexible printed cables have a large flat conducting area, they can handle larger currents than conventional wire cable. Table 10.3 shows the approximate current capacities and derating values when a multiple number of conductors are laminated within a common cable.

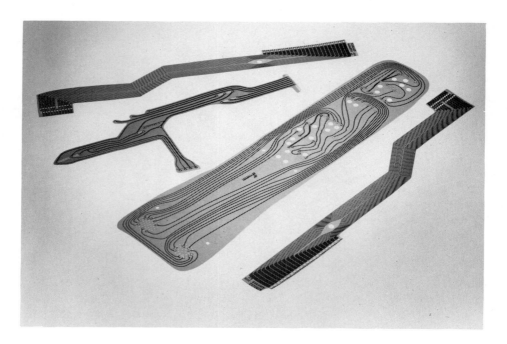

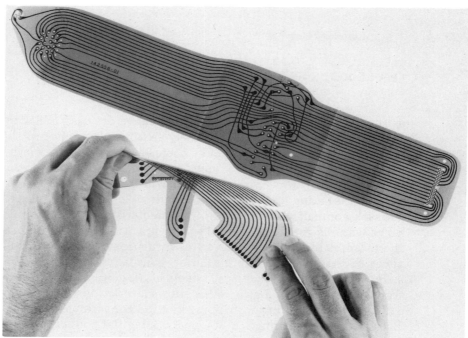

FIGURE 10.4. Flexible printed circuit cables. Courtesy of DuPont Company.

TABLE 10.1

Insulation Characteristics for Flexible Printed Cables.
(Courtesy of Insulfab Plastics Inc.)

	TFE FLUORO-CARBON	TFE GLASS CLOTH	FEP FLUORO-CARBON	FEP GLASS CLOTH	POLYI-MIDE	POLY-ESTER	POLY-CHLORO-TRIFLUORO-ETHYLENE	POLY-VINYL FLUORIDE	POLY-PROPYLENE	POLY-VINYL CHLORIDE	POLY-ETHYLENE
Specific Gravity	2.15	2.2	2.15	2.2	1.42	1.395	2.10	1.38	0.905	1.25	.93
Square inches of 1 mil film per pound	12,800	13,000	12,900	13,000	19,450	21,500	12,000	20,000	31,000	22,000	30,100
Service Temp. Deg. C (Minimum)	−70	−70	−225	−70	−250	−60	−70	−70	−55	−40	−20
(Maximum)	250	250	200	250	+250	150	150	105	125	85	60
Flammability	Nil	Nil	Nil	Nil	Nil	Yes	Nil	Yes	Yes	Slight	Yes
Appearance	Translucent	Tan	Clearbluish	Tan	Amber	Clear	Clear	Clear	Clear	Translucent	Clear
Thermal Expansion X 10 inches/inch/deg. F	70	Low	50	Low	11	15	45	28	61	–	–
Bondability with Adhesives	Good	Good	Good	Good	Good	Good	Good	Good	Poor	Good	Poor
Bondability to itself	Good	Poor	Good	Good	Poor	Poor	Good	Good	Good	Good	Good
Tensile strength PSI 77° F	3,000	20,000	3,000	20,000	20,000	20,000	4,500	8,000	5,700	3,000	2,000
Modulus of Elasticity PSI	80,000	3	70,000	3	430,000	550,000	200,000	280,000	170,000	–	50,000
Volume Resistivity ohms-cm	2 x 10	10	10	10	10	1 x 10	1 x 10	3 x 10	10	1 x 10	1 x 10
Dielectric Constant 10-10 cycles	2.2	2.5/5	2.1	2.5/5	3.5	2.8-3.7	2.5	7.0	2.0	3-4	2.2
Dissipation factor 10-10 cycles	.0002	.0007/.001	.0002	.0001/.001	.002/.014	.002-.016	.015	.009-.041	.0002/.0003	.14	.0006
Dielectric strength (5 mils thickness) volt/mil	800	650/1600	3,000	650/1600	3,500	3,500	2,000	2,000	.125 in thk 750v/mil	800	1,500
Chemical Resistance	Excellent	Excellent	Excellent	Excellent	Excellent	Excellent	Excellent	Good	Excellent	Good	Excellent
Water Absorption, %	0	.10/68	0	.18/30	3	0.5	0	15	.01	.10	0
Sunlight Resistance	Excellent	Excellent	Excellent	Excellent	Excellent	Fair	Excellent	Excellent	Low	Fair	Low

TABLE 10.2

Comparison Chart to Select Most Appropriate Stripping Method:
Size of Black Sector Indicates Effectiveness of Stripping Method

METHOD OF STRIPPING

FLAT CABLE TYPE	DIELECTRIC	BOND STRENGTH	Knife Edge — Sharp — Room Temperature — NASA (1)	Knife Edge — Sharp — Room Temperature — Gore (2)	Knife Edge — Sharp — Hobel (3)	Knife Edge — Dull — Hot — NASA (4)	Knife Edge — Dull — Hot — Hughes (5)	Friction Wheel — Rush Viking Carpenter (6)	CHEMICAL (7)	Hot Air (8)
UNSHIELDED	Polyester	High				●	◑(7)	○	●	◔(7)
UNSHIELDED	Polyimide-amide	High							◑	●
UNSHIELDED	Polyimide FEP	High			◔			◔	◑(6)	◔(6)
UNSHIELDED	Polyimide FEP	Low	●		◔			◔	◑(6)	◔(6)
UNSHIELDED	TFE	Very Low	●	●				◔		
SHIELDED	Polyester	High			◔	◔(7)	◔(7)	◔(7)	●	◔(7)
SHIELDED	Polyimide-amide	High			◔			◔	●	
SHIELDED	Polyimide FEP	High			◔			◔	◑(6)	◔(6)
SHIELDED	Polyimide FEP	Low	◔		◔			◔(6)	◑(6)	◔(6)

◕ SIZE OF BLACK SECTOR INDICATES RATING

187

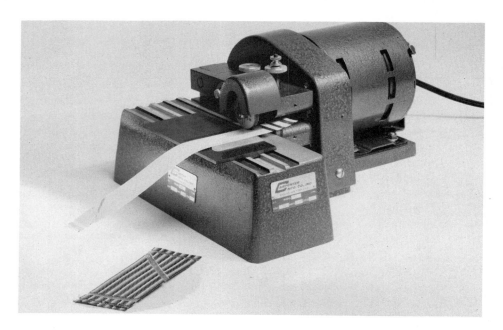

FIGURE 10.5. Friction wheel stripper. Courtesy of Carpenter Manufacturing Company.

TABLE 10.3

Conductor Temperature Rise and Current Derating for Flat Cables: (1) Rise in Temperature above Ambient as a Function of Current; (b) Approximate Derating Curve for Flat Cable

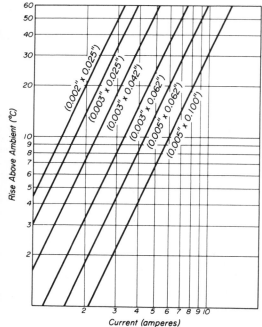

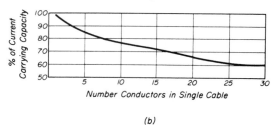

(b)

(a)

10.2 INSULATING BASE MATERIALS FOR PRINTED CIRCUIT BOARDS

Printed circuit boards of the rigid variety will be discussed exclusively in the remaining sections of this text since they are by far the most common. The insulating or resin material for these boards is either a *thermosetting plastic* or a high-temperature-resistant *thermoplastic polymer* that provides the support for the conductor pattern and components. The discussions will be limited to the thermosetting plastics since they are used almost exclusively in rigid boards.

Base materials for printed circuit boards are laminates containing a *reinforcement* such as *paper, glass,* or *fabrics,* including *cotton cloth, asbestos, nylon cloth,* and *glass cloth.* These laminates are formed under pressure and heat. The thermosetting plastic resins common to the manufacture of printed circuit boards are *phenolic, epoxy, melamine, silicone,* and *Teflon.* Silicone-glass is mechanically inferior and more expensive than either phenolic, epoxy, or melamine resin boards. Teflon is used primarily for microwave applications and is approximately fifteen times more expensive than phenolic and at least five times as costly as the epoxy and melamine resin types. The phenolic types are less expensive than the epoxy resins and for many applications are both electrically and mechanically suitable. In general, the epoxy types possess superior electrical and mechanical characteristics than phenolic or melamine. In most cases, the reliability of a particular type of board will by far outweigh the cost factor.

To aid the technician in selecting the most suitable base material for a specific application, there follows a description of some of the electrical, mechanical, and chemical qualities of the more common base materials. The classifications are in accordance with NEMA (National Electrical Manufacturers Association) standards. These standards designate phenolic resin printed circuit boards as either *X, XX,* or *XXX,* which represent the amount of resin present, the amount increasing with the number of *X's.* In general, the more resin content, the more improved are the electrical and mechanical characteristics of the board. A *P* suffix together with the *X* designation denotes that the phenolic is capable of being punched *but* only at elevated temperatures, whereas a *PC* suffix indicates cold-punch qualities.

Grade XX. This grade of paper-base phenolic possesses fair electrical characteristics and good mechanical properties. It machines well but is not recommended for punching. It offers little abrasive action to drills and, as a result, does not dull these bits easily. This grade is used by many manufacturers as a reference for determining the costs of other laminate types. For example, if the price index of nylon cloth phenolic is 3.5, this indicates that its cost is that much more than grade *XX,* which has a cost index of *1* (see Table 10.5).

Grade XXP. This type is similar in electrical and mechanical properties to grade *XX* and it can be punched at increased temperatures of between 200 and 250°F (93 to 121°C).

Grade XXX. This material possesses some improved electrical and mechanical properties over the *XX* type and is recommended for use at radio frequencies. It is available in a *self-extinguishing* form, meaning that it will not support combustion if ignited. This grade possesses fair *dimensional stability* (i.e., it maintains its size within close tolerances under environmental changes such as temperature or moisture).

Grade XXXP. This grade is recommended for most general-purpose applications and is similar to grade *XXX*, but possesses an additional advantage in that it can be punched at elevated temperatures. Frequencies up to 10 MHz are permitted with this grade.

Grade XXXPC. This is similar in electrical and mechanical properties to grade *XXXP* board and has higher insulation resistance and lower water absorption. These qualities make this type of board particularly useful for high-humidity applications. In addition, this material can be punched at temperatures of from 70 to 120°F (21 to 49°C). For these reasons, *XXXPC* printed circuit boards are widely preferred over grades *XX* and *XXP*. However, when punching and moisture are not considerations, grade *XXXP* is generally selected.

Grade G-5. This is a melamine-glass base material with excellent electrical and mechanical properties. It is the least expensive glass cloth base but is one of the most abrasive grades and, as a result, is very seldom used for general-purpose printed circuit applications.

Grade G-10. Manufactured with a glass-cloth reinforcement and epoxy resin, this grade possesses superior electrical and mechanical properties when compared to phenolic boards. Because of its high flexibility, low dielectric losses and moisture absorption, and high bond strength between the base and the conducting foil, this laminate is widely used and is recommended for frequencies up to 40 MHz.

Grade G-11. Also employing a glass-cloth reinforcement with epoxy resin, this grade is similar in characteristics to grade *G-10* but is more flexible at higher temperatures. In addition, it is more resistant to acids, alkali, and heat than *G-10*, but is more difficult to machine. The cost of both of these grades is the same.

Grade FR-3. This grade is designed for optimum balance of electrical and mechanical properties and is manufactured with paper reinforcement impregnated with an epoxy resin. It is considered superior to grade

XXXPC because of its higher arc resistance, lower moisture absorption, and improved dimensional stability qualities. It is used almost exclusively as a self-extinguishing type in the higher frequency ranges, up to 10 MHz.

The information presented for these various types and grades of laminates is included with additional information in Tables 10.4, 10.5, and 10.6. These tables provide for a complete comparison of a wider range of electrical, mechanical, and chemical properties of the described grades, in addition to many other base materials and their properties. All of these properties should be evaluated when selecting the most suitable base material for a particular application.

Single- and double-sided copper-clad laminates are available in sheets that range from 26 to 36 inches wide by 36 to 48 inches long. Smaller, precut dimensions are also available at extra cost. The thicknesses of the laminates, including the copper foil, range from $\frac{1}{32}$ to $\frac{1}{4}$ inch in increments of $\frac{1}{32}$ inch.

A characteristic of all printed circuit boards is their tendency to *warp* and *twist*. These conditions are the result of the two dissimilar materials (base and copper foil) bonded together to form the laminate. After etching, however, the boards will significantly straighten out because of the large amount of copper loss. Warp and twist describe the deviation of the board from a straight line fixed by two points at the extremities of the board. To determine the degree of warp, this line is taken along the longest edge. Twist is evaluated along a diagonal line between opposite corners of the board. These values of warp and twist are expressed in percentage of board or diagonal length, with the thinner boards tending to warp more than the thicker ones. Deviations range from approximately 12% for a thickness of $\frac{1}{32}$ to 5% for a $\frac{1}{4}$-inch thickness. Double-sided boards reduce these values by at least a factor of 2. These deviations of the boards become important considerations when designing for plug-in-type circuits, wherein only extremely small amounts of warp or twist can be tolerated.

10.3 CONDUCTING FOIL

Copper is the principal type of foil used in the manufacture of printed circuits. For special applications, nonstandard materials such as *aluminum*, *steel*, *silver*, and *tin* foils are used. Copper foil has the advantages of high conductivity, excellent soldering characteristics, low cost, and ready availability in a variety of widths. The copper used for printed circuits is as least 99.5% pure in order to maintain its high-conductivity properties.

The manufacturers of copper foil for printed circuit applications employ two basic production methods. These methods are termed *rolled foil* and *electrodeposited foil*. Rolled foil is formed from refined blocks of pure copper. The foil thus produced is largely free of pinholes and im-

LAMINATED THERMOSETTING SHEETS
PRINCIPAL GRADES, USES AND PROPERTIES

These are AVERAGE and not GUARANTEED properties.
Guaranteed values are NEMA or applicable MIL-P specifications.

DESCRIPTION CHARACTERISTICS — APPLICATIONS	NEMA GRADE	Conforms to Military Spec. MIL-P	COLOR*	Standard Thicknesses (others available)	MECHANICAL TIMES 1000				
					Compr. Str. (Lb./Sq. In.) ½" Face	Flexural Str. (Lb./Sq. In.) 1/16" Face—Lengthwise	Flexural Str. (Lb./Sq. In.) 1/16" Face—Crosswise	Tensile Str. (Lb./Sq. In.) 1/16" Lengthwise	Tensile Str. (Lb./Sq. In.) 1/16" Crosswise
Paper/Phenolic — General Purpose. For panels, contactors, terminal blocks — high strength, low cost	X		N.B.C.	.010—2.0	38.0	28.0	20.0	20.6	16.0
Paper/Phenolic — For applications requiring good mechanical strength and dim. stability with elec. properties secondary	XP		N.B.C.	.020—2.0	35.0	30.0	25.0	20.0	16.0
Paper/Phenolic — Excellent cold punching, high strength	XPC		N.B.C.	.015—.250	42.0	23.0	20.0	17.0	12.7
Paper/Phenolic — An economy grade offering excellent shear and punch characteristics at room temp.	XPC		N.B.C.	.031—.125	32.0	24.0	20.0	12.0	10.0
Paper/Phenolic, good mechanical, electrical and machining properties	XX	3115, PBG	N.B.	.010—2.0	34.0	22.0	19.0	16.0	13.0
Paper/Phenolic — Excellent punching, high mechanical and high electrical. Best all-purpose	XXP		N.B.C.	.015—.250	41.0	23.0	18.0	16.3	13.8
Paper/Phenolic — Excellent electrical grade used where dim. stability or min. cold flow is essential	XXX	3115, PBE	N.B.	0.10—2.0	35.0	20.0	17.0	13.1	10.5
Paper/Phenolic — Best quality paper base. High insulation resistance, low loss material	XXXP	3115, PBEP	N.B.	.031—.250	28.0	19.0	17.5	12.0	9.5
Paper/Phenolic — High quality, cold punching material. High insulation resistance	XXXPC	3115, PBE-P	N.B.	.031—.50	30.0	19.0	15.0	14.0	11.0
Cotton Cloth/Phenolic — General purpose. Excellent mach. and punching. For panels, wedges, gears, wear strips	C	18324 15035, FBM	N.B.	.031—10.0	41.0	22.0	18.0	12.1	9.80
Cotton Cloth/Phenolic — Best canvas base electrical grade. Excellent acid resistance. For panels, steam valve discs	CE	15035, FBG	N.B.	.031—3.0	39.0	21.0	19.0	13.5	9.80
Cotton Cloth/Phenolic. A finer weave than Grade C. Better for fine punching, close machining. Mechanical usage	L	15035, FBI	N.B.	.020—4.0	39.0	21.5	18.0	11.9	7.9
Cotton Cloth/Phenolic — Best insulation of cloth grades. Excellent resistance to moisture, mild acids, alkalies	LE	15035, FBE	N.B.	.010—3.0	38.0	22.0	17.0	15.0	10.6
Fine Weave Cotton Cloth/Phenolic, with graphite added. Non-insulator, used for thrust bearings, pump bearings	Linen Graphite	5431 (AER.)	Green-Black	.008—6.0	37.0	20.0	16.0	15.0	11.0
Asbestos Paper/Phenolic — Suitable for continuous use at 150°C. Good mechanical, poor electrical strength	A	8059 (USAF)	N.B.	.015—1.0	45.0	28.0	21.0	18.0	13.0
Asbestos Cloth/Phenolic — High heat resist., dimensional stab., low water absorp., low thermal expan., good impact resist.	AA	8059 (USAF)	N.	.047—4.0	40.0	30.0	19.0	9.2	8.1
Nylon Cloth/Phenolic — Excellent electrical and good mechanical properties under humid conditions	N1	15047-NPG	N.	.015—1.125	40.0	13.5	10.5	10.0	6.0
Glass Cloth/Phenolic — Used in applications where stability of elec. and mech. properties is essential	G3		N.	.031—2.0	55.0	20.0	18.0	23.0	18.0
Glass Cloth/Melamine — Excellent mechanical and electrical properties. Good arc resistance. flame retardance	G5	15037, GMG	Gray-Brown	.008—3.5	70.0	50.0	43.0	39.0	36.0
Glass Cloth/Silicone — Suitable for high temp. (Class H)) applications. Low loss electrical properties	G7	997, GSG	White	.010—2.0	62.0	38.0	32.0	33.0	31.0
Glass Cloth/Melamine — An arc and flame resistance laminate with outstanding elec. prop. to meet MIL-P15037C, GME	G9	15037C, GME	Gray-Brown	.008—3.5	62.0	83.0	61.0	48.0	35.0
Glass Cloth/Epoxy — High ins. resist., very low absorp. Highest bond strength of glass laminates. High stability in humidity	G10	18177, GEE	Green	.010—2.0	50.0	70.0	55.0	49.0	38.0
Glass Cloth/Epoxy — High flexural strength retention at elevated temperatures. Extreme resistance to solvents	G11	18177, GEB	Green	.010—2.0	58.0	70.0	65.0	48.0	37.0
Paper/Epoxy — Designed for optimum balance of mechanical and electrical properties, with excellent machinability	FR-3	22324, PEE	Ivory	.031—.50	32.0	26.5	22.0	12.0	9.0
Glass Cloth/Epoxy — A flame resistant epoxy laminate with excellent machining and electrical properties	FR-4	18177, GEE	Green	.010—.125	55.0	82.0	75.0	48.0	37.0
Glass Cloth/Epoxy — High strength retention at elevated temperatures. Self-extinguishing	FR-5	18177, GEB	Green	.010—1.5	50.0	58.0	52.0	49.0	36.0

Sheet Sizes: Range from 36" to 48" wide x 36" to 120" long, depending on Grade and Thickness.
Consult nearest Insulfab office for additional information.

* Colors: Symbols N, B, or C refer to Natural, Black, or Chocolate. "N" ranges from light tan to dark brown. Other referenced colors are "Natural for grade." MIL-P specifications approve Natural colors only.

Table 10.4

Insulating Material: Grades, Uses and Properties (Courtesy of Insulfab Plastics Inc.)

Mod. of Elas. (Lb./Sq. In.) 1/16" Lengthwise ($\times 10^6$)	Mod. of Elas. (Lb./Sq. In.) 1/16" Crosswise ($\times 10^6$)	Izod Impact (Ft. Lb./In. Notch) 1/16" Build-Up Edge—Lengthwise	Izod Impact (Ft. Lb./In. Notch) 1/16" Build-Up Edge—Crosswise	Bonding Strength 1/2" (lbs.)	Rockwell Hardness M Scale	Diel. Str. (V/Mil) Perp. To Lam. 1/16" S/S @ 23°C	Diel. Breakdown KV Par. To Lam. 1/16" S/S @ 23°C	Dissipation Factor @ 1 Mc. 1/16" As Received	Dissipation Factor @ 1 Mc. After 24 Hr. In Water 1/16" @ 23°C	Diel. Constant @ 1 Mc. 1/16" As Received	Diel. Constant @ 1 Mc. After 24 Hr. In Water 1/16" @ 23°C	Continuous Operation Temp. Limit Deg. C	Coef. Of Expansion In./In./C° Lengthwise $\times 10^{-5}$ 1/16"	Coef. Of Expansion In./In./C° Crosswise $\times 10^{-5}$ 1/16"	Deformation Under Load 1/16" (Percentage) @ 70 C	Thermal Conductivity BTU/Hr./Sq. Ft./In. Thick/F°	Water Absorp. % 24 Hr. @ 23°C 1/16" Thick	Specific Gravity	NEMA GRADE
1.70	1.40	1.00	0.80	1030	102	580	45	0.042	0.047	5.4	5.6	120	2.0	2.5	0.98	1.6	1.70	1.38	X
1.50	1.10	0.70	0.60	1100	110	550	62	0.054	0.080	6.5	7.5	120	2.0	2.5	0.60	1.6	3.00	1.44	XP
1.00	0.80	0.65	0.60	——	93	470	65	0.046	0.078	5.1	6.5	120	1.6	3.0	1.69	1.7	5.00	1.36	XPC
1.40	1.00	0.90	0.75	——	85	550	73	0.051	0.073	5.1	5.8	120	1.6	3.0	2.0	1.7	2.6	1.37	XPC
1.26	0.88	0.60	0.50	900	102	625	60	0.040	0.046	5.0	5.4	120	1.1	2.6	0.93	1.7	1.53	1.35	XX
1.10	0.80	0.80	0.70	——	105	555	65	0.036	0.044	5.0	5.3	120	1.5	1.7	1.59	1.7	1.25	1.36	XXP
1.36	0.90	0.60	0.50	960	110	745	68	0.034	0.036	5.0	5.4	120	1.7	3.5	0.56	1.8	0.78	.35	XXX
0.62	0.47	0.5	0.5	——	98	740	90	0.028	0.030	3.8	3.9	120	1.06	1.21	1.64	1.7	0.40	1.27	XXXP
0.70	0.60	0.60	0.50	1050	104	740	70	0.028	0.032	4.1	4.3	120	1.1	1.2	1.67	1.7	0.40	1.28	XXXPC
1.05	0.79	2.80	2.40	2100	103	310	40	0.053	0.089	5.2	5.8	120	1.8	2.2	1.00	2.3	2.00	1.33	C
1.14	1.00	2.50	2.20	1800	104	400	60	0.045	0.050	5.0	5.6	120	1.8	2.2	1.30	2.3	1.52	1.35	CE
1.00	0.80	1.70	1.40	1820	105	300	50	0.055	0.070	5.8	6.5	120	2.2	3.5	1.20	2.2	1.60	1.33	L
0.91	0.75	2.00	1.60	1820	108	430	60	0.045	0.050	5.6	6.0	120	2.2	3.5	0.94	2.2	1.20	1.33	LE
0.84	0.70	1.87	1.15	1700	105	—	——	——	——	—	——	120	2.2	3.5	0.96	2.9	2.00	1.40	Linen Graphite
1.30	1.10	3.00	2.80	1250	100	120	5	——	——	—	—	150	1.0	1.3	1.06	2.3	1.00	1.56	A
1.42	1.06	5.00	3.50	2100	99	90	10	——	——	·	—	170	0.9	1.4	1.30	3.5	1.20	1.64	AA
0.40	0.35	4.00	2.50	1500	85	330	70	0.030	0.040	3.7	3.8	120	3.0	5.0	3.40	2.2	0.50	1.17	N1
1.70	1.50	8.00	6.00	1010	103	500	60	0.018	0.090	4.9	8.0	150	1.0	1.5	0.30	1.8	2.10	1.73	G3
2.10	1.90	13.00	10.00	1800	123	395	34	.011	.065	6.1	7.2	150	1.0	1.1	.60	2.4	2.2	1.9	G5
1.80	1.60	16.00	16.0	850	102	400	70	0.002	0.003	4.1	4.2	250	1.0	1.0	0.25	1.4	1.77	0.10	G7
3.20	2.20	15.00	13.00	2100	115	400	65	0.012	0.015	7.0	7.2	150	1.5	1.8	0.50	2.4	0.68	1.92	G9
1.90	1.80	10.00	8.00	2300	105	510	65	0.021	0.022	4.5	4.6	130	1.0	1.5	0.26	1.8	0.20	1.80	G10
2.50	2.40	11.00	9.00	2100	114	600	60	0.018	0.022	4.8	5.0	150	1.0	1.4	0.10	1.8	0.20	1.77	G11
1.00	0.90	0.70	0.65	——	95	550	65	0.031	0.032	4.3	4.4	120	1.3	2.5	1.50	1.6	0.40	1.42	FR-3
2.80	2.50	11.00	9.00	2100	107	500	65	0.016	0.018	4.2	4.4	130	1.2	1.5	0.25	1.8	0.20	1.93	FR-4
2.50	2.40	13.00	9.00	2000	109	490	60	0.013	0.019	4.5	4.6	150	1.0	1.5	0.10	1.8	0.13	1.95	FR-5

TABLE 10.5

Insulating Material Comparison Chart (Courtesy of Insulfab Plastics Inc.)

Use only for general selection. Consult complete data sheets for final decision. Price comparisons are only a guide. Properties improve in quality from 1 to 10.

PHE: Phenolic — MEL: Melamine — SIL: Silicone — POLY: Polyester.
*Indicates Non-significant Values.

PROPERTIES	X	XP	XPC	XX	XXP	XXX	XXXP	XXXPC	C	CE	L	LE	A	AA	N1	G3	G5	G7	G9	G10	G11	GPO1	GPO2	FR2	FR3	FR4	FR5
ELECTRICAL																											
Insulation Resistance	5	2	3	7	5	7	7	10	1	2	1	3	1	*	10	1	2	2	6	10	10	1	1	10	10	10	10
Dielectric Str. Perp. to Laminations	9	6	8	10	8	9	8	10	1	6	1	6	2	*	8	8	8	5	8	10	10	6	5	10	10	10	10
Dielectric Str. Par. to Laminations	1	6	2	6	8	8	8	10	1	4	1	6	1	*	10	2	4	6	9	6	6	5	5	9	9	6	6
Dielectric Losses Radio Frequency	*	1	1	2	2	3	3	5	*	1	1	1	*	*	6	2	5	10	4	6	6	5	5	5	5	6	6
Dielectric Losses Power Frequency	1	1	1	2	2	3	3	4	1	2	1	3	*	*	9	2	6	10	3	10	9	6	6	4	4	10	10
ARC Resistance	1	1	1	1	1	1	1	1	*	1	1	1	*	*	1	1	8	10	9	5	5	3	2	5	5	5	5
Dielectric Properties Stability	1	3	2	5	5	7	6	8	1	3	1	4	*	*	10	3	5	6	9	8	8	3	3	8	8	8	8
MECHANICAL																											
Mechanical Strength	5	2	2	3	3	2	4	2	4	3	4	3	2	4	2	6	8	4	9	10	10	8	8	2	3	10	10
Impact Strength	3	3	2	1	2	1	2	1	4	3	3	2	2	5	4	8	8	7	9	10	10	8	8	1	1	10	10
Bond Strength	1	3	*	1	1	2	2	*	9	9	7	7	9	9	9	1	9	1	7	10	8	6	6	*	*	10	10
Water Absorption	1	2	2	3	5	6	5	8	3	5	3	5	4	4	9	4	4	9	7	10	10	6	6	8	9	10	10
Dimensional Stability Due to Moisture	1	2	2	3	4	5	3	6	1	3	1	3	5	7	5	7	8	10	8	10	10	6	6	6	6	10	10
Heat Resistance	3	3	2	2	2	2	2	2	3	3	3	4	7	8	3	8	9	10	9	6	9	5	6	2	3	6	9
Dimensional Stability Due to Temp.	5	2	5	3	3	2	2	2	4	1	4	2	8	9	1	10	9	9	10	7	8	5	5	2	3	7	8
PHYSICAL																											
Specific Gravity	1.36	1.36	1.36	1.36	1.36	1.32	1.33	1.34	1.36	1.35	1.36	1.35	1.55	1.58	1.15	1.67	1.90	1.73	1.95	1.80	1.80	1.74	1.75	1.38	1.42	1.90	1.93
Machinability	4	4	6	5	6	5	4	5	7	7	8	7	4	1	8	2	2	3	1	4	3	3	3	5	7	4	4
Punching	4	10	9	9	7	3	7	8	8	7	10	8	3	2	8	3	3	3	2	3	2	3	3	8	9	5	5
Flame Retardency	1	1	1	1	1	1	1	1	1	1	1	1	5	5	1	1	10	10	10	1	2	7	7	7	8	10	10
CHEM																											
Acid Resistance	*	*	*	2	2	2	2	1	4	6	4	6	*	2	*	4	2	8	2	4	6	4	4	1	1	4	4
Alkalai Resistance	*	*	*	2	2	1	1	1	1	2	2	2	1	3	*	3	4	2	4	5	7	3	3	1	1	5	5
Res. to Oxidizing Agents	*	*	*	2	2	1	2	1	1	1	1	1	1	1	*	2	4	8	4	4	7	3	3	1	1	4	4
GENERAL																											
NEMA GRADE	X	XP	XPC	XX	XXP	XXX	XXXP	XXXPC	C	CE	L	LE	A	AA	N1	G3	G5	G7	G9	G10	G11	GPO-1	GPO2	FR2	FR3	FR4	FR5
Military Type		PBG		PBG	PBE	PBE	PBEP	PBEB	FBM	FBG	FBI	FBE			NPG		GMG	GSG	GME	GEE	GEB				PEE	GEE	GEB
Base	Paper	Paper	Paper	Paper	Paper	Paper	Paper	Paper	CRS Fab.	CRS Fab.	Fine Fab.	Fine Fab.	ASB Paper	ASB Paper	Nylon Cloth	Cont Glass Cloth	Cont Glass Cloth	Cont Glass Cloth	Cont Glass Cloth	Cont Glass Cloth	Cont Glass Cloth	Mat Glass	Mat Glass	Paper	Paper	Cont Glass Cloth	Cont Glass Cloth
Type Resin	PHE	PHE	PHE	PHE	PHE	PHE	PHE	PHE	PHE	PHE	PHE	PHE	PHE	PHE	PHE	PHE	MEL	SIL	MEL	EPOX	EPOX	POLY	POLY	PHE	EPOX	EPOX	EPOX
A.I.E.E. Insulation Class	A	A	A	A	A	A	A	A	A	A	A	A	B	B	A	B	B	H	B	B	F	B	B	A	A	B	F
Comparative Price	.8	.9	.9	1.0	1.0	1.2	1.3	1.4	1.3	1.3	1.7	1.7	.9	2.6	3.5	2.5	2.2	5.8	2.4	3.1	3.1	.7	.8	1.1	2.9	3.1	3.6

TABLE 10.6

Copper-clad Properties

		Bond Strength Lbs./Inch Width				Hot Solder Resist. @500°F. Sec-Min
Nema Grade	Mil-P Type	1 oz. Ave.	Min.	2 oz. Ave.	Min.	
XXXP	13949B PP	10	9	12	10	10
XXXP		10	9	12	10	10
FR-2		10	9	12	10	10
XXXP		10	9	12	10	15
FR-3	22324 PEE	10	9	12	10	25
G-10	13949B GE	11	10	13	12	40
FR-4	13949B GF	11	10	13	12	20
G-11	13949B GB	9	8	10	9	40
FR-5		9	8	10	9	30

perfections and has a smooth surface on both sides. Made in widths not exceeding 24 inches and thicknesses less than 0.001 inch, this rolled film has excellent tensile-strength properties. Because of its smooth surfaces, however, rolled foil requires special bonding treatments to enhance the *wetting* between the adhesive and the foil and to remove contaminants.

Electrodeposited foil, used almost exclusively in the manufacture of printed circuit boards, is produced by the *plating* of the film from solutions of *copper sulfate* or *copper cyanide* on a revolving stainless steel drum from which the foil is continuously stripped. The inner surface of the resulting film exhibits a smooth finish whereas the outer surface is coarse, thereby promoting improved bonding with the increase of surface area. The thickness of the foil is controlled by the solution concentration and the electrical and mechanical parameters of the plating process. Thicknesses of less than 0.001 inch and widths in excess of 5 feet are obtainable by this method.

Foil thicknesses, regardless of the manufacturing process, are specified for printed circuit boards in *ounces* of foil per *square foot*. Foils of 1 ounce per square foot have an approximate thickness of 0.0014 inch whereas 2- and 3-ounce foils have thicknesses of 0.0028 and 0.0042 inch, respectively. Other available thicknesses are shown in Table 10.7, which gives film thickness with the thickness tolerance.

The selection of a thickness of copper for a particular circuit application is determined principally by the amount of current that it must be able to handle. This current-carrying capability depends primarily upon the thickness and width of the conductor path, in addition to the tem-

TABLE 10.7

Copper Foil Thickness Tolerances

Weight oz/ft^2	Normal Thickness in Inches	Tolerance in Inches
½	0.0007	±0.0002
1	0.0014	+0.0004 −0.0002
2	0.0028	+0.0007 −0.0003
3	0.0042	±0.0006
4	0.0056	±0.0006
5	0.0070	±0.0007

TABLE 10.8

Maximum Recommended Current-carrying Capacity for Various
Conductor Widths and Thicknesses

Conductor width in inches	Current in Amperes			
	½ oz Foil	1 oz Foil	2 oz Foil	3 oz Foil
0.005	0.13	0.50	0.70	1.00
0.010	0.50	0.80	1.40	1.90
0.020	0.70	1.40	2.20	3.00
0.030	1.00	1.90	3.00	4.00
0.050	1.50	2.50	4.00	5.50
0.070	2.00	3.50	5.00	7.00
0.100	2.50	4.00	7.00	9.00
0.150	3.50	5.50	9.00	13.00
0.200	4.00	6.00	11.00	14.00

perature. Owing to the larger radiating conductor surface area, the current capacities of printed circuit boards can easily approach 50 amperes. Table 10.8 shows the maximum current capacities for several conductor width and foil thicknesses to aid in the proper weight selection.

10.4 BONDING

The manufacturers of printed circuit boards choose the proper adhesives to bond the conducting foil to a particular base material in order to produce copper-clad laminates that will suit the needs of a variety of applications. The characteristics of *bond strength*, *hot solder resistance*, and *chemical resistance*, as listed in the tables in Sec. 10.3, largely depend on the particular adhesive used by the manufacturer for bonding and therefore will influence the technician's choice of board. Of these characteristics, bond strength is one of the most important. This affects the reliability of a printed circuit board. If the foil should lift from the base during manufacture or in service, repair is virtually impossible. The bond must withstand the stresses involved in processing, fabricating, and servicing the printed circuit. These stresses include chemical attack from etching solutions; physical forces introduced through shearing, bending, twisting, shock, and vibration; and thermal shock during the soldering process. When shearing large sheets of single-sided copper-clad laminate into smaller blanks, the foil side should be facing *upward* with the insulating base material against the shear table. By cutting through the foil side first, there is less tendency of tearing the foil away from the edge of the blank. In addition, care must be exercised that the shear blade be allowed to return *slowly* to its normal position as it passes the newly cut edge of the board. If the blade were allowed to spring back quickly, it might snap the edge of the board just enough to crack the base material and upset the bond along that edge.

The two most common types of adhesives used in printed circuit bonding are *vinyl-modified phenolics* and *modified epoxys*. Although both adhesives display excellent bonding qualities, the epoxy resins are superior in this respect. The manufacturing process of bonding involves the application of a uniform thickness of adhesive on the foil, after which it is forced-air-dried and laminated to the base material in a press. Once the adhesive has cured, the excess foil is trimmed around the edges of the board. Finally, the foil is cleaned to remove all films and oxides resulting from the manufacturing process.

10.5 BOARD SELECTION

Selecting the most appropriate printed circuit board grade for a particular application depends primarily on the mechanical and electrical requirements of the circuit. The information contained in Tables 10.4, 10.5, and

10.6 will dictate which grade of board is most suitable for a specific application. An evaluation of the various board characteristics will be made with reference to selecting the most appropriate grade of base material for the amplifier's printed circuits. This example will serve as a guide to illustrate the important factors that need to be considered for proper board selection:

Example: A printed circuit board grade is to be selected to meet the amplifier circuits' requirements as follows:

Type of construction — prototype

Maximum continuous current — 1.6 amperes

Maximum voltage — 36 volts

Frequency — audio range (maximum 20 kHz)

Board mounting — corner securing by machine screws to metal chassis

Hardware mounting — swaged terminals and machine screws

Machine operations — drilling and shearing only

Soldering — hand

Grades of printed circuit boards containing silicone or Teflon resins are not considered because of their higher cost and high-frequency characteristics that are far beyond the requirements of the amplifier. Melamine resin boards are also not considered because of their extremely abrasive characteristics, which make them difficult to drill.

One important consideration for comparison and selection is *hot solder resistance.* This characteristic expresses thermal bond strength as a function of time and temperature. Since the boards will be hand-soldered, it is difficult to accurately control soldering time and temperature. Therefore, a board grade with a high value of hot solder resistance should be selected. Table 10.6 compares some common boards. Grades *G-10* and *G-11* are far superior to the *FR* and phenolic grades listed.

Bond strength also needs to be considered in terms of the mechanical stresses that result from the fabrication of the board. Table 10.5 shows that the epoxy-glass resins (*G-10*, *FR-4*, and *FR-5*) have a higher bond strength than all the other grades.

Warp and twist are of no great concern, since all the boards in the amplifier circuit are to be secured to the chassis element with machine screws and nuts.

The *machinability* of the board is another important mechanical consideration. As shown in Table 10.5, the phenolics can be drilled easier than the epoxy-glass types. Finally, the base material's *mechanical strength* must be taken into account. Mechanical strength and machinability are inversely related. The epoxy glass is stronger than the phenolics. Since mounting holes will be formed in the corners of each board and terminals will be swaged to serve as circuit interconnections, fracturing of the base material is a necessary consideration. Although the machinability of the glass-base epoxy materials is inferior to that of the phenolics, this characteristic is not significant since it refers to punching and the abrasive action of tools. Because no punching or machining other than drilling will be performed on any of the boards used in the amplifier, the mechanical superiority of glass-base epoxy outweighs the small difference in machinability characteristics.

In considering the electrical characteristics, as shown in Tables 10.4, 10.5, and 10.6, *all* the board materials listed for the amplifier circuits have parameters which are in excess of those required. The parameters of *dielectric constant* and *dissipation factor* are the most important for high-frequency applications. Therefore, since the amplifier will operate in the low-frequency (audio) range, these parameters need not be considered here. For dc to low-frequency applications, the parameters that must be considered are *arc* and *tracking resistance, insulation resistance*, and *dielectric breakdown voltage.*

Arc resistance is the ability of a base material to prevent undesired conducting paths created by an arc among portions of the existing conductor pattern. Tracking resistance describes the formation of undesired conductor paths *through* the insulating material. These characteristics need not be considered at low power and low voltage applications, such as the 36 volts maximum of the amplifier.

Insulation resistance, expressed in megohms, is the measure of the total *leakage* resistance among adjacent conductors. Because of the extremely high ohmic values, as listed in Table 10.6, this parameter does not become a consideration for this application.

Dielectric breakdown voltage is a function of the spacing among conductors and, as such, need not be considered here for purposes of board material selection. This property will be discussed under conductor spacing in Chapter 11.

Based on the mechanical and electrical requirements of the amplifier circuit, grades *XXXP*, *G-10*, and *G-11* provide a realistic compromise of the listed requirements. Of these three, *G-10* is the grade selected. Grades *G-11* and *XXXP* are eliminated because their bond strength and machinability are lower. One final consideration that influenced the selection of the *G* series over the phenolic types is the ease of tracing circuits on the completed printed circuit board. As shown in Table 10.4, the phenolics are available in opaque *natural, brown,* or *black* colors. The brown color is used almost exclusively with printed circuit boards. The type *G* grades are generally *translucent* with a *green tint*. This translucent characteristic is extremely helpful in inspecting circuits. It is difficult to compare component and conductor patterns with a schematic diagram when the base material is opaque. By holding the translucent board up to a strong light, the relationship among components and conductor pattern is easily seen.

With the base material selected, the final step is the choice of the most suitable base thickness and copper foil weight. As mentioned previously, warp and twist present no problem in this amplifier package. Since all boards will be firmly secured to the chassis element, no other mechanical parameters need to be considered. For these reasons, $1/16$-inch base material will be used. For the smaller boards, $1/32$-inch material may be used; but for the sake of consistency and to avoid board fracture, especially around swaged terminals and mounting holes, all boards in this package will be $1/16$ inch.

In selecting the foil thickness, the major consideration is the amount of current that the circuit is expected to handle. The maximum circuit current is 1.6 amperes. From Table 10.8, for a typical conductor width of 0.030 inch, the maximum current capacity for 1-ounce copper foil is approximately 1.9 amperes. This standard 1-ounce foil, which will easily meet the circuit current demands, is therefore selected.

In summary, the solution to the problem of board selection for the amplifier package is *$1/16$-inch 1-ounce G-10 single-sided copper-laminated printed circuit board.*

The printed circuit board having been selected, the technician must next prepare the necessary drawings and artwork for transforming the schematic diagrams of the various circuits into conductor patterns that will be later processed. The following chapter develops the techniques for producing printed circuit conductor pattern artwork.

PROBLEM SET

10.1 Describe the meaning of the NEMA suffixes *X, P,* and *C.*

10.2 Using Table 10.4, select the laminate that provides the optimum balance of mechanical and electrical properties.

10.3 Referring to Table 10.5, match the grades of insulation independently in terms of the least and most desirable characteristics of cost, machinability, punching quality, water absorption, bond strength, and arc resistance.

10.4 Using Table 10.8, select the minimum conductor width for $1/2$-ounce, 1-ounce, 2-ounce, and 3-ounce copper foil if the maximum dc current rating is to be 2.5 amperes.

10.5 Referring to the tables provided in this chapter, determine the optimum pc insulating material and minimum conductor thickness and width to meet the following requirements: 1-ampere dc current; a frequency of 20 kHz; drilling and shearing operations only; wire-wrap terminals and finger-type connectors for lead and conductor path connections.

11

Single-Sided Printed Circuit Artwork Design and Layout

11.0 INTRODUCTION

After the most suitable type of printed circuit board has been selected, the next phase in the fabrication of a printed circuit is the design of the conductor pattern *artwork master*. A properly designed and constructed master will become a precision instrument for faithfully reproducing the foil configuration in processing the printed circuit board. The artwork constructed must not only accurately form the circuit's electrical connections but must also provide for the space required for placing the components on the board.

There are several methods of producing quality artwork. However, the use of commercially available precut tapes is one of the more popular techniques and will be discussed exclusively. Included for discussion in this chapter will be (1) selection of a proper grid system, (2) reduction requirements, (3) development of component and conductor pattern layout drawings, (4) master artwork preparation and tolerances, and (5) solder and marking masks.

11.1 GRID SYSTEMS

To facilitate the placement of components, in addition to providing the proper spacing among the intended conductor paths, some type of uniform spacing system must be used. For most printed circuit artwork applications, *grid systems* having grid spacings of 0.050, 0.100, and 0.125 inch are generally used. Grid systems are uniform patterns or arrangements of horizontal and vertical lines that aid in simplifying printed circuit board layout. The thickness of the grid lines is approximately 0.003 inch. Depending on the system selected, every second, fourth, fifth, eighth, or tenth line is *accented* to a thickness of approximately 0.006 inch. These accent lines aid in laying off dimensions and establishing a tabulated coordinate system to identify specific component or hole positions at the intersections of particular horizontal and vertical lines. Selecting the most suitable grid system depends on factors such as (1) scale reduction, (2) required tolerances, and (3) capacity of the reducing equipment.

Whenever possible, artwork masters should be made *four times* larger than the actual size of the board pattern (4:1 scale) to maintain reasonable tolerances. This scale is preferable since any error in component positioning or tape placement made on the master will be reduced by a factor of 4 on the processed board. Another advantage of the 4:1 scale over smaller available scales is that, from a practical standpoint, this larger size facilitates handling. Although tapes, device outlines, and other layout materials are available for 4:1, 2:1 and 1:1 scales, the artwork should not be smaller than 2:1 if reasonable tolerances are to be met. If the photographic processing equipment will not make 4:1 reductions or accept a 4:1 enlargement of a printed circuit design, the technician is forced into using a 2:1 scale artwork layout. If no reduction equipment is available, only 1:1 scale artwork can be made. Under these conditions, the technician must exercise extreme care in placing the tapes to realize even a most marginally acceptable board.

Artwork prepared for 4:1 reductions on 0.125-inch grid systems are common whereas a 0.100-inch system is used for close tolerance work. For 2:1 reductions, 0.100-inch systems are standard, with 0.050-inch grids used for the closest tolerances. Artwork for a 1:1 scale requires a 0.050-inch grid system as a minimum.

One grid system that yields excellent results consists of 0.005-inch thick *filmed translucent polyester* (Mylar) lined with black rectangular coordinates. Sheet polyester is preferred since a high degree of dimensional stability, in the order of ±0.002 inch in 48 inches, can be maintained. For this reason, commercial artwork makes use of polyester film grid systems almost exclusively. When dimensional accuracy as provided by the polyester system is not required, simple quadruled paper, with intercepts of 0.100 or 0.050 inch, can be used with satisfactory results. Figure 11.1 shows several common grid patterns. Depending on the application and tolerances required, it may be acceptable to sacrifice the high

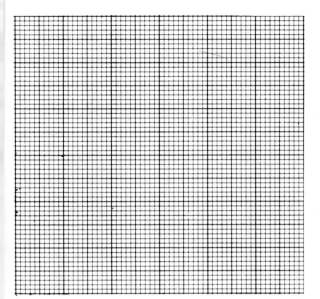

(a) Grid patterns common to printed circuit layouts (0.050 inch).

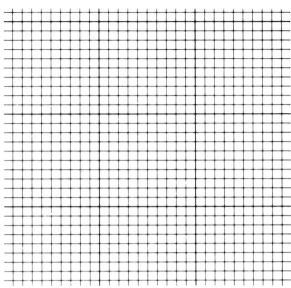

(b) Grid patterns common to printed circuit layouts (0.100 inch).

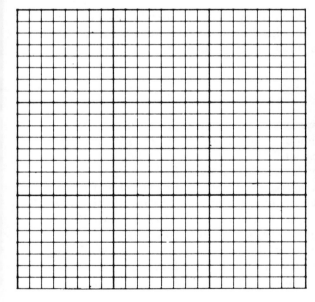

(c) Grid patterns common to printed circuit layouts (0.125 inch).

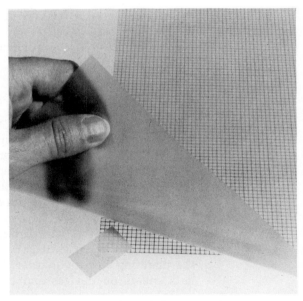

(d) Overlay method used for artwork master construction.

FIGURE 11.1. Grid systems and initial setup for artwork construction. Grid patterns common to printed circuit layouts are shown in (a), (b), and (c). Overlay method used for artwork master construction is shown in (d).

degree of dimensional stability obtainable from a preestablished grid system on polyester to that of simple ruled paper. A sizable saving in cost would be realized by the latter selection.

The polyester sheets must be handled carefully to keep these materials in a workable state. They should always be stored flat. If these sheets are rolled in storage for any length of time, they will not lie flat on the work surface. If rolled after the artwork tapes have been applied, the tape will buckle or stretch. This would upset the artwork tolerance because of altering of the original tape position. Sheets should be stored in a cool, dry area to minimize dimensional alterations from temperature and humidity changes. If stored under extreme conditions, the polyester sheets should not be used until they have been allowed to stabilize for at least 24 to 36 hours in a 68 to 75°F (20 to 24°C) temperature and 50% humidity environment. These precautions are necessary for work involving extremely close tolerances, which in turn requires materials possessing high dimensional stability.

To begin preparing the artwork, the grid system sheet is securely fastened with masking tape to a work surface. A drafting board acts as an excellent work surface for this application. A sheet of 0.005-inch-thick *clear* polyester is then taped over the grid system only on one edge so that it may be lifted away. This clear polyester sheet will eventually become the artwork master backing and the grid system will later be used to facilitate component positioning. See Fig. 11.1d for this arrangement. This *overlay* procedure is economical since the grid paper can be reused.

11.2 SPACE ALLOTMENT AND COMPONENT POSITIONING

After the most suitable grid system that satisfies the reduction factor, photographic reduction equipment, and required tolerance has been selected and taped to the work surface, the printed circuit layout may begin. The clear polyester sheet is first lifted away from the grid system to allow a sheet of *vellum* (tracing paper) to be taped over the grid. The component layout will be drawn on the vellum with the aid of the underlying grid system. All the components and hardware that will eventually be mounted to the printed circuit (pc) board should be made accessible. The circuit schematic diagram and preliminary sketches developed during the planning stages (Chapter 2) as well as any manufacturers' mechanical data sheets on devices and hardware should also be collected and kept available.

The dimensions of the openings in the amplifier's chassis under which the pc boards will be mounted are first obtained from the chassis layout drawings. These dimensions are next multiplied by the reduction factor being employed and transferred, in pencil, to the vellum. These pencil marks will serve as a *component border* (i.e., the components to be positioned on the pc board must fall within this border). In cases in which

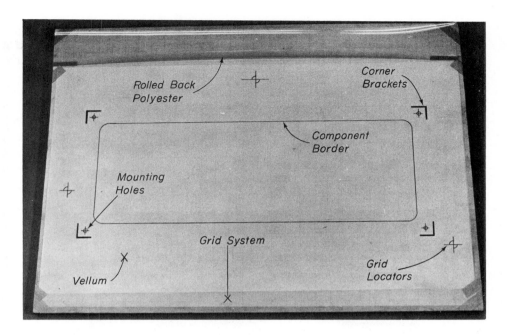

FIGURE 11.2. Initial component layout.

the boards are to be secured to the chassis, as in the design of the amplifier, an *overall board outline*, indicated by penciled *corner brackets* on the vellum, is drawn. This information is obtained from the engineering drawings. Outside the overall board outline, three staggered *grid locator marks* should be drawn that will be used for aligning the pad masters. Figure 11.2 shows this initial layout.

The technician can now indicate on the vellum the placement of significant locations, such as input and output terminals as well as power connections and grounds, in accordance with the information specified in the planning stages and represented on the engineering sketches. External connections are usually in the form of terminals that are positioned along component border edges. These external terminal positions should be considered first in layout. Depending on component positioning, all input and output connections may be positioned along one end of the pc board. This arrangement would be typical of modular design in which boards are used with the pc connectors for the purpose of plug-in-type systems. If the plug-in design is not required, wiring between boards is simplified by positioning input connections along one end with the output connections at the opposite end of the board. End-positioned external connections are vital to prevent leads from crossing the pc board, which could impair service.

Two types of aids in developing component layouts that are readily available are *layout templates* and *precut component outlines*. Figure 11.3a shows several common template styles. These templates, constructed of plastic, provide component outlines such as resistors, capacitors,

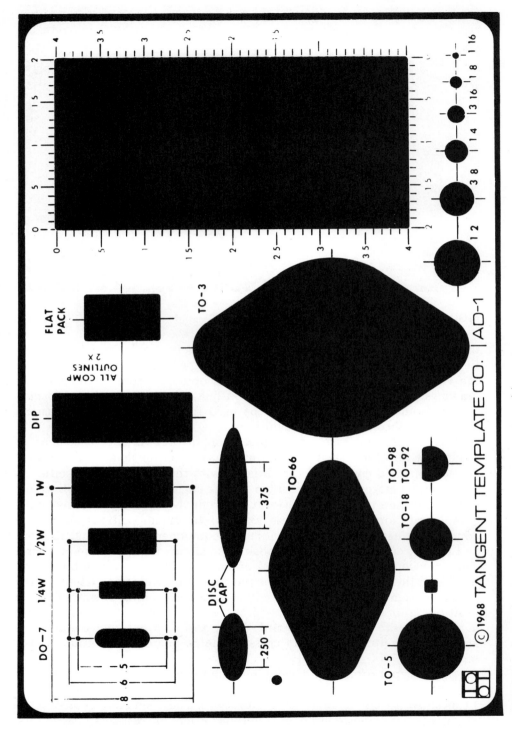

FIGURE 11.3. Layout aids for printed circuit drawings. (a) Layout templates. Courtesy of Tangent Template Co., San Diego, California.

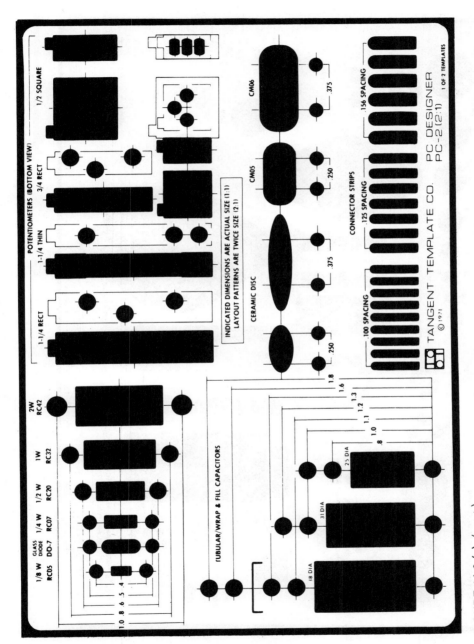

FIGURE 11.3.(a) (cont.)

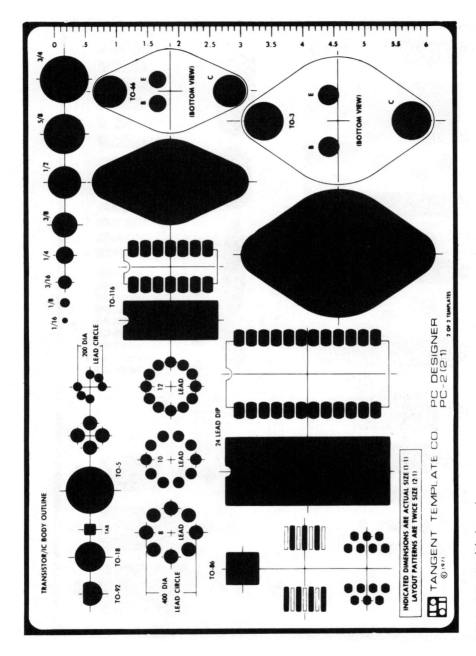

FIGURE 11.3.(a) (cont.)

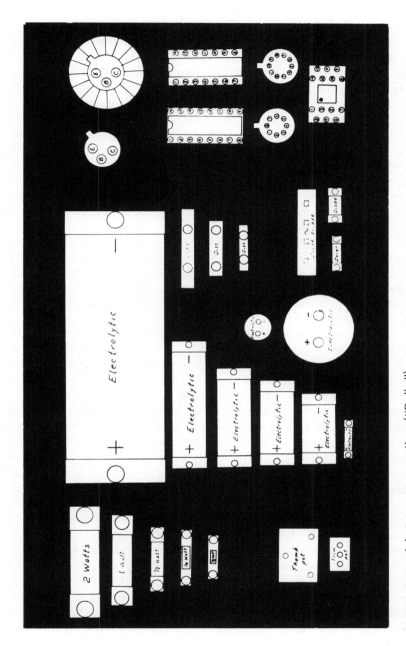

FIGURE 11.3.(b) Component outlines ("Dolls.")

transistor pin arrangements, and common integrated circuit patterns, thus eliminating the need, to some extent, of measuring individual terminal positions. These templates are available in scales of 1:1, 2:1, and 4:1 to further aid the technician in layout work.

Component outlines, such as those shown in Fig. 11.3b, can easily be constructed from thin cardboard with the use of a scaled template, pencil, and scissors in any of the desired device and component sizes and shapes. These *dolls* provide an excellent means of establishing component placement, lead orientation, and component density within the component border of the layout. The technician should realize when preparing dolls that the leads of components such as transistors or integrated circuits will be inserted into the *insulated* side of the pc board. The orientation of these leads must therefore be sketched on the dolls as if viewed through the *top* of the device and properly labeled (i.e., *E* to designate emitter and *1* to designate pin number 1 on integrated circuits). If this technique is not strictly adhered to, errors that cannot be corrected are certain to occur in the conductor pattern.

The cut component outlines are somewhat more desirable to use than templates since they allow for rapid movement and rearrangement about the grid system. The use of the template requires the technician to draw the desired component outline on the vellum. Any later change in position of any component or device means that the original outline will have to be erased.

Dolls used to establish the approximate component layout followed by the component outlines transferred to the vellum with the aid of layout templates allows for the most rapid and accurate results. Both of these layout aids are demonstrated later in this section.

The procedure for developing optimum component placement involves such considerations as size, weight, and heat-dissipation properties as well as the electrical criteria. This task, then, becomes quite complex and requires a meaningful approach for initiating layout. Three methods for solving this problem have been proposed. These are the *schematic viewpoint*, the *central viewpoint*, and the *peripheral viewpoint*, which will be discussed here.

Since considerable care is taken by the draftsman in developing a schematic drawing to minimize crossovers and to present a well-balanced circuit layout, it seems logical that the first phase of component positioning would be to place the cut component outlines on the vellum in the approximate positions they occupy in the schematic diagram. Once the dolls have been positioned, the conductor pattern is drawn between components on the vellum to see that all electrical connections can be made without *conductor crossovers*. These crossovers would result in undesired electrical connections that do not appear in the schematic diagram. Conductor paths can be drawn on the vellum *through component body outlines* because the components will be mounted on the *opposite* side of the board from the conductor pattern when the pc board is assembled.

Figure 11.4a shows the initial component placement as a result of employing the schematic viewpoint for the complementary-symmetry power amplifier board. Notice that some rearrangement is necessary because components and hardware exceed the component border and result in poor component density. When a component layout follows a schematic diagram exactly, the conductor pattern would be similar to that of the schematic. However, if even one component is rearranged, the conductor pattern should be checked and, if necessary, corrected in order to meet the electrical requirements of the schematic.

Components that present electrical or thermal problems are considered for repositioning first. For the example shown in Fig. 11.4, diode D_1 must be repositioned so as to be mounted to the heat sink supplied for the complementary-pair transistors. This technique is to compensate for thermal instability. Mechanical problems are introduced in the initial positioning of this heat sink. Adequate heat dissipation of the output transistors will require a 2- by $3\frac{1}{2}$-inch sheet of aluminum. If mounted as shown in Fig. 11.4a, the width of this heat sink is greater than the opening provided on the chassis for mounting the board. From the chassis drawings, the allowable clearance between the top of the chassis and the enclosure is under 3 inches. This eliminates the possibility of the $3\frac{1}{2}$-inch

FIGURE 11.4. Use of schematic viewpoint for component positioning. (a) Doll positioning using schematic viewpoint. (b) Necessary rearrangement of components because of physical limitation of component border. (c) Final component repositioning with conductor paths.

(a)

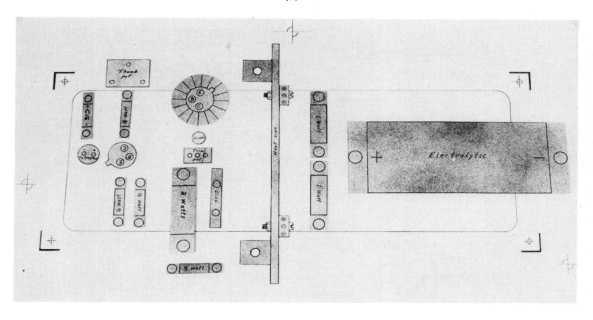

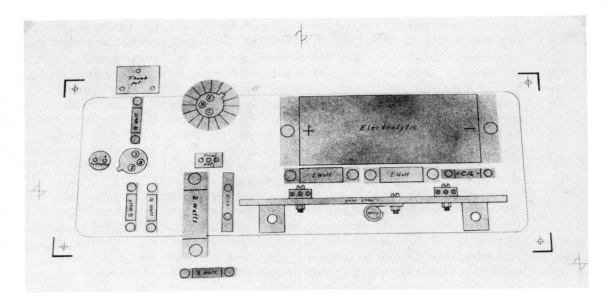

(b)

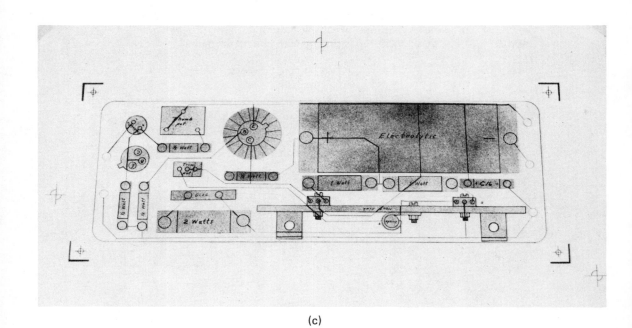

(c)

FIGURE 11.4. (cont.)

vertical position. The solution to this problem is to reposition the heat sink on the board as shown in Fig. 11.4b. The height of the sink remains 2 inches above the pc board and will mount easily within the enclosure. The repositioning of the heat sink means that the large output coupling capacitor and the two 0.47-ohm resistors must be repositioned parallel to the long edge of the pc board. Figure 11.4b shows the component arrangement up to this point. The smaller components can now be rearranged. The completed component positioning and conductor path check is shown in Fig. 11.4c, which uniformly utilizes all the available pc board space and results in a *balanced* component arrangement. Notice that all resistors and capacitors are mounted either parallel or perpendicular to an edge of the pc board and that uniform spacing between components, whenever possible, is maintained. As will be seen later, positioning groups of components such as resistors and capacitors either parallel or perpendicular to the edge of a pc board will aid significantly in laying out the final conductor pattern.

To complete the component layout, *light* position marks are made on the vellum around the dolls. With the dolls removed and with the aid of the underlying grid system, the drafting template is used to obtain accurate component size, position, and lead hole locations. Figure 11.5 shows the completed component layout drawing and final conductive paths for the power amplifier circuit. At this point, all components that require specific orientation or assembly on the finished board are *keyed* on the layout drawing. This technique involves the use of symbols such as a plus (+) sign to indicate capacitor polarity for electrolytics, (E) to represent the emitter lead of a transistor that could be positioned improperly, (1) to represent the number *one* pin of integrated circuits, as well as numerical and literal symbols to indicate external connections. By using this method of keying, once the corresponding end of a component or a particular pin or lead in a group is identified and positioned into the board, all other component leads associated with the key will be oriented in the correct location.

The second method of approaching the problem of component placement, termed the central component viewpoint, might be chosen over the schematic viewpoint if a *single* predominant component such as a transistor, integrated circuit (IC), or relay exists on the circuit schematic. The predominant component is first centered within the component border established on the vellum, using the grid as a guide. The support components, such as resistors and capacitors, are then oriented on the vellum, continually checking for correct electrical conducting paths. Figure 11.6a shows an *initial* component arrangement for the IC preamplifier board using dolls. It should be pointed out that high-gain amplifiers require short input lead connections to eliminate possible noise from the input of the system. For this reason, the initial layout must be rearranged. Figure 11.6b illustrates this component rearrangement.

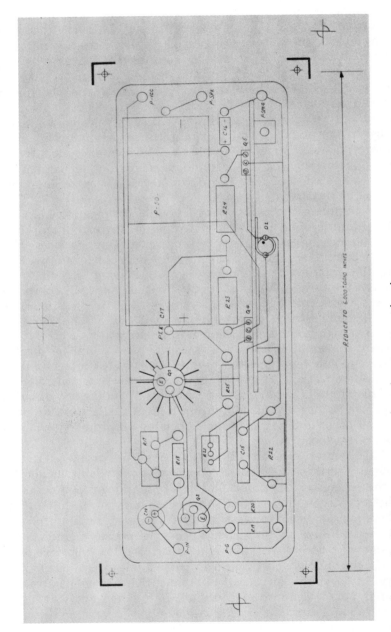

FIGURE 11.5.　Finished component and conductor pattern drawing.

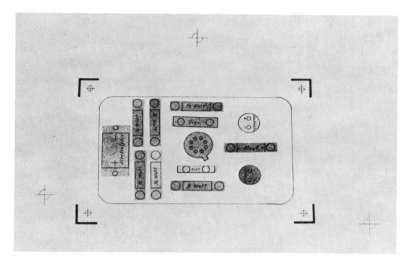

(a)

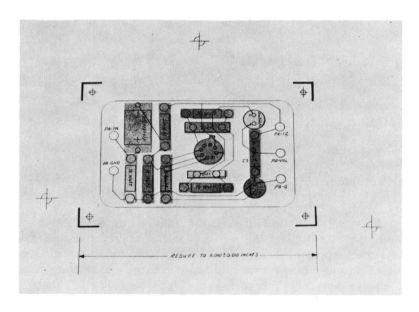

(b)

FIGURE 11.6. Use of central viewpoint for component and conductor pattern layout. (a) Doll positioning using central viewpoint. (b) Final component repositioning with conductor paths.

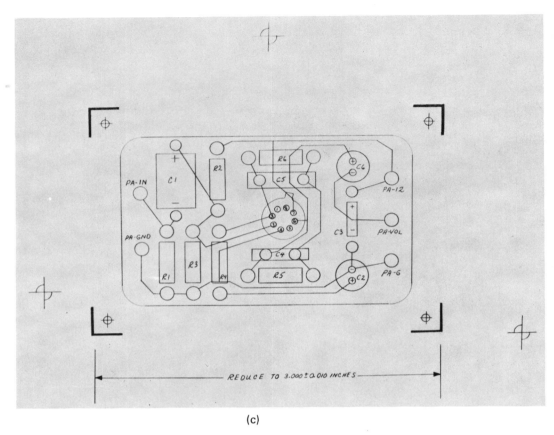

(c)

FIGURE 11.6. (cont.) (c) Finished component and conductor pattern drawing.

Finally, the remaining support components are rearranged to balance and make uniform the spacing around the IC. Figure 11.6c shows a completed component layout for the IC preamplifier board together with all necessary keying information drawn on vellum.

The third method of component placement, termed the peripheral viewpoint, is based on the initial planning stages of a package. If many of the circuit's components will not be located on the pc board but instead will be mounted either on a chassis or front panel, this approach should be chosen for the pc board layout. The tone control circuit exemplifies this method. In this circuit, two potentiometers will be mounted on the front panel. Wiring from these panel-mounted components to the board can be greatly simplified if all pc terminal connections for these external leads are positioned near the potentiometers. To a great extent, the component placement on the pc board will be predetermined by these external connections. This is a common situation wherein the conductor pattern on a pc board is designed for input and output terminations at one end of the board to align with conventional plug- or receptacle-type printed circuit connectors. (See Chapter 14 for further discussion of connectors.) Figure 11.7 shows the completed component layout for the tone control circuit board. Notice that all the terminals for potentiometer connections are along one edge of the board and that much of the closely associated circuitry is positioned close to these terminals.

The three methods of approaching the problem of component placement just examined are intended only as a guide to help minimize what may appear to be a formidable problem. With experience, the technician will find his or her own approach to laying out a pc board.

Upon completion of the component and conductor pattern layout, the following observations should be noted:

1. The component and conductor pattern composite drawing enables the technician to readily observe *both* sides of the pc board simultaneously and therefore determine the effects of any layout modification on either side of the board.

FIGURE 11.7. Resulting layout drawing using peripheral viewpoint.

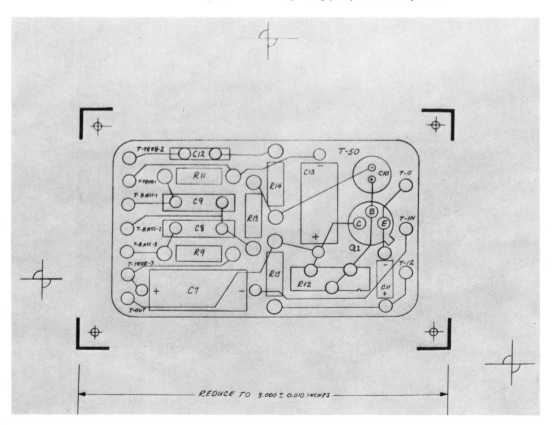

2. Even though the component layout appears balanced and the conductor paths are electrically correct, there may be conductor pattern crowding. This *crowding* occurs when component placement forces many conductor paths to lie close together, resulting in nonuniform conductor density (i.e., making inefficient use of the total foil area). Although this condition may in no way impair circuit operation, the technician may choose to slightly modify components and/or conductor path arrangement to achieve conductor pattern balance. Proficiency in obtaining both component and conductor path balance *simultaneously* requires considerable layout experience.

3. Component spacing must be consistent with any specified spacing tolerances. For this, the grid system used to align and orient components is an immeasurable aid. Use of the intercepts of the grid system to position leads for components and external connections is also extremely helpful in minimizing misalignment of the eventual taped conductor pattern artwork.

FIGURE 11.8. Component layout drawing for the amplifier's power supply board.

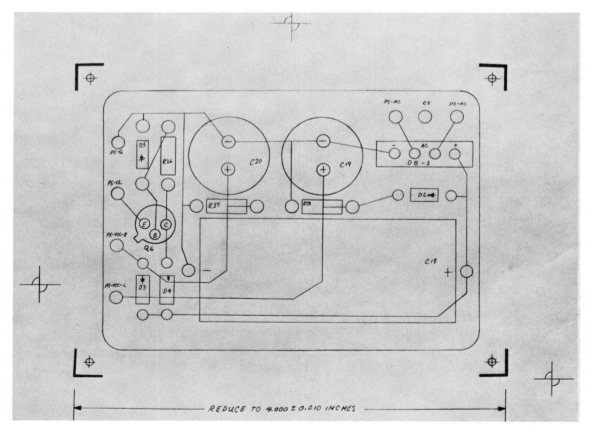

4. Spacing among similar components, such as resistors, should be uniform to enhance the appearance of the board and to result in a more systematic drilling sequence.

It cannot be overemphasized that the appearance of any completed pc board is directly dependent on proper layout procedures. It is therefore imperative that this phase of design be undertaken with utmost care.

The final component and conductor path layout for the power supply board is shown in Fig. 11.8. This final layout completes all required pc drawings for the amplifier.

11.3 DRAFTING AIDS FOR PRINTED CIRCUIT ARTWORK

The basic aids used for printed circuit artwork produced commercially are precut pressure-sensitive adhesive tapes and conductor pattern shapes. These aids provide a quick and accurate method of developing artwork masters. Tapes normally used for single-sided pc board applications are available in opaque *black* and vary in thickness from 0.002 inch to 0.006 inch and in widths from 0.015 inch to 2 inches. Figure 11.9 shows some common tape sizes used in pc artwork design. Although the selection of tape width will depend on such factors as current, voltage, type of board insulation, board protective coating, and thickness of copper foil, some minimum specifications will be presented to serve as a guide in tape selection.

For general-purpose pc work, $\frac{1}{16}$ (0.0625)-inch conductor width with a minimum spacing of $\frac{1}{32}$ (0.031) inch is normally used with a 1:1 scale. Conductor widths that are made or reduced to $\frac{1}{32}$ (0.031) inch with a minimum spacing of $\frac{1}{32}$ (0.031) inch often are used for small signal circuits with low power characteristics. For 4:1 and 2:1 reductions, tape widths of 0.125 ($\frac{1}{8}$) inch and 0.0625 ($\frac{1}{16}$) inch, respectively, are employed to obtain a final reduced conductor width of 0.031 ($\frac{1}{32}$) inch. These larger tape widths are easier to handle and yield a more accurate artwork master. It is recommended that the beginner not select a tape width of less than $\frac{1}{32}$ (0.031) inch because of the difficulty in handling and the care that must be exercised to manipulate these narrow widths. For 1:1 artwork masters, tape widths of $\frac{1}{32}$ (0.031) inch are preferable when dexterity and electrical characteristics permit this size. Otherwise, a maximum of $\frac{1}{16}$ (0.0625) inch should be used. Drafting aids for lead patterns associated with transistors and integrated circuits are more readily available in 2:1 and 4:1 scales, so it is preferable to work in either of these two scales. A scale of 1:1 should be limited to circuits not employing intricate wiring associated with transistor and integrated circuit devices. A 5:1 scale is only used when an unusually high degree of accuracy is required. From the standpoint of economy, the 2:1 scale is preferable to the 4:1 scale.

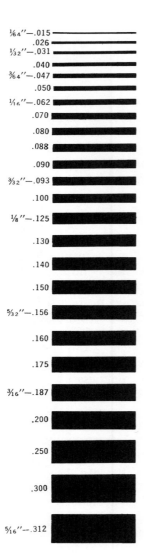

1⁄64″—.015	
.026	
1⁄32″—.031	
.040	
3⁄64″—.047	
.050	
1⁄16″—.062	
.070	
.080	
.088	
.090	
3⁄32″—.093	
.100	
1⁄8″—.125	
.130	
.140	
.150	
5⁄32″—.156	
.160	
.175	
3⁄16″—.187	
.200	
.250	
.300	
5⁄16″—.312	

FIGURE 11.9. Common tape widths used for printed circuit artwork designs.

The foregoing information does not necessarily conform to any particular military or industrial specifications but presents realistic and practical tape sizes that are suitable for general-purpose applications. For more detailed information relative to tape width as a function of current capacity, refer to Table 10.8.

Many precut tape configurations as well as tapes for conductor paths are available. Figure 11.10 shows many of the most common of these outlines. A brief description of each with their application follows.

Terminal pads, also called *teardrops*, are available in all common conductor widths. They are filleted at the terminal ends and are available in *single-end entry*, *double-end entry*, and *elbow angles* to facilitate the layout work. Lead hole diameters should be selected to be no greater, when reduced, than the component lead or wire they are to accept. The radius of the conductor pad should be at least three times the lead hole diameter. In addition, the entry width (dimension *A* in Fig. 11.10) should be the same width as the tape used for the conductor path.

If smooth curves cannot be made by simply bending the straight tape during application, precut *elbows* with angles of 30, 45, 60, and 90 degrees, or *universal circles*, may be used. These aids are shown in Fig. 11.10. Universal circles provide not only various bend radii, but also an infinite number of angles by cutting the appropriate circle to the desired angle. If narrow tape is used, conductor path direction can be changed by forming the tape into a smooth radius. Each size and type of tape has a minimum bend radius. Exceeding this minimum narrows the tape width.

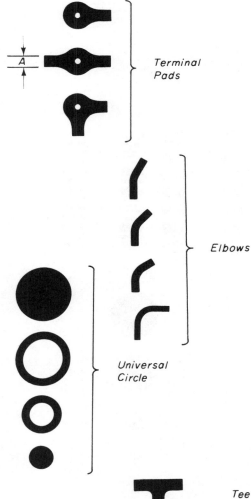

FIGURE 11.10. Precut tape configurations. Courtesy of Bishop Graphics, Inc.

221

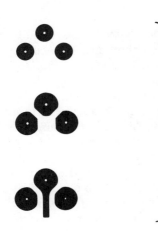

Transistor Pads

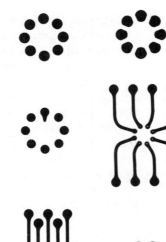

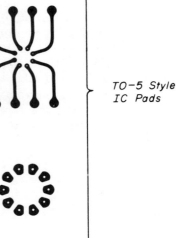

TO-5 Style IC Pads

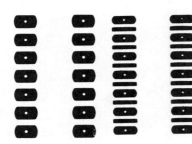

Dual In-line IC Pads

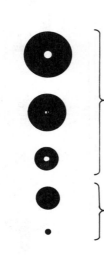

Donuts

Solid Donuts

Flatpack IC Pads

FIGURE 11.10. (cont.)

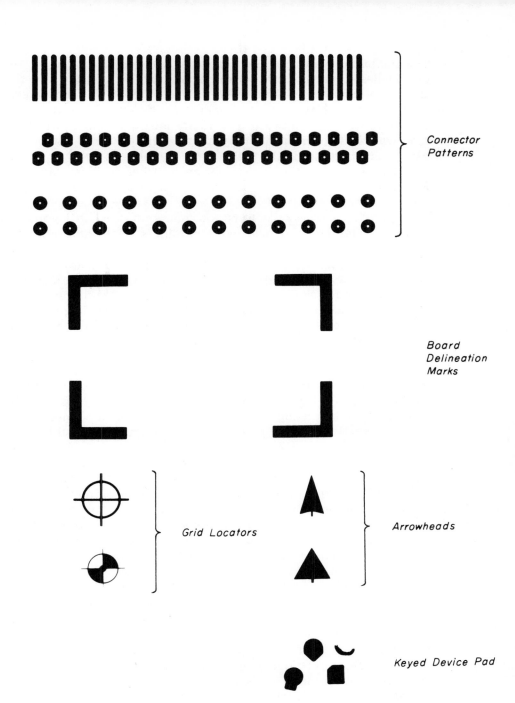

Connector Patterns

Board Delineation Marks

Grid Locators

Arrowheads

Keyed Device Pad

FIGURE 11.10. (cont.)

223

The minimum radius for each tape size can be determined by observing the smallest bend that can be formed without causing the tape to *buckle* along the inside edge of the bend or *stretch* along the outside edge. The minimum bend radius is approximately the width of the tape used. Stretching the tape between two points should also be avoided since the tape will tend to creep, thereby altering the position of the conductor path or creating gaps in the path.

Conductors that meet at right angles to each other may be filleted. For this purpose, *tees* with the same entry widths as the conductor are used.

Terminal circles or *donuts* are commonly used in layout work for locating holes and are selected according to the same criteria as are terminal pads.

Special shapes for mounting the common transistor and integrated circuit case configurations are also shown in Fig. 11.10. Included are terminal pads for the TO-5 3 lead transistor, the TO-5 8 and 10 lead IC, the 14-pin dual-in-line IC, and the 14-lead flatpack. Many more precut drafting aids are available. All of these shapes are mounted on a pressure-sensitive adhesive backing ready for application.

Connector patterns, also shown in Fig. 11.10, are of three basic types: *spaced-edge* (finger), *staggered*, and *in-line*. The use of these aids saves considerable time by eliminating repetitive taping. The spaced-edge pattern, often referred to as *terminal connector strips*, is used with insertion-type board-edge receptacle connectors. If a connector plug that mounts directly onto the pc board and has contact pins either swaged or soldered to the conductor pattern is used, a staggered or in-line type donut configuration is required, the choice of which will depend on the type of contact pin arrangement on the plug. (See Chapter 14 for a discussion of various connectors.) The connector patterns, shown in Fig. 11.10, are accurately scaled for standard printed circuit connectors or plugs. These precision-spaced aids are available on a transparent pressure-sensitive adhesive acetate film for group unit application. Strips of up to 18 pads (donuts) or 22 edge connectors are available, which may be cut to the desired number.

Board delineation marks are used to outline the board's overall dimensions and to position the reduced artwork accurately onto the board for processing. These aids are positioned along the outside edges of the grid lines representing the four corners of the board. These marks are used as guides only and are not processed onto the finished pc board.

Grid locators (targets or bulls-eyes) are used to align one master accurately over another when developing solder masks, marking masks, or double-sided board layouts. These targets are positioned *outside* the board delineation marks and are not an integral part of the finished pc board.

Arrowheads are used in conjunction with numerals that designate artwork reduction dimensions. A whole number with the appropriate tolerance is positioned in the center between two vertical pieces of tape on the artwork. The tape is separated by a distance approximately as great as that which is between the delineation marks. The arrowheads are positioned with their points against the inside edges of the vertical tape. This technique is shown in Fig. 11.17.

The precut tapes discussed are an illustrative selection and do not constitute a complete listing. Nonstandard-device outlines, such as the keyed-type transistor pattern shown in Fig. 11.10, are available from some manufacturers on special order but are uneconomical unless large quantities are used. For additional standard drafting aids, appropriate manufacturers' catalogs may be consulted.

11.4 PRELIMINARY CONSIDERATIONS FOR ARTWORK CONSTRUCTION

In preparing to construct the artwork master, several considerations need to be noted to avoid improper tape arrangements. These considerations are:

1. A separate terminal pad must be provided for each component lead to be soldered to the conductor pattern. In cases in which large conductor areas are used instead of individual conductor pads, each component lead must still be provided with its own access hole.

2. Whenever possible, conductor paths should be along the vertical and horizontal lines of the grid system. However, irregular paths are acceptable only if they simplify the conductor pattern.

3. Attention should be focused on the conductor width and spacing. The cross-sectional area of the conductor path is determined in part by the amount of current it will be expected to handle. As discussed in Chapter 10, once the foil thickness has been selected, the minimum conductor width allowable is determined as outlined in Sec. 10.3 with the use of Table 10.8. Once this width has been established, the spacing between conductors must next be determined. Table 11.1 lists recommended conductor spacings for boards that will not be encapsulated (sealed) when completed. These spacings indicated are the *minimum* for a wide range of ac or dc voltages and take into consideration the *dielectric breakdown* voltage characteristics. These characteristics refer to that voltage at which flashover (arc) among conductor paths occurs. For optimum circuit reliability, the components should be positioned in such a manner that the conductor width and spacing are not less than the minimums specified throughout the entire conductor pattern. In this phase of layout, the design must not be compromised. Once these minimum values have been determined, any reduction in either one for the purpose of preferred component repositioning for convenience or for the sake of appearance is unacceptable.

TABLE 11.1

Recommended Minimum Conductor Spacing

Voltage between conductors DC or AC peak (volts)	Minimum spacing
0-150	0.025 inch
151-300	.050 inch
301-500	.100 inch
Greater than 500	.0002 (inch per volt)

(a) Sea Level to 10,000 Feet.

Voltage between conductors DC or AC peak (volts)	Minimum spacing
0-50	0.025 inch
51-100	.060 inch
101-170	.125 inch
171-250	.250 inch
251-500	.500 inch
Greater than 500	.001 (inch per volt)

(b) Over 10,000 Feet.

An example that illustrates the use of Tables 10.8 and 11.1 for the purpose of selecting an appropriate conductor width and spacing follows:

Specifications
Maximum continuous current — 1.6 amperes
Maximum voltage — 36 volts

Table 10.8 shows that a conductor width of 0.030 inch will safely handle 1.9 amperes using 1-ounce copper foil. This width allows an acceptable margin of safety. The spacing for the specified voltage requirements can be determined from Table 11.1, which shows that the minimum spacing for a 0- to 150-volt range is 0.025 inch. Therefore, no conductor path on the finished board can be closer to any other path than this minimum. The solution to this problem, therefore, is:

Conductor width — 0.031 inch
Conductor spacing — 0.025 inch

Notice that these specifications apply to the *finished* pc board. A 4:1 taped artwork master would therefore employ 0.125-inch tape width with a minimum spacing between conductor paths of 0.100 inch. A 2:1 taped artwork master would require a 0.062-inch tape width and 0.050-inch minimum spacing in order to achieve the desired results.

4. Finally, terminal area shapes and conductor path configurations must be evaluated so that only recommended artwork forms will be employed. Figure 11.11 shows various terminal shapes and path layouts with the preferred pattern configurations, along with those that should be avoided. The use of those configurations that are not recommended may introduce problems such as undesirable voltage distribution along ground paths, nonuniform solder flow onto the conductor pattern (especially at terminal areas), and weakening of the foil bond.

FIGURE 11.11. Recommended pattern configurations.

Recommended		Not Recommended	
Pattern Design	Comments	Pattern Design	Comments
	Uniform distribution of solder around lead access hole		Non-symmetrical pad area around lead access hole will result in non-uniform solder flow
	Double filleted entries result in symmetrical solder fillets		Non-uniform pad area will result in a non-uniform solder fillet
	Allows symmetrical solder fillets		Uniform solder flow is prevented around outside edges of pad
	Pattern allows for symmetrical solder fillets		Non-uniformity in solder flow due to large center area of pattern
	Properly taped entry		Inferior entry results in sharp conductor path edges detracting from appearance and hampering good solder flow
	Contour elbow maintains good foil bond		Weakened foil bond due to sharp exterior edge of elbow which could create problems during etching
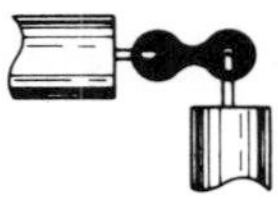	For symmetrical soldering an individual access hole for each component lead is required		Excessive pad area is removed to accommodate multiple leads resulting in foil bond weakening
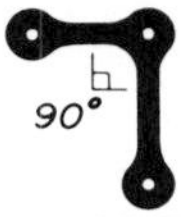	Shortest possible path length	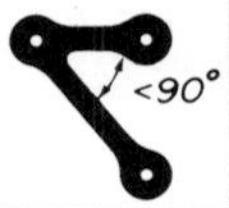	Excessive path length and sharp angle at entries
	Lower voltage drops due to heavy buss		High voltage drops along thin conductor paths

11.5 TAPED ARTWORK CONSTRUCTION

To begin constructing the artwork master for a single-sided board, a sheet of polyester is placed over the component and conductor pattern drawings previously constructed and is firmly taped to the drawing board. This *pad master* will become the support for the taped artwork.

With a scalpel or other sharp instrument, the terminal pads (teardrops) are peeled from their protective paper backing and transferred to the polyester sheet. These aids should be positioned by using the underlying conductor pattern drawing and centered with the aid of the grid system intercepts of the component layout. The entry of the teardrops should be oriented to align with their associated conductor path direction (Fig. 11.12a). A single-entry teardrop will properly accommodate only one termination of a path with a filleted entry. If another conductor is terminated at the same terminal pad, it is necessary to provide an additional filleted entry. Figure 11.12b shows an unfilleted conductor termination. Figure 11.12c shows one method of obtaining a second filleted entry by overlapping two single-entry teardrops. When this method of filleting is used, tape buildup results. This is undesirable since it may yield poor line definition during the photographic reduction process. The preferred method is to use the proper double-entry teardrops, as shown in Fig. 11.12d. The technique of overlapping is necessary only if conductor path entry angles are other than 90 or 180 degrees apart. These angles are standard with the double-entry teardrops, as shown in Fig. 11.10.

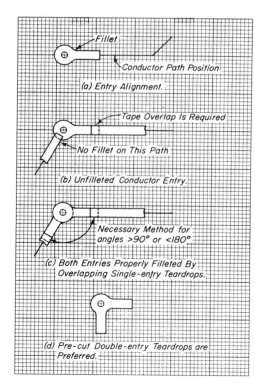

(a) Entry Alignment.

Fillet

Conductor Path Position

(b) Unfilleted Conductor Entry.

Tape Overlap Is Required

No Fillet on This Path

(c) Both Entries Properly Filleted By Overlapping Single-entry Teardrops.

Necessary Method for angles >90° or <180°

(d) Pre-cut Double-entry Teardrops are Preferred.

FIGURE 11.12. Methods of conductor filleting. (a) Entry alignment. (b) Unfilleted conductor entry. (c) Both entries properly filleted by overlapping single-entry teardrops. (d) Pre-cut double-entry teardrops are preferred.

(a)

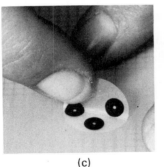

(c)

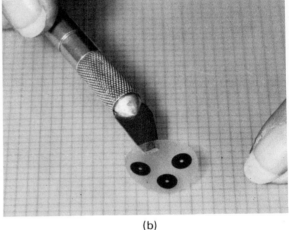

(b)

FIGURE 11.13. Recommended method of positioning group units. (a) A knife is used to remove unit from backing. (b) Hold unit on blade to guide into desired position. (c) Firmly press entire unit into position. Courtesy of Bishop Graphics, Inc.

Complex device outlines, such as those for transistors, integrated circuits, or connector patterns, are next transferred to the pad master, as indicated by the component and conductor layout drawings. The positioning of these group units is accomplished by the technique shown in Fig. 11.13a and b. These group units must be firmly pressed into place once positioned to ensure that the entire adhesive backing makes positive contact with the surface of the pad master (Fig. 11.13c).

The pad master for the complementary-symmetry amplifier board completed to this point is shown in Fig. 11.14.

With all device outlines and terminal pads properly positioned, the conductor paths can next be formed. The end of the tape is positioned to overlap the entry of the terminal pad and pressed firmly into place. As it is unrolled, the tape is guided along the conductor path and pressed down securely against the pad master until the desired termination point is reached (Fig. 11.15a). The conductor layout and the grid system are used as guides for positioning the tape. The tape width should be positioned *approximately* over the single lines representing conductive paths on the component layout drawing viewed through the pad master. During this procedure, minor changes in tape direction or spacing can be made from the plan originally laid out in the component layout drawing if necessary (Fig. 11.15b). A sufficient amount of tape should be unrolled to reach between termination points. Once the tape has been positioned and pressed into place, the index finger should be firmly run over the entire length of tape a second time to affix it securely to the pad master and to ensure an even surface. Once the second termination point is reached, the tape is cut. *Caution:* Do not attempt to cut *down through* the tape with the scalpel. Instead, the blade is rested flat against the tape and positioned along the intended cut. The tape is then pulled back over the blade edge. This technique is shown in Fig. 11.15c. This eliminates the possibility of cutting through both the tape and terminal pad, which would result in an open circuit in the conductor pattern at that point.

FIGURE 11.14. Pad master with all device outlines and teardrops positioned.

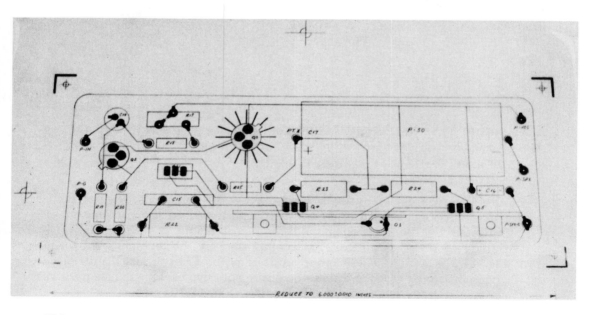

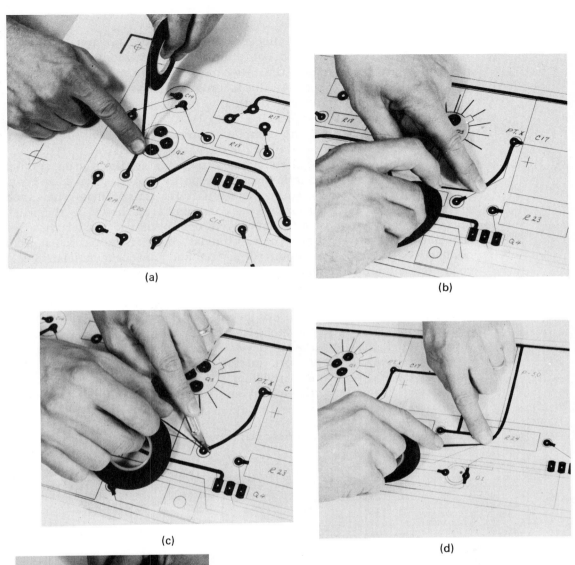

(a)

(b)

(c)

(d)

(e)

FIGURE 11.15. Taping techniques. (a) Tape should overlap teardrop entry. (b) Minor conductor repositioning for improved contour. (c) Proper knife placement for tape cutting. (d) Method of contouring sweeping bends. (e) Tee to fillet conductors at right angles.

When a conductor path changes direction with a sweeping bend, the tape is formed to the desired contour by using the index finger of one hand to apply pressure and the other hand to uniformly guide the tape direction. Figure 11.15d shows this technique. When conductors intersect at right angles and fillets are desired, tees are positioned as shown in Fig.11.15e.

The pad master for the complementary-symmetry amplifier board with the conductor paths completed is shown in Fig. 11.16.

Upon completion of the conductor path taping, the routing should be checked with the schematic diagram to ensure that no errors have been made in the conductor patterns. These patterns must provide the exact electrical interconnections among the components as shown on the schematic diagram. The most efficient way to check the pad master with the schematic diagram is to have one technician read off the particular lead connections made by the conductor pattern and another technician check each corresponding connection on the schematic. A colored pencil aids in marking over the wiring and connections on the schematic as they are compared. These marks are essential in discovering errors in the pad master. Remember that any error not detected during this check will result in a corresponding error in the etched pattern. Continual checking for error through every phase of artwork development will ensure accurate results.

To complete the artwork master, border delineation marks, grid locators (targets) positioned in alignment with the points on the drawing, and artwork reduction dimensions are added. The completed artwork

FIGURE 11.16. Completed taped pad master.

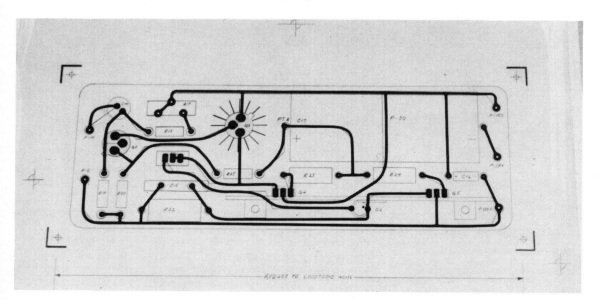

should contain keyed information if a marking mask is not to be used. (As will be seen later, a marking mask identifies completely all components on the insulated side of the board.) In addition, all components that might be positioned incorrectly on the pc board should be keyed similarly to the layout drawings. Finally, a descriptive identification number is positioned on the master.

Notice that all literal and numerical designations on Fig. 11.17 are *reversed*. The reason for this is that the original artwork master is generated from the *component* side of the board. Since a photographically reduced negative of this tape layout will be needed to process the *foil* side of the board, a *physical reversal* of this negative will be necessary, resulting in the proper orientation of literal and numerical designations as well as correct device lead and conductor patterns. Figure 13.3 shows the foil side of the tone control circuit board being processed with the correct negative orientation. Reversed literal and numerical designations are available for keying pad masters generated from the component side of the pc board. An alternative and more economical method of keying is to apply dry transfers (discussed in Chapter 9) *directly* to the side of the pad master *opposite* to that of the taped conductor pattern.

The completed artwork masters for the amplifier's preamp, tone control, power amplifier, and power supply boards are shown in Fig. 11.17.

FIGURE 11.17. Completed conductor pattern artwork masters. (a) Power amplifier.

(a)

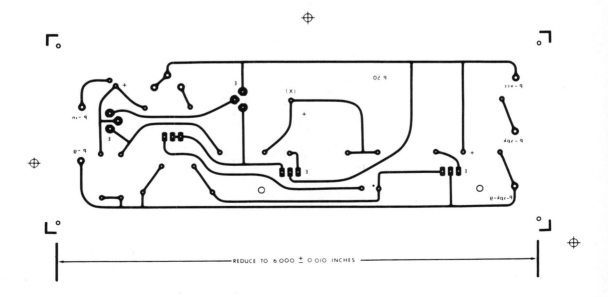

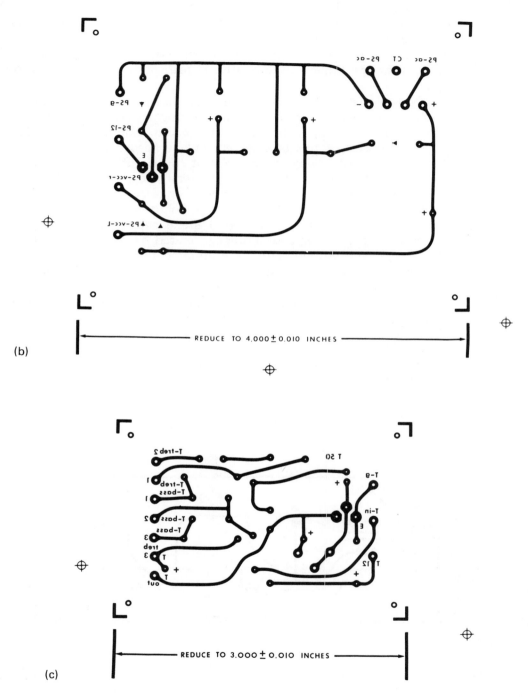

(b)

(c)

FIGURE 11.17. (cont.) (b) Power supply. (c) Tone control.

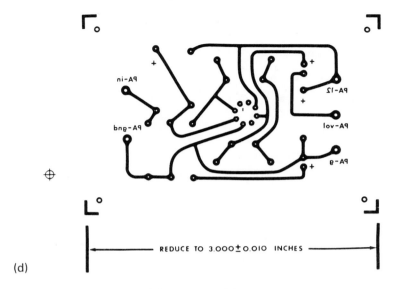

REDUCE TO 3.000±0.010 INCHES

(d)

FIGURE 11.17. (cont.) (d) IC preamplifier.

11.6 SILK-SCREEN SOLDER AND MARKING MASKS

The procedures and techniques thus far discussed for designing component layouts, conductor pattern drawings, and pad master artwork represent the minimum requirements for producing pc boards. To provide for more efficient soldering, a *solder mask* may be employed.

The artwork for the solder mask is prepared on a sheet of clear polyester securely taped to the pad master alone. Duplicate size solid donuts, as shown in Fig. 11.10, and device outlines are positioned on this polyester sheet directly over those areas on the underlying conductor pattern that will eventually be the soldered terminal pads on the processed board. Corresponding targets and board delineation marks must also be added to aid in all necessary artwork and board alignment. This solder mask artwork is a *positive* pad master.

To form the solder mask over a conductor pattern, a *silk screen* is necessary. This silk screen is prepared from a reduced negative of the positive pad master. The silk screen is then produced by using the techniques discussed briefly in Chapter 9 and to be considered in detail in Chapter 13. All the areas seen as clear on the positive solder mask artwork will be clear on the silk screen. The opaque terminal areas will be filled with the film gelatin when the screen is processed. The silk-screen solder mask must have the same reduction size as the conductor pattern artwork master. When the conductor side of the board is positioned under the silk screen, the application of an appropriate solder resist through the

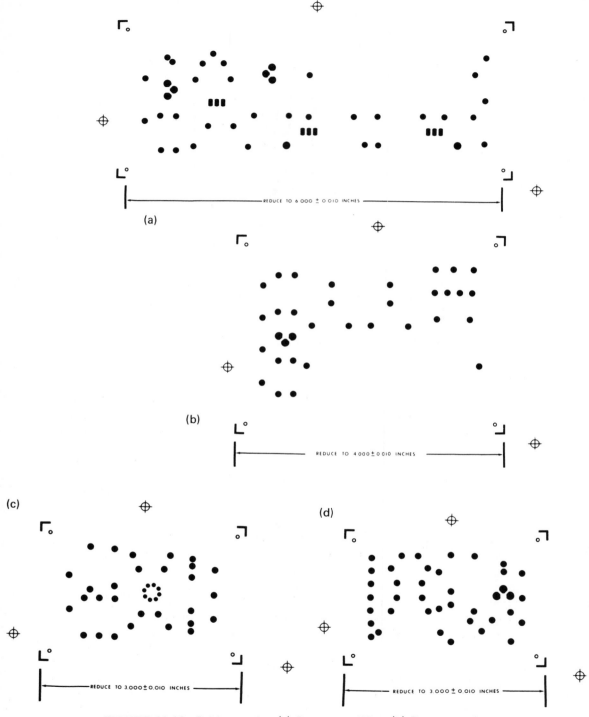

(a)

REDUCE TO 6 000 ± 0 010 INCHES

(b)

REDUCE TO 4 000 ± 0 010 INCHES

(c)

REDUCE TO 3.000 ± 0.010 INCHES

(d)

REDUCE TO 3.000 ± 0.010 INCHES

FIGURE 11.18. Solder masks. (a) Power amplifier. (b) Power supply.
(c) IC preamplifier. (d) Tone control.

screen will coat all areas of the board, including the conductor pattern with the resist *except* those terminal areas to be soldered.

Completed solder masks prepared from the preamplifier, tone control, power supply, and power amplifier pad masters are shown in Fig. 11.18. Notice the targets positioned for alignment purposes.

The solder mask will protect the conductor side of the board during the soldering process. It is employed *after* the board has been etched. Only the exposed terminal areas will allow solder to alloy with the conductor pattern and provide sound electrical and mechanical connections for component leads. A solder mask is utilized almost exclusively for automatic soldering methods such as *dip* and *wave* soldering (described in Chapter 15). Since only the terminal areas are exposed, less solder is required, thereby reducing weight and cost. Without the mask, the entire conductor pattern would be alloyed with the solder.

A *marking mask* is employed to aid in component assembly. This mask shows component outlines, positioning, and identification numbers. When processed into a silk screen, the mask will allow this information to be printed onto the *insulated* side of the pc board. For the purpose of preparing marking masks, precut outlines of the type shown in Fig. 11.19 are used. To construct the marking mask, a piece of clear polyester is taped over the component layout. With a scalpel, the component outlines are transferred to their appropriate positions. This procedure is shown in Fig. 11.20. Component identifications, such as R_1, R_2, C_1, C_2, Q_1, etc., should be in accordance with the schematic diagram and are easily produced with the aid of dry transfers (see Chapter 9). The marking mask artwork's grid locations and reduction scale should coincide with that of the master artwork.

The marking mask artwork is then processed into a silk screen with the screen becoming a *negative* marking mask. The application of a suitable epoxy paint will produce the desired information on the insulated side of the board. The completed marking masks for the preamplifier, tone control, power amplifier, and power supply boards are shown in Fig. 11.21. At this point, a differentiation between the information generated by a marking mask and the symbols produced from the conductor pattern pad master on the *copper* side of the pc board should be noted. As previously mentioned, the painted information on the insulated side of the board is used to simplify component assembly. The symbols on the foil side are for quick reference for testing purposes. However, as is often the case, the total information found on both sides of the board is helpful in identifying components and in rapidly locating test points.

Because of the complexities and difficulties inherent in the design of printed circuits, a *flow chart* is provided in Fig. 11.22 to aid the technician by showing the logical sequence of procedures necessary to generate all the drawings and artwork for pc boards. Inspection of this flow chart will show that it is a complete sequential review of this chapter in outline form.

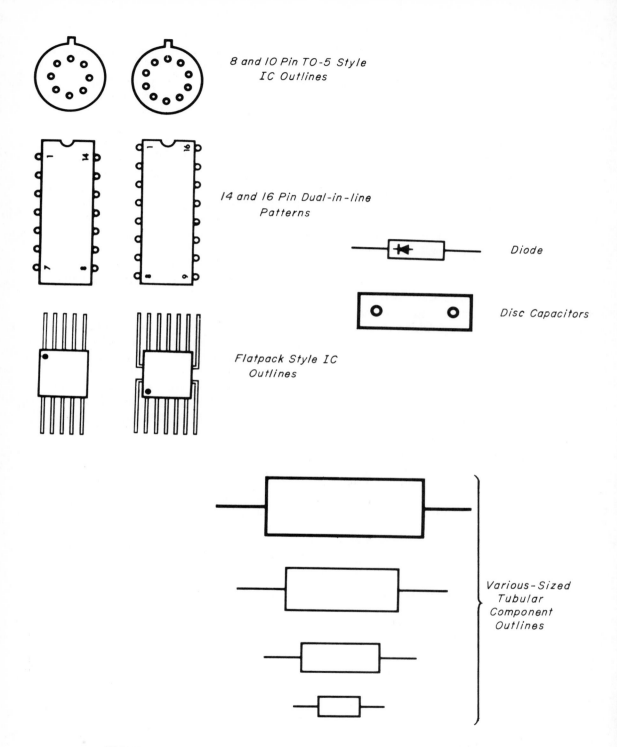

8 and 10 Pin TO-5 Style
IC Outlines

14 and 16 Pin Dual-in-line
Patterns

Diode

Disc Capacitors

Flatpack Style IC
Outlines

Various-Sized
Tubular
Component
Outlines

FIGURE 11.19. Device and component outlines. Courtesy of Bishop
Graphics, Inc.

FIGURE 11.20. Construction of marking mask pad master.

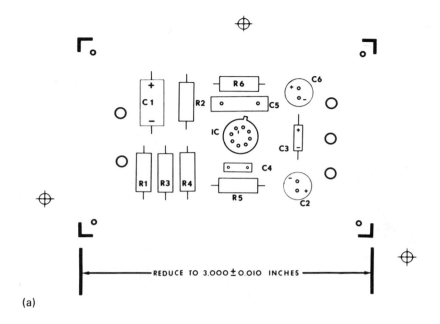

(a)

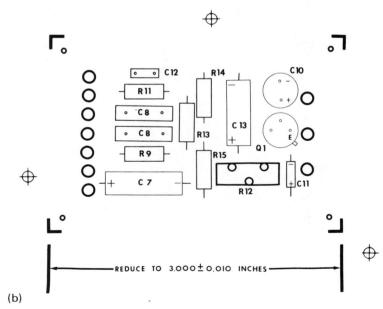

(b)

FIGURE 11.21. Marking masks. (a) IC preamplifier. (b) Tone control.

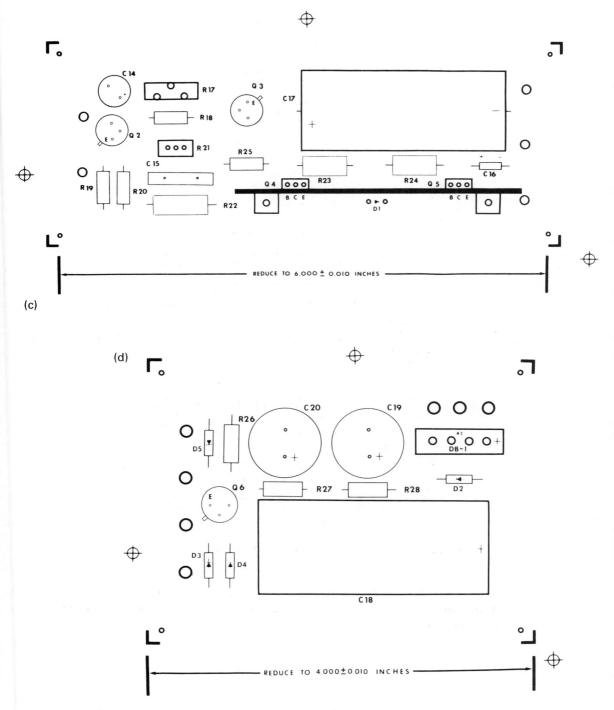

FIGURE 11.21. (cont.) (c) Power amplifier. (d) Power supply.

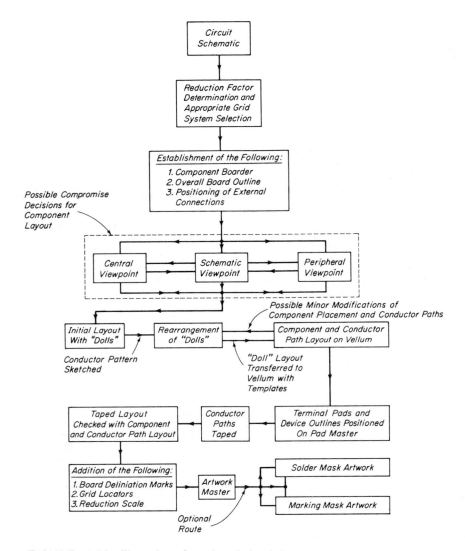

FIGURE 11.22. Flow chart for printed circuit layouts.

The next chapter is devoted exclusively to the processes required in producing a pc board from the drawings and artwork masters developed in this chapter.

PROBLEM SET

11.1 Using the appropriate templates and poster paper, trace and cut 2:1 and 4:1 scale drafting aids (dolls) for all resistor and capacitor sizes determined and tabulated for Problem 3.4. Also construct dolls for transistors, diodes, and integrated circuits of the 8- and 10-pin TO-5 style, and the 14- and 16-pin dual-in-line configurations with reference to Fig. 11.3b and Appendix III.

11.2 Referring to the 14-pin dual-in-line test jig configuration shown in Fig.11.23, construct the test jig conductor pattern drawings to a 1:1 scale and tape the artwork masters for the 8-pin TO-5, 16-pin dual-in-line, and TO86 flatpack configurations. The conductor pattern for the TO86 flatpack test jig is to be designed for soldered finger connections requiring the double 90-degree bend configuration shown in Fig. 14.32c. Each printed circuit board for the test jigs must be no larger than $3\frac{1}{2}$ by $4\frac{1}{2}$ inches. The pc boards are fabricated in Problem 13.3. Drilling terminal pads and installing terminal socket pins are performed in Problem 14.2. Tinning of the conductor pattern and soldering of terminals and socket pins are done in Problem 15.2. A test jig stand is constructed in Problem 16.4.

FIGURE 11.23.

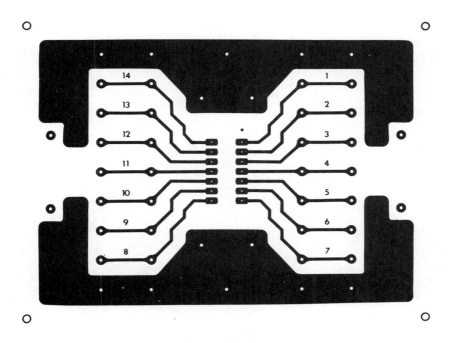

11.3 Draw the component layout and conductor pattern to a 2:1 scale for the circuit diagram shown in Fig. 11.24a. This circuit is a seven-fixed-interval UJT timer. The pin arrangement of the timer selector switch used in this circuit and shown in Fig. 11.24b is identical to that of the 14-pin dual-in-line integrated circuits shown in Appendix III. L1, S8, and the battery are to be chassis-mounted. External connections may be made to any board edge. The component layout design must be confined within a 2½-by 4½-inch border. Choose a suitable chassis, indicating lamp, and battery and battery clip styles for the design. An optional design may be considered that uses individual chassis-mounted time-interval selector switches. The seven time intervals for the circuit range from approximately 5 to 35 seconds. The pc board is fabricated in Problem 13.5 and assembled and soldered in Problems 14.3 and 15.3, respectively. Installation of the pc board into the chassis fabricated in Problem 8.5, along with the mount-

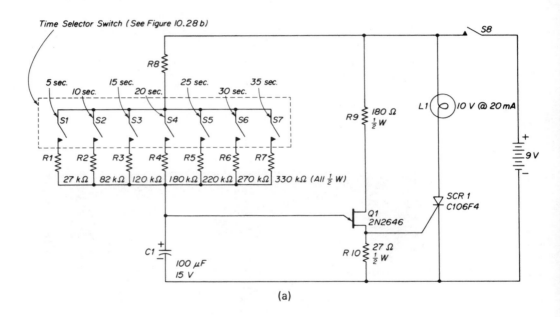

(a)

FIGURE 11.24.   (b)

ing of the on-off switch, indicating lamp, and battery clip, is accomplished in Problem 18.5. All interconnections among the chassis-mounted components and hardware and the pc board are soldered in Problem 19.5.

11.4 Construct a 4:1 scale component layout and conductor pattern for the series feedback regulator circuit shown in Fig. 11.25. All input and output connections are to be located on one edge of the pc board and designed for finger-type connection. The maximum size of the pc board is to be 3 by 3 ½ inches. The printed circuit is fabricated in Problem 13.6 with component assembly and soldering performed in Problems 14.4 and 15.4, respectively.

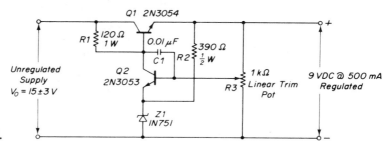

FIGURE 11.25.

11.5 Design 2:1 scale component and conductor pattern layouts for the tachometer circuit given in Fig. 11.26. Maximum pc board size is to be 1½ by 2 inches with provision for a suitable pc connector on one of the 2-inch sides of the board. Construct 2:1 artwork masters and marking and solder mask layouts for this design. All components and external connections are to be identified and keyed where necessary. The silk screens are constructed in Problem 13.4 and employed in Problem 13.7. Printed circuit board component assembly and soldering are undertaken in Problems 14.5 and 15.5, respectively. The pc board and meter are mounted in Problem 18.6. The harness for interconnections is constructed in Problem 20.3. An optional double-sided printed circuit design may be considered for this problem.

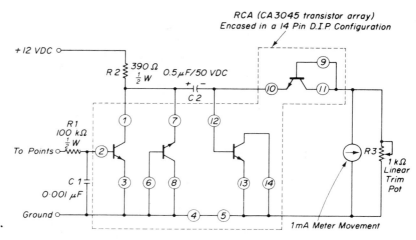

FIGURE 11.26.

245

12

Printed Circuit Photographic Processing

12.0 INTRODUCTION

When the conductor pattern artwork masters and the solder and marking mask artwork have been completed, they are ready to be processed photographically into working tools that will be used to process the finished pc board.

The topics to be discussed in this chapter include copy camera adjustments for photographic reduction of the artwork master, making an exposure guide, and one-to-one contact printing. Chemicals, bath preparations, and procedures used in the photographic processes are also discussed in detail.

12.1 PHOTOGRAPHIC REDUCTION OF THE ARTWORK MASTER

Commercial photocopy reduction cameras for producing either positives or negatives are capable of accepting artwork masters in excess of 36 by 42 inches. These masters may be reduced by a factor of 5 to 1 (5:1), 4 to 1 (4:1), or 2 to 1 (2:1), depending on the type of lens used. When desired, 1 to 1 (1:1) negatives of the master may also be made. These artwork reductions can be produced within a tolerance of ±0.001 inch. The exposures taken with such equipment have high-quality line resolution and high contrast, with negligible imperfections such as light spots or pinholes in opaque areas of the exposures.

The equipment mentioned above is expensive and is not normally necessary unless extremely accurate pc work is required. Less expensive equipment for photographic reduction of artwork masters is available that will produce positives or negatives of a quality acceptable for many pc applications. An example of a common reduction method is available school and office equipment that produces scaled transparencies for visual aids for overhead projections. The *Xerox No. 4 Copier*, shown in Fig. 12.1 can be used to reduce pc artwork masters. This camera, used with the appropriate film, will yield good quality exposures ideally suited for pc work.

FIGURE 12.1. Xerox No. 4 camera.

The Xerox No. 4 Copier will be discussed to illustrate photographic procedures and camera operation. Other cameras used for pc photography involve similar operating procedures but may be distinctive enough to require the technician to consult the manufacturer's instructions for any variations. The Xerox No. 4 Copier has eight basic parts. These are: (1) *copy board*, (2) *cover glass*, (3) *four 500-watt reflector flood lamps*, (4) *lens*, (5) *bellows*, (6) *film holder*, (7) *percentage scale*, and (8) *timer*. To operate this camera for reducing artwork masters, the procedure is as follows. First, the vertical copy board is unlatched from the cover glass frame. The copy board is then lowered to the horizontal position (Fig. 12.2a). The register guide lines on the copy board are used for center

(a)

(b)

FIGURE 12.2. Loading pad master into camera's copy board. (a) Lowering copy board. (b) Engaging latches to press cover glass against artwork.

positioning of the artwork master. Once the master is centered, the cover glass is replaced and the *locking latches* on the top edge of the copy board are engaged by gently pressing together the upper edges of both the copy board and the cover glass frame. This is shown in Fig. 12.2b. With the cover glass locked to the copy board, the artwork master will be held firmly in place. Raising the counterbalanced copy board to the vertical position will reestablish the correct position for exposure. This copier can produce 1:1 and 2:1 reductions and can extend its capabilities to 4:1 and 5:1 with the appropriate lenses from a maximum copy board capacity of 18 by 23 inches. To adjust for the desired reduction, the two scale-reduction *inner locking wheels* are rotated counterclockwise. Two sets of these wheels are located over the percentage scale on the lower edges of the front and rear sections of the camera with scale pointers immediately to the right of each set of adjustment wheels (Fig.12.3a). Both scale pointers (front and rear) are positioned over corresponding numbers on the percentage scale by turning the front and rear *adjustment wheels* located to the outside of the locking wheels. By turning the adjustment wheels in opposite directions, the camera bellows may be expanded or compressed. This causes the distance between the lens and film holder to change. To *decrease* the size of the image, the bellows should be compressed to move the lens closer to the film holder. The reverse procedure will *increase* the image size.

FIGURE 12.3. Camera adjustment for artwork reduction. (a) Percentage scale adjustment assembly. (b) Focus adjustment test setup.

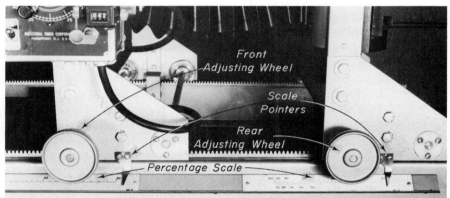

(a)

(b)

To serve as an example, following is the adjustment procedure to yield 2:1 reductions on the Xerox No. 4 Copier. The front and rear scale pointers are positioned on the numerical *50* on the percentage scale (50% reduction). This adjustment is secured by tightening the inner locking wheels. The correct distance of lens-to-artwork and film-to-lens is now set. To test the focus adjustment, a ground glass *focusing plate* is inserted into the rear of the camera directly behind the bellows (Fig. 12.3b). The lens is set to the widest aperture opening (lowest f-stop number). This will allow maximum light transmission to the focusing plate. By turning on the flood lamps with the switch located in the center of the camera bed frame and engaging the toggle switch on the timer to "constant output," the artwork image will appear on the focusing plate for inspection. (Most cameras are equipped with a protective lens cover that must first be removed to observe the image.) If adjustment for a sharper image is necessary, the rear inner locking controls are loosened and the rear outer focusing wheel is rotated. This wheel should be first turned slightly clockwise and then counterclockwise until the image is sharp and accurately reduced. For accuracy of reduction adjustment, a scale is held with its graduations *perpendicular* against the rear face of the focusing plate and aligned with the image of the scale reduction factor from the artwork master. The dimension thus obtained should agree with the dimension between the corresponding marks on the artwork master. If the reduced dimension is not accurate, the rear focusing adjustment wheel is again released and the final fine adjustment is made. The rear inner locking wheels are again tightened and the timer toggle switch is brought to the "off" or "timed output" position. The focusing plate can now be removed. Even though the flood lamp switch is left in the "on" position, the lamps will not be illuminated in the "timed output" mode until the timer is depressed. With the fine focusing completed, the camera is set for film loading. Under safe light conditions the film to be exposed is placed in a *film holder*. The film holder is then inserted into the rear of the camera with the sensitized side of the film toward the lens. The lens is adjusted to the appropriate f-stop (see Sec. 12.2) and the film is ready for exposure.

Exposure is controlled by an electrical timer that is synchronized with the shutter mechanism. The timer pointer, as shown in Fig. 12.4, is set to the desired number of seconds and activated by depressing the timer start button. Immediately, the lights are illuminated and the shutter is opened. The film is exposed for the preset time, at the end of which the timer activates the switching that turns off the lamps and closes the shutter. After each exposure, the timer will recycle to the preset position in preparation for another timed exposure. The mechanical counter on the timer assembly records the number of exposures taken. Exact exposure time and the method for determining this time will be discussed in the next section.

FIGURE 12.4. Timer unit.

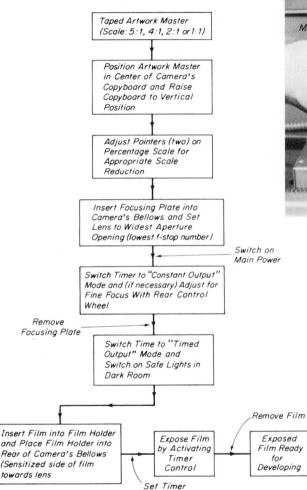

Taped Artwork Master
(Scale: 5:1, 4:1, 2:1 or 1:1)

Position Artwork Master
in Center of Camera's
Copyboard and Raise
Copyboard to Vertical
Position

Adjust Pointers (two) on
Percentage Scale for
Appropriate Scale
Reduction

Insert Focusing Plate into
Camera's Bellows and Set
Lens to Widest Aperture
Opening (lowest f-stop number)

Switch on
Main Power

Switch Timer to "Constant Output"
Mode and (if necessary) Adjust for
Fine Focus With Rear Control
Wheel

Remove
Focusing Plate

Switch Time to "Timed
Output" Mode and
Switch on Safe Lights in
Dark Room

Insert Film into Film Holder
and Place Film Holder into
Rear of Camera's Bellows
(Sensitized side of film
towards lens

Remove Film

Expose Film
by Activating
Timer
Control

Exposed
Film Ready
for
Developing

Set Timer

FIGURE 12.5. Flow diagram for photographic reduction of artwork master.

Once the film has been exposed the film holder is removed from the camera. The film is now ready to be developed. This process is also discussed in the next section.

A flow diagram of the photographic reduction procedure just discussed is shown in Fig. 12.5 to aid the technician in following the proper sequence.

12.2 EXPOSING AND PROCESSING PHOTOGRAPHIC FILM

The Xerox No. 4 Copier or similar camera for reduction of artwork masters requires the use of suitable film. This film must possess high line contrast, reasonably fast processing time, and excellent dimensional stability. Several manufacturers have produced sheet film with these qualities, among them Kodak Kodalith Ortho Type 3 and Dupont Cronalith film. For purposes of illustration, the Kodak Ortho 3 film and its associated chemicals will be discussed. If other film is used, the specific manufacturer's literature on film processing must be consulted.

The Kodak negative film is processed in the following sequence: *exposure, develop, stop, fix, wash,* and *dry.* Each of these steps is described below.

Exact exposure time cannot be stated specifically, because of the many variables involved, such as temperature, lighting, and f-stop setting. Therefore, a method that accounts for these factors and allows for the determination of exact exposure time is presented. Under *safe-light* conditions (Wratten 1-A or equivalent), a sheet of Kodak negative film is removed from its light-tight box. Under normal lighting, this film would appear pale *red* but appears as *light gray* under the safe lamp. The *emulsion* side of the film (as opposed to the 0.004-inch-thick *triacetate* support backing side) is determined by its *lighter* color. If extremely high dimensional stability and durability are required, 0.007-inch film is available at slightly higher cost. Graphic art film often has one corner notched to aid in identifying the sensitized emulsion side. Most films are arranged with the emulsion side toward the operator when the long length is vertical. Again, the manufacturer's instructions should be consulted.

The film is secured in the film holder so that the emulsion side is directed *toward* the lens. A thin metal light shield is inserted between the film holder and the lens (Fig. 12.6). An artwork master consisting of several horizontally taped lines on a polyester backing is positioned onto the copy board, as shown in Fig. 12.7. The line width should be at least $\frac{1}{8}$ inch to $\frac{1}{4}$ inch. With the percentage scale adjustment wheels and pointers set for the intended reduction, the lens aperture is set to "*f/8*" and the timer pointer to 4 seconds. The light shield is extracted to expose approximately 1 inch of film to the lens and the timer start button is depressed. When the exposure is completed, the light shield is extracted an additional inch from the film carrier. (The light shield now extends 2 inches from the outside edge of the film holder.) Again the timer is activated for a 4-second exposure. This second segment of the film will have been exposed for 4 seconds whereas the first 1-inch strip will have been exposed for 8 seconds. This procedure of exposing one strip of film at a time will result in approximately seven verticle 1-inch strips, each exposed sequentially from 4 seconds (the last exposure) to 28 seconds (the first exposure). The number of exposed 1-inch strips will, of course, depend on the size of film

FIGURE 12.6. Insertion of light shield between bellows and film holder.

FIGURE 12.7. Test artwork to process exposure guide.

used. When developed, this will provide the technician with a guide for determining exposure time under his exact working conditions. Additional exposure guides can be developed for various f-stops, temperatures, and reduction scales, thus furnishing the technician with a complete record of exposure guide films to aid in determining desired line contrast for the factors considered. A sample exposure guide film prepared as outlined above is shown in Fig. 12.8.

Kodak film is developed by the use of two concentrated chemicals: *Kodalith Liquid Developer Solution A* and *Kodalith Liquid Developer Solution B*. Since these chemicals are in liquid form, they require no special mixing equipment. For best results, they should be mixed with water, which is at a temperature of between 65 and 70°F (18 to 21°C). (This water temperature is recommended for mixing all film baths.) The following five steps outline the preparation of the developer.

1. Determine the volume of developer required.
2. Divide this total in half to determine the amount of diluted developer A and diluted developer B required.
3. Divide this figure by 4 to find the amounts of both concentrates A and B needed.
4. Add three equal parts of water to concentrate A as determined in step 3; repeat for concentrate B.
5. Mix diluted A to diluted B to yield the total volume as determined in step 1.

FIGURE 12.8. Four-second-interval exposure guide at f/8.

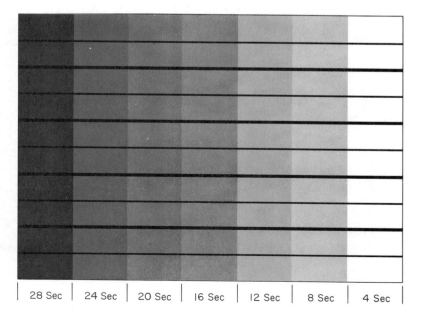

| 28 Sec | 24 Sec | 20 Sec | 16 Sec | 12 Sec | 8 Sec | 4 Sec |

The concentrates A and B are never added together. They must first be diluted before mixing.

An example for the preparation of the developer will aid in understanding the various steps. In this example, it is required that *1 quart of Kodalith Developer be prepared.* Following are the five steps just presented:

1. Total volume required — 1 quart (32 ounces).
2. Subvolume — 32/2 = 16 ounces of diluted A and 16 ounces of diluted B.
3. 16A/4 = 4 ounces of A concentrate; 16B/4 = 4 ounces of B concentrate.
4. Four times 3 = 12 ounces of water. Four ounces of A plus 12 ounces of water = 16 ounces of diluted A solution; 4 ounces of B plus 12 ounces of water = 16 ounces of diluted B solution.
5. Sixteen ounces of A diluted solution added to 16 ounces of B diluted solution yields 32 ounces (1 quart) of working developer.

The prepared solution is placed in a plastic developing tray. Under safe-light conditions, the exposed film with the emulsion side *up* is placed in the tray and constantly agitated until the film is developed completely. At a developing solution temperature of 68°F (20°C), the development time is between 2 and 2½ minutes, the time decreasing as the temperature increases. At 74°F (23°C), the developing time is 1¼ to 2 minutes and at 80°F (27°C) reduces to 1 to 1½ minutes. A chart of development time in minutes for specified temperatures is shown in Fig. 12.9. Developing time is critical and strict adherence to the specified time must be observed to reduce the possibilities of under- or overdeveloping. The negative film of the tone control circuit pc board being developed is shown in Fig. 12.10.

Bath Temperature	Developing Time
68° F	2 min. 15 sec.
69° F	2 min. 10 sec.
70° F	2 min. 7 sec.
71° F	2 min. 3 sec.
72° F	1 min. 58 sec.
73° F	1 min. 52 sec.
74° F	1 min. 49 sec.
75° F	1 min. 45 sec.
76° F	1 min. 40 sec.
77° F	1 min. 37 sec.
78° F	1 min. 32 sec.
79° F	1 min. 26 sec.
80° F	1 min. 22 sec.
81° F	1 min. 18 sec.
82° F	1 min. 15 sec.

FIGURE 12.9. Table of developing time as a function of bath temperature.

255

FIGURE 12.10. Developing sequence for reduced negative.

An explanation of the developing process will aid in understanding the reason for paying close attention to the required time. When the film is initially exposed in the camera, certain portions remain underexposed because the black conductor pattern on the copy board absorbs light whereas surrounding white area reflects light. As a result, the underexposed film portions form a latent image of the conductor pattern. When the film is developed, it yields a white conductor pattern image on a black background. During the developing process, the total emulsion surface is softened and feels oily to the touch. As a result, extreme care needs to be exercised to avoid scratching this surface. If the film is allowed to remain in the developer for an excessive amount of time, the entire emulsion surface will become totally black. The film negative should be agitated in the developer long enough for a clear-cut image to appear with the desired contrast. Optimum contrast can be gauged by continual visual inspection of the device outline configuration. Attention should be focused on the smallest lead holes to be developed. When they appear as distinct sharp black dots, the developing process should be terminated *immediately* by quickly placing the film into a *stop bath*. This stop bath is made up of 1 part of *glacial acetic acid* to 32 parts of water. To serve as an example of this mixture, to prepare 330 milliliters of stop bath, 10 milliliters of glacial acetic acid is mixed with 320 milliliters of water.

The developer is completely neutralized in the stop bath when the oily characteristics of the film emulsion no longer exist. This neutralization process will take approximately 1 minute. There is no visible change in the image. The negative is gently agitated in this stop bath for approximately 1 minute and then is placed in a tray containing *Kodak Rapid Fixer (Solution A)* with a *hardener (Kodak Solution B bath for fixer)* at 68°F (20°C) for approximately 1 to 2 minutes. The black areas of the film are insoluble in the fixer but the white emulsion of the conducting pattern image will dissolve in the solution. The negative must be agitated in the fixer solution for a sufficient period of time for all the emulsion in the conducting pattern area to be dissolved, exposing the underlying transparent backing. A black tray should be used because it permits the complete dissolving of the white emulsion to be more easily detected. The visual impression is a totally black film. The hardener is included in the fixer solution to stabilize the remaining emulsion.

To prepare 1 gallon of fixer bath, 1 quart of solution A is added to 2 quarts of water at 65 to 70°F (18 to 21°C). Three and one-half fluid ounces of solution B (hardener) are stirred into the solution. Finally, sufficient water is added to make the total solution equal to 1 gallon.

FIGURE 12.11. Final film rinse and drying. (a) Water rinse. (b) Sponge used to remove excess water.

(a)

(b)

After the fixing process, the film negative is washed in a spray of water at approximately 70°F (21°C) for several minutes to remove all traces of the fixing solution (Fig. 12.11a). Once the film has been washed, it can be allowed to dry by simply suspending it with a clamp attached to an overhead line. This method is not recommended, however, since the surfaces of the negative tend to water-spot. To prevent this spotting a squeegee or photographic sponge may first be used to remove the excess water and then the negative should be allowed to dry. Figure 12.11b shows a negative being squeegeed to remove excess water. The film is now completely processed into a negative of the printed circuit pattern. The negative should be handled carefully because it will scratch easily. Because of the involved nature of the photographic process, a flow diagram is provided in Fig. 12.12 to guide the technician through the individual steps.

The completely processed negatives for all the printed circuit conducting patterns for the amplifier system together with the reduced solder and marking mask films are shown in Fig. 12.13.

FIGURE 12.12. Flow diagram for developing negative film.

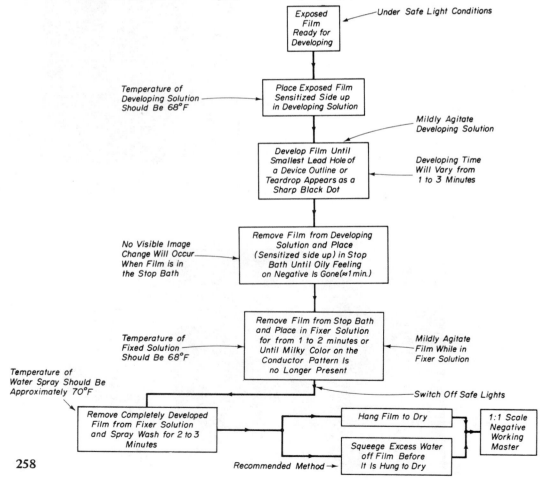

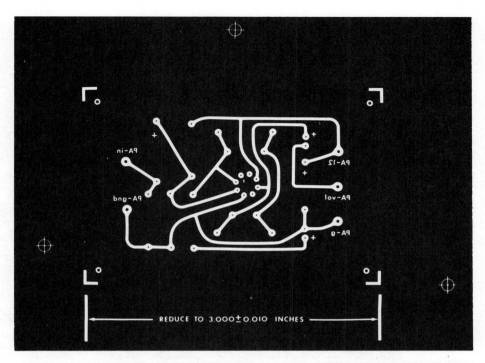

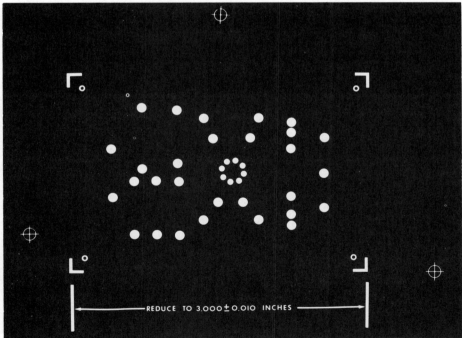

FIGURE 12.13. Processed negatives for the complete amplifier system. (a) IC preamplifier.

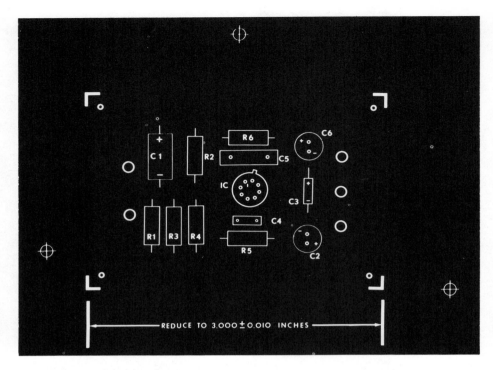

FIGURE 12.13.(a) (cont.)

FIGURE 12.13. (cont.) (b) Tone control.

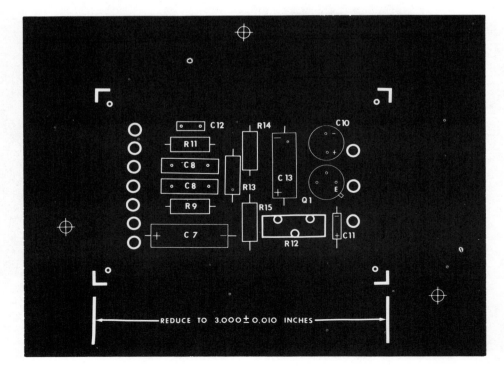

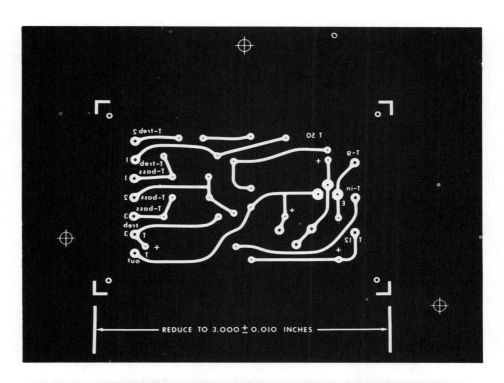

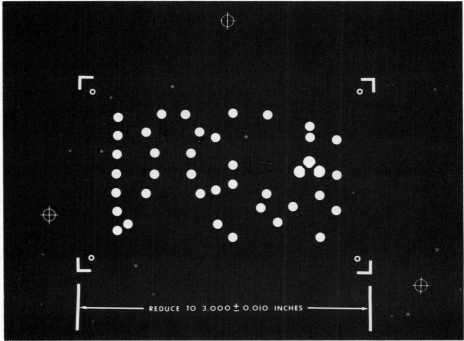

FIGURE 12.13.(b) (cont.)

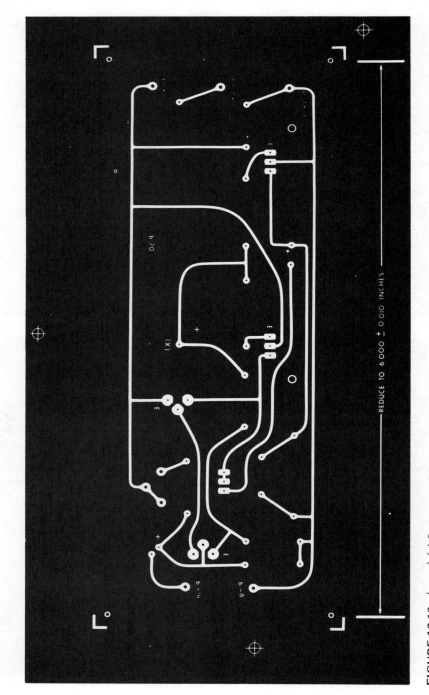

FIGURE 12.13. (cont.) (c) Power amplifier.

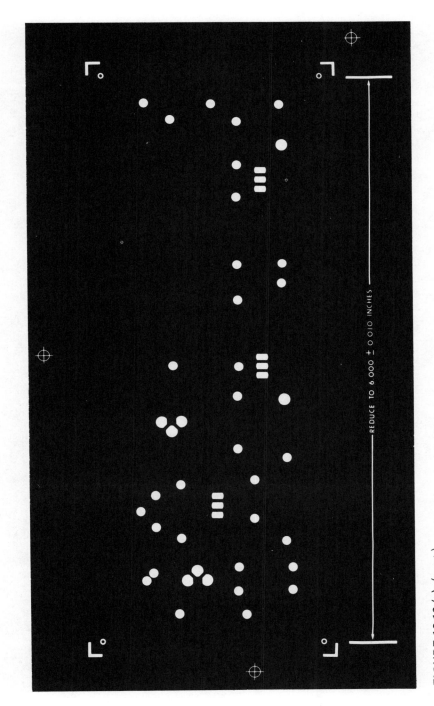

FIGURE 12.13.(c) (cont.)

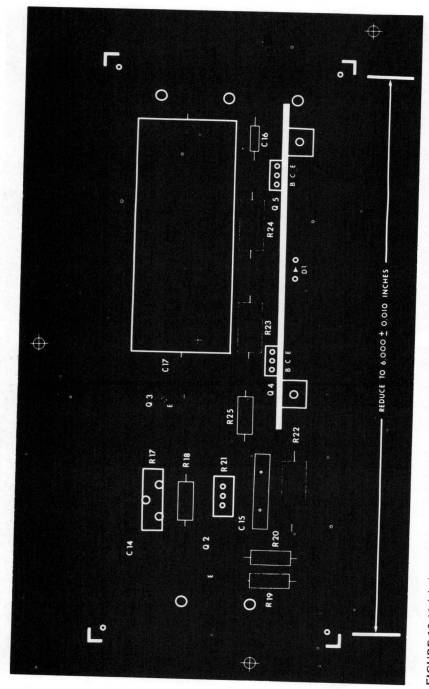

FIGURE 12.13.(c) (cont.)

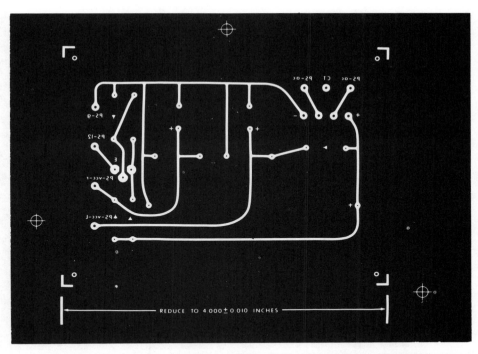

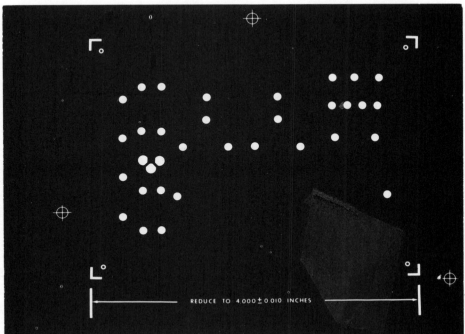

FIGURE 12.13. (cont.) (d) Power supply.

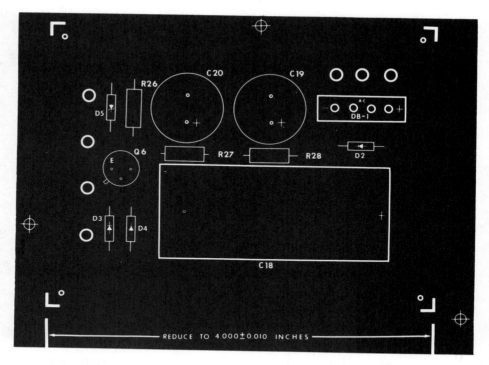

FIGURE 12.13.(d) (cont.)

12.3 CONTACT PRINTING

When the artwork masters have been photographically reduced into negative transparencies, the required 1:1 scale *positives* of the marking mask negatives can now be processed. If the film used in the initial reduction process yields a negative transparency, one more operation will generate the positive transparencies. To produce a positive from a reduced artwork negative is to in fact obtain a positive transparency of the original artwork master that is reduced to the proper 1:1 size scale for processing a pc board marking mask. (The process of converting a positive transparency of a marking mask into a silk screen will be discussed in detail in Sec. 13.6.)

The processing of a 1:1 scale positive from the same-size negative can best be accomplished by contact printing using the same Kodak Kodalith Ortho Type 3 film used for the reduction of the original artwork masters. Under safe-light conditions, the film is placed onto a sheet of white paper with the sensitized side up toward the light. The reduced negative is placed onto the film and this combination is then covered with a sheet of ¼-inch heat-treated plate glass. The negative is thus held in direct contact with and flat against the film by the weight of the glass. Directly above the glass, a 275-watt sun lamp is placed at a distance of between 5 and 7 inches. It is very important that the light source be provided with a shield that directs the light onto the work and at the same time protects the eyes from exposure to light rays. Light in the ultraviolet range is extremely dangerous to the eyes and prolonged exposure must be avoided. If a light shield is not available, protective glasses should be worn.

An arrangement to contact-print Ortho 3 film for the tone control circuit marking mask is shown in Fig. 12.14. Commercially available contact printing systems feature vacuum frames and adjustable-intensity light sources. However, they are relatively expensive and satisfactory results can be obtained from the unit shown in Fig. 12.14, which was constructed from common materials.

To expose the film, the sun lamp switch is turned on and off as quickly as possible. (The exposure time is approximately $1/4$ second.) The sun lamp will not have reached its full intensity, but the emitted light is sufficient to completely expose the film.

After exposure, the film is removed from the frame and, still under safe-light conditions, processed through the same developer, stop bath, fixer, and water rinse processes as outlined in Sec. 12.2. The flow diagram of Fig. 12.12 used to process negative film can also be used for contact print processing.

With the completion of the photographic processes, the next phase in pc board fabrication will be to process the board into a functional circuit. Single-sided board processes are discussed in Chapter 13 and double-sided layout and processing are discussed in Chapters 16 and 17, respectively.

FIGURE 12.14. Exposure unit and film arrangement for contact printing.

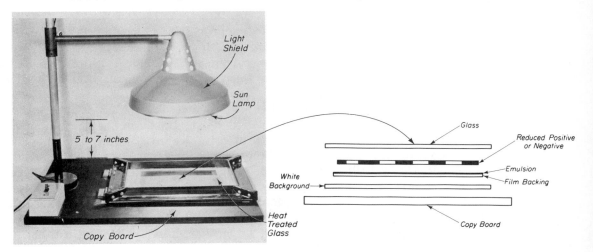

13

Single-Sided Printed Circuit Board Processing: Print-and-Etch Technique

13.0 INTRODUCTION

Of the many methods available for the processing of single-sided pc boards by the simple print-and-etch technique, the use of liquid photo-resist (wet film) is perhaps the most economically feasible for prototype work. This technique requires inexpensive materials and equipment, thus making it ideally suited for this purpose.

The topics discussed in this chapter include detailed discussions on circuit board cleaning, photosensitive resist application, etching, immersion tinning, conductor pattern protective sealing, drilling, and the silk-screen process for preparing marking masks.

In the wet-film technique of circuit board processing, certain methods are more suitable for prototype work, whereas others lend themselves more readily to mass production applications. The selection of a specific process will depend on such factors as costs, quantity of boards to be fabricated, and labeling and solder mask requirements. The use of the various chemicals and materials for processing are included to aid in the selection of the most appropriate method. Also included is information on simple construction of some of the equipment necessary for the various phases of board fabrication.

13.1 PHOTOSENSITIZING THE PRINTED CIRCUIT BOARD

For developing the conductor pattern onto the foil side of the copper-clad pc board, a negative of the original conductor pattern artwork master is required. This negative becomes one of the basic elements used to process either one board for prototype work or hundreds of boards for mass-production requirements.

The surface of the foil must first be treated so that selected areas of the copper can be protected from the etching solution, which will remove all the copper except the conductor pattern. The copper may be treated by several methods to accomplish selective foil pattern protection. The method to be considered is *photosensitizing*. In this process, the foil is treated with a *photosensitive resist*, exposed through a 1:1 negative of the conducting pattern, and developed. The result will be a 1:1 positive conductor pattern produced onto the surface of the foil, which is etchant resistant, whereas the undesired areas of the foil are unprotected and will be removed by the etching process.

There are several chemicals readily available to implement this photosensitized technique. The chemicals and procedures discussed in this section will be those of the *Dynachem Corporation*. Chemicals of other manufacturers may require processing and exposure times that are different from those discussed.

The process begins by cutting the necessary-size copper-clad board from sheet stock. (See Chapter 10 for information on shearing pc boards.) The work must then be thoroughly cleaned before the photo-resist is applied. This cleaning is essential in order to remove all contaminants, such as grease or copper-oxide film, which readily forms on exposed copper. Unless these contaminants are removed, they will adversely effect the adhesion of the resist with the copper surface. Copper foil can be cleaned by scouring the surface with wet *pumice* and a stiff bristle brush. This technique is shown in Fig. 13.1. If pumice is not available, any gentle abrasive household cleanser that contains no bleach may be substituted.

FIGURE 13.1. Copper foil cleaning method.

Coarse abrasives should not be used because they would deeply scratch the soft copper foil. Proper cleaning removes contaminants without causing surface imperfections in the foil. Cleaning is sufficient when water flushed over the surface of the copper "sheets" and no water "beads" remain. Remaining grease contaminants will appear as dry spots from which the water tends to pull away. To complete the cleaning process, the foil is thoroughly rinsed with water and swabbed with a lint-free cloth to remove any remaining deposits of pumice or cleaner. Once the copper is completely cleaned, all surface moisture must be removed. This is accomplished by first patting dry with a paper towel. Since moisture severely affects resist adhesion, a final drying operation is important. Total drying will be achieved if the board is placed in an oven for approximately 10 minutes at a temperature of 120°F (49°C). After the board is removed from the oven and allowed to cool to room temperature, the application of the photo-resist can begin. The board must be handled carefully because fingerprints will contaminate the copper. The cleaning and drying process appears time-consuming and tedious, but it is only through a thoroughly clean and dry surface that a quality printed circuit will result.

The photo-resist is applied by *spraying, dipping, rolling,* or *spinning.* Spraying is preferable for prototype work because special equipment is necessary for rolling and spinning and large vats of resist, which can easily become contaminated, are needed for dipping. The spraying application of *DCR* (Dynachem Photo-Resist) is simplified since it is readily available in 16-ounce aerosol cans.

This photo-resist should be used in a ventilated spray booth under subdued light. Although in its liquid state DCR photo-resist is not extremely sensitive to light, direct sunlight should be avoided. The board to be coated should be positioned vertically in the spray booth. It is important that a uniform thickness of resist be applied if exposure time is to be controlled. After shaking the can, the spray should be tested for 1 or 2 seconds *away* from the board to ensure complete atomizing of the resist. A partially clogged nozzle will prevent uniform spraying and may cause large droplets to splatter the work surface. Once the nozzle is cleared, it is held approximately 8 to 10 inches from the surface of the board. The spraying is started at the bottom of the board, making horizontal passes that overshoot the edges while moving the can slowly upward to complete the coating. This procedure of spraying from bottom to top is to minimize photo-resist sagging (Fig. 13.2). The valve button on the aerosol can should be held firmly depressed during the entire spraying process to ensure a continuous and even flow of resist. Immediately after spraying, the board should be shifted to a horizontal position with the treated side on top for approximately 1 minute. In this position, the wet resist will uniformly level itself over the entire copper surface, thus minimizing the possibility of *pinholes* forming in the coating. Pinholes are minute voids

FIGURE 13.2. Application of photo-resist with aerosol can.

through the resist which expose areas of the underlying copper. They will cause portions of the conductor pattern to be removed during the etching process.

Optimum drying of the photo-resist is best accomplished in an oven set at between 100 and 115°F (38 to 46°C) for 10 to 15 minutes. This form of drying is preferable to room temperature drying to minimize *solvent entrapment* between the copper surface and the photo resist, which interferes with complete wetting and bonding.

When the photo-resist hardens, it becomes sensitive to light. Consequently, care must be exercised to protect the surface from exposure to *any* intense light when removed from the oven to cool. The board should be allowed to cool to room temperature before exposing it to a light source. Large-diameter 16-mm film cans are often used to store sensitized boards for short periods of time.

The light-sensitive photo-resist can be exposed with several types of light sources. Since the maximum sensitivity of DCR photo-resist is at 3600 angstroms, a black light or sun lamp would be ideally suited. Exposure time will vary depending on such factors as light intensity, type of source, distance of the source from the surface, and the thickness of the resist. Optimum exposure can best be determined by trial and error. For this type of photo-resist, a 275-watt sun lamp at a distance of 5 to 7 inches above the work for approximately 4 minutes yields proper exposure.

The pc board can best be exposed by placing it on a flat surface with the sensitized side *up* in subdued light. The reduced negative of the conductor pattern is positioned directly onto the sensitized surface. The border delineation marks on the negative must be aligned with the corners of the board. In addition, the negative is positioned so that all literal and numerical designations are readable. Care must be exercised to ensure that the sensitized surface will be exposed through the correct side of the negative. If an orientation error is made and the board is later etched, it will be impossible to correct the conductor pattern. The correct orientation of the tone control circuit negative on the sensitized copper surface is shown in Fig. 13.3.

With the negative held in direct contact with the sensitized surface by the weight of a sheet of heat-treated glass, the light source is positioned and an exposure is made. Refer to Fig. 12.14 for a similar exposure technique.

Those photo-resist areas exposed to the light through the negative will become insoluble to the developer and thus remain on the copper. Those areas under the opaque portions of the negative are not exposed and remain soluble to the developer. This photo-resist will wash away, leaving a photo-resist mask on the surface of the copper that duplicates the desired positive conductor pattern.

To develop the board, it is immersed in a developing tray of *Dynachem Developer* at room temperature with the resist side *up* for 30 to 40 seconds. The solution should be gently agitated by raising and lowering one end of the tray. It is important that developing time does not exceed 1 minute lest the solution begins to dissolve the resist pattern. At the completion of the developing process, the resist pattern cannot be

FIGURE 13.3. Correct alignment of negative on sensitized copper.

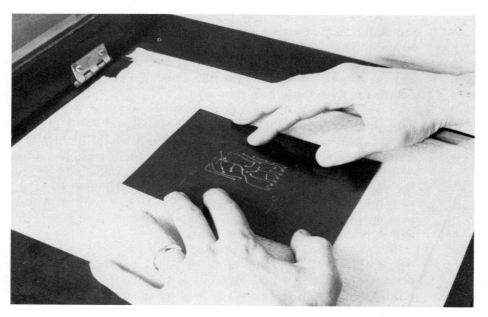

detected because it is transparent. However, before the board is etched, the pattern should be carefully inspected for quality and for the detection of any imperfections. The pattern is made visible by applying *Dynachem Dye* from an aerosol dispenser immediately after the board is removed from the developer. The entire surface of the board is coated with the dye and allowed to stand for 10 to 15 seconds (Fig. 13.4a). The dye will penetrate and adhere to all photo-resist remaining on the surface of the board (pattern). The board is then rinsed in flowing water at high

FIGURE 13.4. Conductor pattern inspection is aided by the use of dye. (a) Dye application. (b) Water spray rinse.

(a)

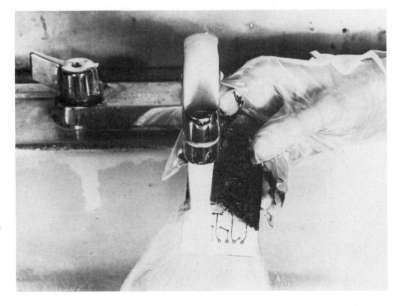

(b)

volume and low pressure (Fig. 13.4b). The resist pattern is now a clearly visible dark-blue color. In this state, the board must be carefully handled because the photo-resist is soft and can be easily removed. The board can now be inspected and any errors or pattern imperfections that may be present can be corrected by using a sharp, pointed knife. If there are any breaks in the photo-resist covering of the desired conductor pattern at this point, they can be repaired by the application of *etchant ink resist* (non-light-sensitive). This resist is carefully applied with a fine artist's brush to close the break. It must overlap the photo-resist on either side of the break while maintaining width. This technique is shown in Fig. 13.5.

When the photo-resist has hardened, the board is ready for etching. To reduce drying time and to ensure the removal of all solvents and improve the durability of the resist to strong etchants, a baking at 200 to 250°F (93 to 121°C) for approximately 30 minutes is recommended.

The photo-sensitizing procedure just discussed is an effective method of preparing a pc board for etching in prototype work. A flow diagram to aid the technician in following the many steps in this process is shown in Fig. 13.6.

All the amplifier boards ready for etching are shown in Fig. 13.7.

FIGURE 13.5. Application of etchant resist to repair defective pattern.

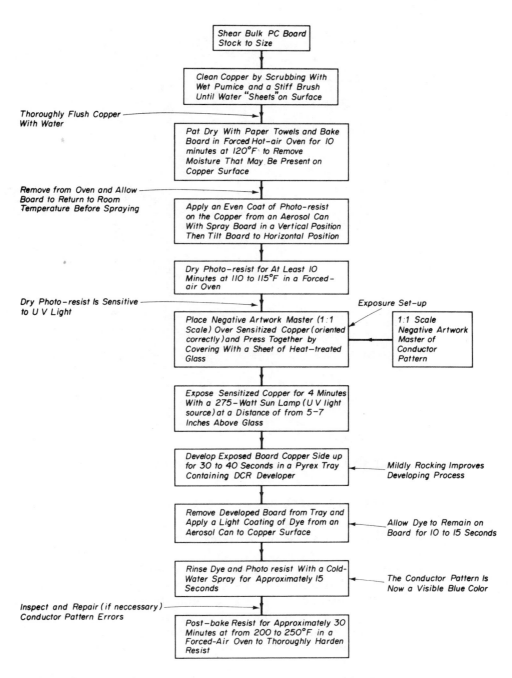

FIGURE 13.6. Flow diagram for sensitizing printed circuit boards with photo-resist.

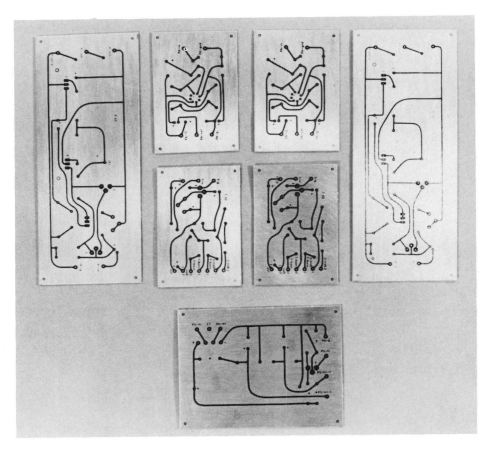

FIGURE 13.7. All pc boards for the amplifier processed and ready for etching.

13.2 CIRCUIT BOARD ETCHING

Etching is the process of attacking and removing the unprotected copper from the pc board to yield the desired conductor pattern. Several commercial etchants and techniques are available for processing pc boards. These materials and methods attack the unprotected copper yet do not affect the adhesive, supporting laminate, and photo-resist. The most common chemicals used by industry as an etchant are *ferric chloride*, *ammonium persulfate*, *chromic acid*, and *cupric chloride*. Of these, ferric chloride is commonly used since it is less expensive and potentially the least dangerous.

Methods of etching include *tray rocking*, *tank etching*, and *spray etching*. Tray rocking is the simplest system, consisting of a tray of etchant attached to a powered rocking table. Tank etching involves dipping the work into a vat containing the etchant, which is maintained at an optimum temperature. Spray etching, although a much faster process, calls for etchant to be pumped under pressure onto the surface of the pc board. This method requires an elaborate and expensive system. These three methods will be described using ferric chloride as an etchant.

Ferric chloride ($FeCl_3$) is an etchant that will react with the unprotected areas of the copper surface and produce copper ions, which are soluble in solution. This chemical is available premixed to the correct concentration. Ferric chloride crystals are also available if liquid storage is a problem but require premixing and concentration testing.

During the etching process, the concentration weakens because the soluble cupric and ferric ions precipitate out of solution in the form of a sludge that tends to settle on the bottom of the etching vat. When the etching time becomes excessively long, the used etchant should be discarded and replaced with a fresh solution.

Ideal etching conditions require that the etchant be heated to a temperature of between 100 and 130°F (38 to 54°C). Temperatures above this range should be avoided because of the fumes that are generated. Although etchant temperature is not critical, the activity of the etching process is increased when the solution is heated above room temperature. With the temperature maintained reasonably constant, the length of time required for etching will be consistent. However, because of the weakening of the solution, etching time will progressively increase.

In addition to heating, efficient etching requires that the etchant be continuously agitated to allow "fresh" solution to flow over the copper. A simple means of achieving this agitation is to use a Pyrex tray filled with etchant and placed on top of a *rocker table*. Rocking motion is imparted to the table by means of a small *motor*, *cam*, *connecting arm*, and off-center *pivot* arrangement. This apparatus is shown in Fig. 13.8. The pc board is placed copper side *up* in the tray. Only one board should be etched at one time since, because of the rocking motion, they tend to come in contact with each other. This could cause some of the etchant resist to be scraped off of a surface. As the table is rocked, a wave is generated in the etching solution that travels back and forth across the tray and board surface, keeping the major portion of any sludge formed off the copper surface. There are two disadvantages in using this arrangement. First, there is no method provided for heating the etchant unless an infrared heat lamp is used, in which case etchant temperature is difficult to control. Second, some sludge is continually carried over the surface of the board, which reduces some of the effectiveness of the etchant. However, outweighing these disadvantages is the fact that this system is relatively simple and inexpensive and suitable for prototype work. A 5- by 5-inch square of 1-ounce single- or double-sided pc board can be completely etched in approximately 30 minutes.

A more elaborate system for agitating the solution is shown in Fig. 13.9, consisting of a *vat* (Pyrex or other suitable nonmetallic material), *heating tape*, *variable transformer*, *air feed*, and *thermometer*. The tem-

FIGURE 13.8. Rocker table.

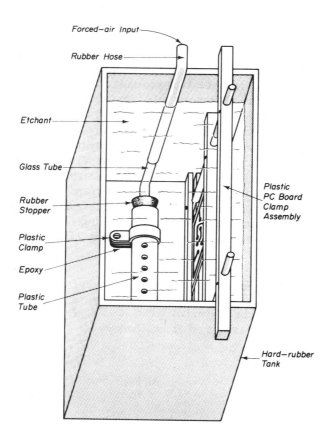

Forced–air Input

Rubber Hose

Etchant

Glass Tube

Rubber
Stopper

Plastic
Clamp

Epoxy

Plastic
Tube

Plastic
PC Board
Clamp
Assembly

Hard–rubber
Tank

FIGURE 13.9. Forced-
air etching system.

perature of the etchant contained in the vat is controlled by varying the voltage level delivered to the heating tape by the transformer. This temperature is monitored by a thermometer placed in the solution in proximity to the boards being etched. The etchant is circulated (agitated) by means of air pumped into the vat under low pressure and escaping through a perforated plastic tube placed along one side of the vat. The sludge (copper and ferric ions) settles to the bottom of the vat, allowing relatively sludge-free etchant to continually circulate across the surface of the work. The boards to be etched are suspended in the solution from noncorrosive rods and clamps with their copper side facing the source of the air bubbles. This method, although more expensive, can etch a 5- by 5-inch 1-ounce board in approximately 15 minutes. Spray etching involves jet spraying the surface of the board with pressurized nozzles. Systems such as that shown in Fig. 13.10 also are available with a heating element to decrease etching time. With such a system, etching times of between 1 and 2 minutes can be realized without difficulty. The etching cycle is controlled by a mechanical timer. Several precautions should be pointed out. The thermostat setting should not exceed the manufacturer's recommended maximum limits to avoid damage to the equipment. In addition, spray etchers of this type may be hazardous if the cover is not properly seated prior to operation to completely contain the etchant within the system.

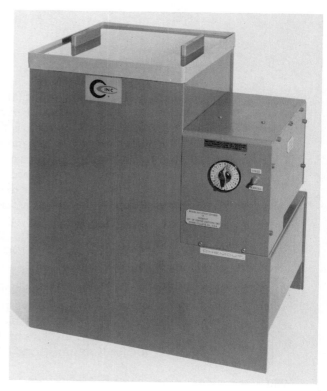

FIGURE 13.10. Industrial etching system. Courtesy of Chemcut Corporation.

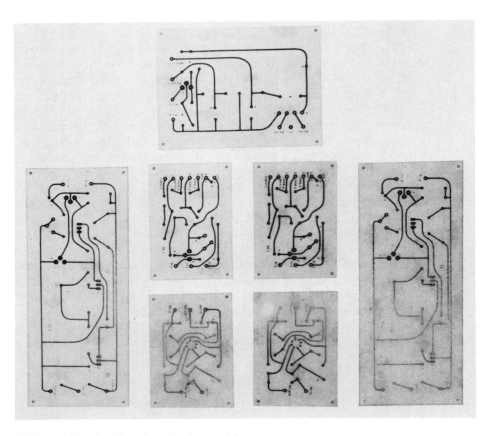

FIGURE 13.11. All pc boards after etching.

When the etching is completed, the board is rinsed under water for approximately 30 seconds and patted dry. This washing process will not remove the photo-resist, which is allowed to remain on the board to protect the copper during drilling and punching operations. Aprons should be worn to protect clothing and tongs used when transferring etched boards from the etchant to the initial rinse. After this washing and drying process, the boards are now prepared for drilling and punching.

The amplifier's pc boards after etching are shown in Fig. 13.11.

13.3 PRINTED CIRCUIT BOARD DRILLING AND PUNCHING

As was discussed in Sec. 6.2, a special wide-shank carbide tip bit is used to drill pc boards. Drilling is always performed from the *copper* side of the processed board for the following reasons: (1) to use the small etched center holes in terminal pads, which aid in bit alignment (it is extremely difficult to locate hole positions for drilling from the reverse side of the board, especially if the insulating base material is opaque), and (2) to minimize the possibility of pulling the copper foil away from the base (Fig. 13.12a). The maximum lead diameter for some common components and devices together with the recommended drill sizes for lead access holes are shown in Table 13.1. After drilling, the holes can be deburred (insulated side) by hand, when necessary, using a slightly oversized drill bit. A $\frac{1}{8}$-inch bit is usually suitable (Fig. 13.12b).

FIGURE 13.12. Printed circuit boards drilling and deburring operations. (a) Wide-shank carbide-tip drill bit used for pc boards. (b) Hand deburring of insulated side.

(a)

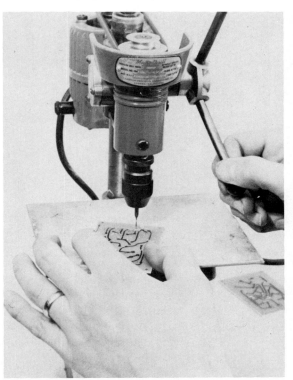

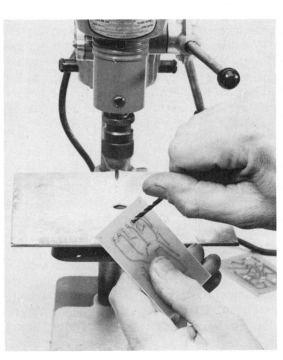

(b)

TABLE 13.1

Lead Hole Drilling Reference for Printed Circuits

Component or Device	Lead Diameter in Inches	Drill Size	Decimal Equivalent of Drill in Inches
1/8 = Watt Resistor	0.016	#75	0.0210
1/4 = Watt Resistor	0.019	#72	0.0250
1/2 = Watt Resistor	0.027	#66	0.0330
1 = Watt Resistor	0.041	3/64	0.0469
2 = Watt Resistor	0.045	#55	0.0520
Disc Capacitor	0.030	#64	0.0360
TO-5 Case Style	0.019	#72	0.0250
TO-18 Case Style	0.019	#72	0.0250
DO-14 Case Style	0.022	#70	0.0280
77-02 Plastic Power Transistor	0.026	#67	0.0320
TO-99 (8 Pin IC)	0.019	#72	0.0250
TO-116 (14 Pin DIP)	0.023	#69	0.0292

Punching should also be performed from the copper side for the same reasons just mentioned. If necessary, the grade of board used may require elevated temperatures for punching (see Sec. 10.2). Where necessary, an oven should be used for heating just prior to the punching operation. The punch and die should be sharp for best results.

Before additional processing can be undertaken, the board must be completely cleaned of all photo-resist, dye, and etchant impurities.

13.4 CONDUCTOR PATTERN SURFACE CLEANING

To prepare the etched copper surface for any additional processes, the photo-resist and dye are first removed simultaneously by applying *Dynachem Stripper* with a soft cloth. The stripper, dye, and resist are flushed from the board in a water rinse for approximately 30 seconds.

Etching in a ferric chloride solution produces contaminants on the copper foil, which will adversely affect soldering or plating. These salts of ferric chloride require an additional cleaning process. The following procedure for their removal is recommended. The contaminating salts remaining on the copper are made water soluble by dipping and mildly agitating in a bath of Lonco Copperbrite No. 48 HT. A final water rinse for 1 minute will remove all residues from the copper surface. After this final rinse, the board is wiped dry with a lint-free cloth. The board is now thoroughly clean and is ready for conductor pattern protective processes.

13.5 CONDUCTOR PATTERN PLATING AND PROTECTIVE SEALING

At this point, the minimum construction procedures have been discussed to fabricate a workable pc board. It is possible to now assemble components and solder directly to the copper foil. However, during the assembly process, the now clean copper will quickly become contaminated through oxidation and handling. To achieve reliable soldering, some means of conductor pattern protection is recommended. If, in addition, the time lapse between conductor pattern processing and component assembly and soldering is extensive (in excess of several hours), the rate of oxidation must be minimized. Two methods of protection that also improve solderability are *immersion tinning* and *rosin-base protective sealers*. These are discussed below.

Immersion tinning is a chemical process that deposits a thin film of pure tin on the copper surface. *Dyna-Tin SN-150* is ideally suited for this purpose. The board to be tinned is suspended into a heated bath of this solution and mildly agitated. The depositing of tin is a chemical reaction rather than an electrochemical plating process.

To make a 5-gallon bath solution for tinning, the procedure is as follows:

1. In a corrosive-resistant container, such as Pyrex, 4 gallons of water are placed and heated to a temperature of approximately 145°F (63°C) with a silica-type immersion heater.

2. Six pounds of Dyna-Tin SN-150 salts are slowly added to the water stirring with a Pyrex ladle until the salts have completely dissolved.

3. To this mixture, 500 cubic centimeters (approximately 16 fluid ounces) of technical-grade sulfuric acid are added. Finally, water is added to make up the 5 gallons of solution. The mixture is thoroughly stirred.

4. The bath temperature is raised to 150 to 160°F (66 to 71°C) for operation.

Prior to immersing the solution, the copper foil is lightly buffed with No. 0000 steel wool to improve the final surface appearance.

The board is totally immersed in the bath and suspended with clamps connected to a motorized shaft for mild agitation. This system is shown in Fig. 13.13. The rate of tin deposit onto the copper will depend on such

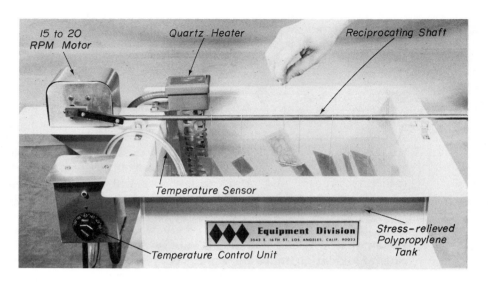

FIGURE 13.13. Immersion tinning system.

Empirical Data Taken on a Fresh
Bath of Dyna–tin SN–150 at 150° F
(rate: 4 millionths of an inch of tin
deposited per minute)

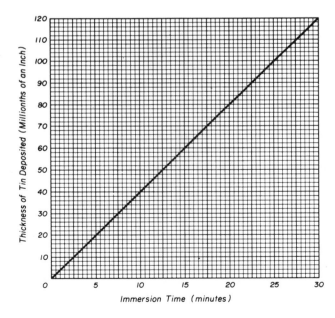

FIGURE 13.14. Thickness of tin deposited as a function of time.

factors as operating temperature of the bath, rate of agitation, and age of the bath. From the manufacturer's literature, the graph in Fig. 13.14 has been developed to aid the technician in determining the amount of tin deposited (in millionths of an inch) as a function of time. Fifty millionths of an inch, requiring from 10 to 15 minutes' immersion, is recommended as the minimum tin thickness that will provide good protection of the copper for several days. The length of time between tinning and soldering will dictate the required thickness.

After the desired thickness of tin has been deposited onto the conductor pattern, the board should be removed from the bath and immediately rinsed in hot tap water at approximately 120°F (49°C) for at least 2 minutes and then dried. The result will be a bright tin luster that will resist oxidation and ensure improved solderability of the foil.

The 5-gallon immersion bath just described is extremely efficient and will deposit a thickness of 50 millionths of an inch of tin surface on approximately 350 to 400 square feet of copper foil before it needs to be discarded. The bath should never be renewed with additional amounts of Dyna-Tin salts or sulfuric acid because this would upset the chemical balance of the bath. However, when necessary, additional water lost to evaporation should be added to the bath to maintain a constant level.

When a plating is not required and an extended period of time between board processing and soldering is expected, an alternative method of conductor pattern protection is available. This method employs a rosin-based protective coating, such as *Lonco Sealbrite No. 230-10*, to keep the copper foil free of contaminants, including moisture, oxides, and greases, and promotes good solderability by entering into the soldering reaction.

Sealbrite No. 230-10 can be applied by dipping, spraying, or roller coating methods. The selection of the most suitable method will depend on the availability of equipment and the intended application. Dipping is the method discussed here to illustrate one technique.

The board is dipped into Sealbrite No. 230-10 for several seconds. One feature of this sealer is that the board need not be dried prior to dipping. As an example, if the board had just been processed with Copperbrite and rinsed, it may be dipped directly since Sealbrite No. 230-10 will completely displace all water from the board surface.

The sealed pc board is then placed into a forced-air oven for approximately 1 minute at 200°F (93°C) to remove all solvents present in the Sealbrite solution. The sealer is then completely dried by forcing cool air from a high CFM fan for approximately 90 seconds downward onto the surface (Fig. 13.15).

Copper protected with Sealbrite No. 230-10 can be stored for as long as 12 months before soldering. It is not, however, recommended for use with a water-soluble flux.

FIGURE 13.15. Forced-air drying technique for Sealbrite.

13.6 SILK-SCREEN PRINTING

A relatively inexpensive method of reproducing identical pc boards is by the silk-screen printing technique. This process utilizes ink resist that is applied through the stencil of the desired pattern that has been affixed to a fine-mesh silk screen. (Nylon or wire mesh can also be used.) Although the silk-screen method is extensively used for processing pc conductor patterns, the illustrative descriptions provided will involve preparing marking and solder masks. The end result will be stencils that will yield acceptable tolerances of line definition.

The necessary materials to produce a silk-screen stencil are a silk or nylon screen mounted on a printing frame with hinged clamps, stencil film, film developer blockout, rubber squeegee, paint or solder resist, and the artwork master processed into a 1:1 scale positive or negative transparency. These materials and the procedures for producing and using a silk screen for the marking mask of the amplifier's tone control circuit board will be discussed as an illustrative example.

The contact films used in the development of silk-screen stencils are normally extremely slow speed types with respect to exposure time and, as a result, do not require darkroom facilities. The film selected for this example is *Ulano Super Prep*. Even though the film speed is slow, it is recommended that this film be used in subdued light avoiding direct sunlight or high-output fluorescent lamps. The 1:1 scale positive transparency, as shown in Fig. 13.16, is used to implement this film.

A piece of plain white paper, larger than the artwork, is first placed flat to serve as a background. The Super Prep film, allowing at least a 1-inch border on all sides, is placed on top of the white sheet with the sensitized (emulsion) side of the film positioned *against* the white paper. (The sensitized side of the film is determined by observing the *dull* finish as opposed to the glossy side of the polyester backing.) The initial arrangement of the film is shown in Fig. 13.17. The 1:1 scale positive marking mask transparency is then placed on *top* of the film and held firmly in place with the glass printing frame. The film is then exposed with a 275-watt bulb, a No. 2 photo-flood lamp, or a Sylvania F20T12/BL ultraviolet lamp mounted at a distance of 15 to 18 inches above and directly over the film. The light source should not be closer than 15 inches from the frame to avoid excessive heating of the film. This is especially true when using photo-flood lamps. The exposure time for the film used in this example is approximately 10 minutes. Exposure times will vary from 5 to 15 minutes, depending on such factors as film type, light source, and exposure distance. Optimum exposure times can be experimentally determined and

FIGURE 13.16. Positive 1:1 scale marking mask of tone control board.

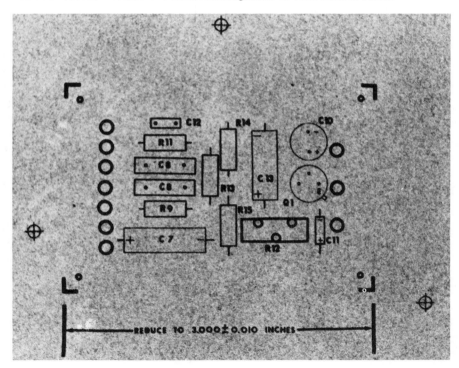

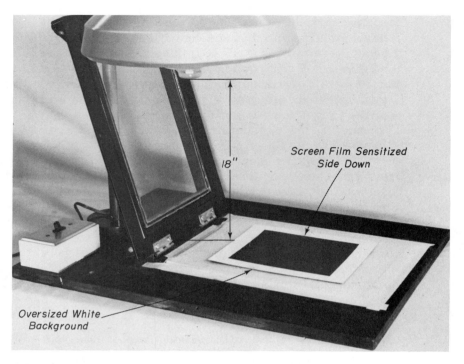

FIGURE 13.17. Exposure arrangement for silk-screen film.

once established, should not vary appreciably. Those areas of the film emulsion exposed to the light source will become insoluble in warm water after the developer bath. Those areas protected from the light by the opaque portions of the positive will dissolve in warm water after the developer bath. After exposure, the film is placed in a tray of developer with the emulsion side *up* for between 1 and 2 minutes. One end of the tray should be slowly raised and lowered to allow the solution to agitate gently over the surface of the film.

The developer for Ulano Super Prep film is available in powder form and packaged in premeasured packets labeled "*A*" and "*B*." To prepare the developing solution, one packet of each of the powders is dissolved together in 16 ounces of water at a temperature of approximately 65 to 70°F (18 to 21°C). This developer is light sensitive and should be prepared in subdued light and stored in an amber-colored bottle. The solution has an effective lifetime of approximately 24 hours, after which time it begins to deteriorate. One indication of this deterioration is the appearance of wrinkles on the polyester support of the film during the developing process. When this occurs, the developer should be discarded and replaced with a fresh mixture.

After the film has been in the developer for the necessary amount of time, it is removed and immediately placed in a tray of water at a temperature of approximately 100°F (38°C). The emulsion side of the film is again facing upward to prevent any smearing of the softened emulsion from the bottom of the tray. Gentle rocking of the water bath will remove the soluble portions of the emulsion (those areas protected during exposure). Approximately 1 minute is all the time required for the water bath to process the film. The film being processed in the water is shown in Fig. 13.18.

The film is next removed from the warm-water bath and rinsed with cold water for about 30 seconds to remove any remaining emulsion from those areas of the negative film that must be completely transparent. *Caution:* Extreme care must be exercised when handling the film after it is taken from the warm-water bath since the remaining emulsion is soft and tacky and is easily smeared. The cold-water rinse is best applied by placing the film on a sheet of glass with the polyester support backing against the glass. The emulsion side can then be gently rinsed under a cold-water tap.

The silk-screen negative stencil of the original artwork master can now be obtained by firmly pressing the soft emulsion of the film against the surface of the silk screen. This process is shown in Fig. 13.19a. After the entire emulsion side has been pressed against the silk screen, both sides of the screen are gently blotted with soft absorbent paper towels to remove any excess water (Fig. 13.19b). A block-out material, such as *Ulano No. 60 Water Soluble Fill-in*, is then spread with a small brush over

FIGURE 13.18. Warm-water used to process film after developing.

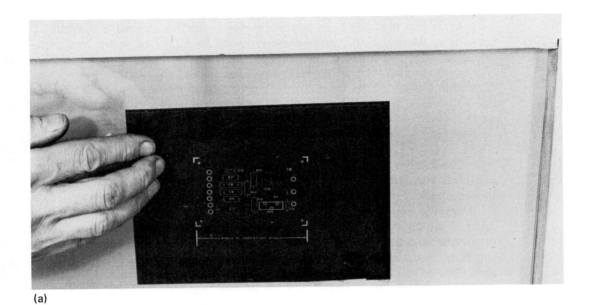

(a)

(b)

FIGURE 13.19. Silk-screen preparation. (a) Initial application of film to screen. (b) Removing excess moisture from both sides of screen.

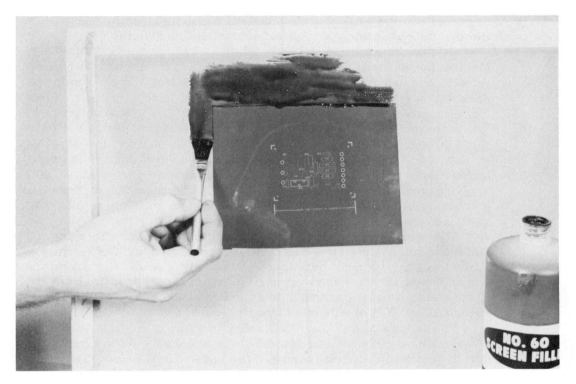

FIGURE 13.20. Block-out applied to unused areas of screen.

the unused areas of the screen bounding the stencil. This block-out confines all excess paint or resist to the upper surface of the screen during printing (Fig. 13.20).

The screen is then allowed to dry thoroughly. If a small fan is available, drying time will only require approximately 30 minutes. After the screen has dried, the transparent backing of the film is peeled from the screen, which is then carefully inspected by tilting it back and forth while holding it up to an intense light. No residual film should remain on those areas of the screen that are not blocked and appear as translucent. This residual film will appear to glisten and, if not removed, will impair smooth transfer of the paint or resist through the screen stencil during its application. This film is removed with standard mineral-oil spirits and a stiff brush. The emulsions forming the stencil or the block-out are resistant to these solvents and will not be affected by this cleaning process. The use of water, however, should be avoided because these emulsions are water soluble.

The stencil just developed is now ready for use. Before proceeding, however, some consideration needs to be given to the type of screen to select for a particular application. In this section, the concern centers around screens for marking and solder masks. Therefore, a moderate degree of control must be maintained on positioning paint or resist flow through the screen in order to effect acceptable line definition.

Screens are classified in two categories, *mesh number* and *durability*. Mesh numbers for silk screens, generally running from No. 1 to No. 25, refer to the number of filaments per inch. A No. 1 mesh screen has approximately 30 filaments per inch and is considered very coarse. Mesh No. 25 represents a fine screen of approximately 196 filaments per inch. The more the filaments per inch, the greater the control of paint or resist flow through the screen and consequently the sharper the resulting pattern. For extremely critical line definition, *monofilament* nylon mesh screens with as many as 283 filaments per inch are also available. For applying epoxy paint for marking masks and solder resist for solder masks, a screen with a mesh number of 20 will produce adequate results. This mesh is also suitable for conductor pattern stencils.

The durability of the screen is a measure of its ability to withstand continued reuse. Durability classifications are *X*, *XX*, and *XXX*, with the latter used only if many reproductions are to be made. For prototype work, screens with a durability classification of *X* are adequate. However, if the screen is to be continually stripped clean and reprocessed for other stencils, *XX*, *XXX*, or wire mesh screens should be considered.

To facilitate the silk-screen labeling process, it is first necessary to provide a base and a set of clamp hinges to accommodate the frame and also to provide positive positioning of the surface to be labeled. These aids are shown in Fig. 13.21. The frame is clamped to the printing base by means of the clamp hinges in such a manner that the stencil is closest to the base as the frame is swung down into the printing position. With the frame properly clamped, the pc board is placed on the printing base under the frame and its corners are aligned with the border delineation marks on the stencil. Once the desired alignment is achieved, three small *register guides* cut from either cardboard or metal, no thicker than the board, are positioned against any three board edges. This procedure is also shown in Fig. 13.21. The register guides are secured in place with masking tape and help hold the work in position during printing. They also allow for rapid positioning of identical work pieces when many boards are to be processed.

Once the work and stencil are properly aligned, a liberal amount of epoxy-type paint is poured across the left-hand margin of the marking mask stencil from top to bottom (Fig. 13.22a). A hard-rubber squeegee, preferably one that will span the entire stencil, is brought down onto the screen behind the pool of paint. With a sweeping motion, the squeegee is drawn across the stencil, spreading the paint as it moves. This technique is shown in Fig. 13.22b. Familiarity with this technique will enable the technician to completely label the entire work surface with a single sweeping motion.

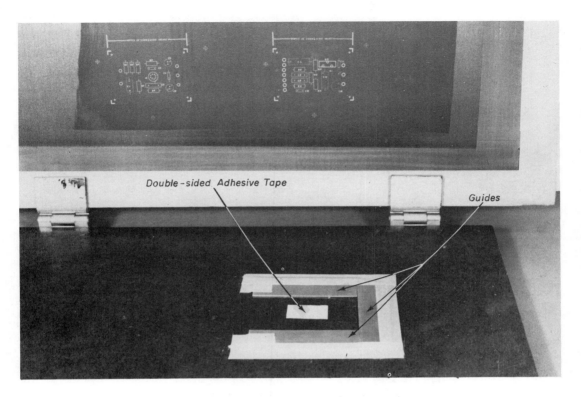

FIGURE 13.21. Register guides to aid in board alignment.

After the stenciling is completed, the screen is raised from the work (Fig. 13.22c). Small boards may tend to cling to the stencil when it is raised. Usually, they will slide slightly before they fall, thus causing the labeling to smear. This problem can be avoided by using double-sided adhesive tape placed on the underside of the work when it is first positioned on the base.

When labeling has been completed, the paint must be allowed to dry thoroughly before handling. Depending on the type of paint used, drying time will vary from 2 hours to a complete curing time of 4 days. Drying time can be reduced by baking according to paint manufacturers' specifications. Certain lacquers and vinyls used in the silk-screen process for labeling will air dry in 5 to 15 minutes. No difficulty will be encountered in obtaining any desired labeling characteristics in terms of color, durability, or finish because there are hundreds of different types of paints, enamels, and lacquers available from graphical suppliers.

The seven labeled pc boards of the amplifier together with the stencils used for each are shown in Fig. 13.23.

FIGURE 13.22. Use of silk screen for board marking. (a) Applying paint along edge of mask. (b) A single stroke of the squeegee labels board.

FIGURE 13.22. (cont.) (c) Results of screening process.

FIGURE 13.23. Marking mask with all boards screened.

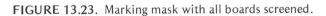

When the labeling process has been completed, the excess paint can be removed from the margins of the stencil with a small spatula and stored. The remaining paint is then washed off with an appropriate solvent. If the existing stencil is no longer to be used, the screen can be reclaimed for further stencil use by washing it in warm water. This will dissolve the water-soluble emulsion and block-out. Difficult areas can be gently scrubbed with a soft-bristle brush.

The discussions above have been limited to silk-screen printing for producing marking masks. This process is also used for processing solder masks. For this purpose, the procedures for developing the stencil are identical except that a 1:1 *negative* is used initially to generate the silk-screen film. Also, instead of using an epoxy paint, a solder resist, such as *Lambert No. 184–10-V green solder resist*, is applied through the screen.

The silk-screen stencils and finished boards with solder masks applied to the amplifier's pc boards are shown in Fig. 13.24.

Because of the complexity of the silk-screen process, a flow diagram is provided in Fig. 13.25 to help the technician follow the various steps.

FIGURE 13.24. Solder mask with all boards screened.

After the pc boards have been processed, the components and hardware must be mounted and leads soldered securely to the conductor pattern. The procedures and equipment necessary for this phase of construction will be discussed in the succeeding chapters.

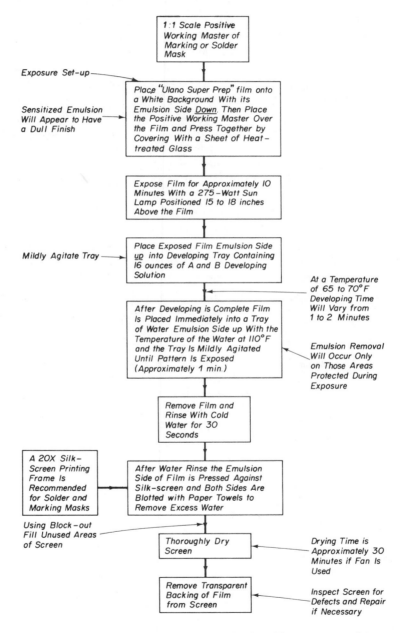

FIGURE 13.25. Flow diagram for developing silk-screen stencils.

PROBLEM SET

13.1 Determine the parts concentration for making three individual pints of photographic developer, stop bath, and fixer with the aid of the example solutions in this chapter.

13.2 Estimate the amount of tin deposited in 17 minutes using the graph in Fig. 13.14 for the immersion tinning process discussed in Sec. 13.5. Also determine the additional time required to increase the tinning thickness by 30 millionths of an inch.

13.3 Fabricate printed circuit device test jigs from the artwork masters designed in Problem 11.2 using $\frac{1}{16}$-inch-thick G-10 board with 1-ounce copper foil. The final etched boards are to be drilled with the appropriate size holes to accommodate suitable turret-type terminals for external interconnections and socket pins for the devices. The terminals and pins are installed in Problem 14.2.

13.4 Prepare silk screens for the marking and solder mask layouts designed for the tachometer circuit of Problem 11.5. (The screens are used in Problem 13.7.) Obtain the meter movement specified for the circuit. Using a commercially available tachometer as a calibration reference, redesign the meter scale and construct a silk screen for the new scale.

13.5 Fabricate a pc board from the artwork master of the seven-fixed-interval UJT timing circuit designed in Problem 11.3 from $\frac{1}{16}$-inch G-10 board with 1-ounce copper foil. Components are assembled in Problem 14.3.

13.6 Construct a pc board from the artwork master for the series feedback regulator circuit designed in Problem 11.4 from $\frac{1}{16}$-inch G-10 board with 1-ounce copper foil. Components are assembled and soldered in Problems 14.4 and 15.4, respectively.

13.7 Fabricate a pc board from the artwork master for the tachometer circuit designed in Problem 11.5 from $\frac{1}{16}$-inch *XXXP* board with 1-ounce copper foil. Using the silk screens fabricated in Problem 13.4, construct the meter scale, mark the pc board, and mask the conductor pattern for soldering. Components are assembled to the pc board in Problem 14.5.

14

Printed Circuit Hardware and Component Assembly

14.0 INTRODUCTION

After a circuit design has been processed into a pc board, two major operations are necessary in order to produce a functional circuit. These are (1) assembling the individual components and hardware that comprise the finished board, and (2) soldering component leads to the etched copper foil conductors. This chapter is devoted exclusively to assembling components and hardware to pc boards.

In recent years, an emphasis has been placed on the development of reliable miniature devices and hardware that are adaptable to pc applications. Consequently, during the planning stages of a pc packaging problem, the technician is faced with the task of selecting the most suitable components and hardware for his specific purpose from an overwhelming variety of types and manufacturers. Coupled with this is the necessity of having available the technical information for mounting these parts onto the pc board. This chapter discusses the applications and assembly techniques of many of the common components and hardware associated with printed circuits.

Specific information for solder terminals and methods for their insertion into the board are considered. Also included are wire-wrap-type terminals and eyelets in addition to information on connectors and associated hardware for board-to-board or board-to-chassis interconnections. Basic techniques and tools necessary for mounting components such as resistors and capacitors are discussed. Included are methods of mounting devices such as diodes, transistors, ICs (integrated circuits), FETs (field-effect transistors), and SCRs (silicon-controlled rectifiers) with and without sockets. In addition, techniques for heat sinking these devices and the thyristor are discussed.

14.1 STANDARD SOLDER TERMINALS

One of the major problems encountered when designing and assembling a pc board is the means by which external connections are made. For this application, the selection of suitable hardware is almost unlimited and is determined by many design factors, some of which were discussed in Chapter 1.

One of the most economical and easily installed terminal types is a metal stud, mechanically and electrically secured to the board, on which clips for testing or wires for final assembly may be attached. These metal studs are termed *solder terminals*. These terminals are mounted through predrilled holes in the pc board's insulating material and copper conductor. They are then *swaged* to form a solid mechanical connection between the insulated side of the board and terminal pad. Finally, they are soldered to the copper conductor to provide a sound electrical connection. Some typical configurations of solder terminals that will satisfy many packaging requirements are shown in Fig. 14.1a. The terminals shown are classified as *turret*, *fork*, *tubular*, *taper pin*, and *minipin* and are available in either single-ended or double-ended styles.

A typical double-turret, single-ended terminal with its associated nomenclature is shown in Fig. 14.1b. These terminals are designated by *post style* and *size* (available in hollow or solid posts), *shank diameter*, and *shank length*. The shank length must fit the pc board thickness for which it is intended. These thicknesses generally range from $1/32$ to $1/4$ inch.

Solder terminals of the type shown are usually formed from tinned brass and are recommended for use with pretinned pc boards. This combination results in the proper flow of solder between the circuit pattern and the terminal, when heated with a soldering iron, to form a sound electrical connection with a minimum amount of solder applied to the joint.

To secure the solder terminals to the pc boards, they are swaged by one of three methods: *hand*, *impact*, or *pressure* staking. All three methods require that a hole be drilled into the pc board that is approximately 0.002 to 0.005 inch larger than the shank diameter of the terminal.

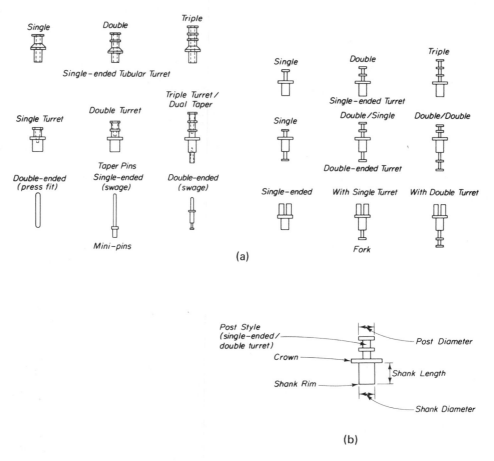

FIGURE 14.1. Solder-type terminals. (a) Typical solder terminal configurations. (b) Solder terminal nomenclature.

For hand staking, an *anvil*, such as that shown in Fig. 14.2a, is held firmly with a *toolholder* secured in a vise. The terminal to be swaged is placed, post *down*, into the anvil with the terminal *crown* seated. The hole in the pc board is positioned onto the terminal shank. This arrangement is shown in Fig. 14.2b. When properly positioned over the terminal, the shank rim should protrude slightly out of the hole through the terminal pad. As can be seen in Fig. 14.2b, a simple toolholder can be fabricated from an appropriate-sized hex nut by first drilling out the threads to accommodate the tool diameter. A second hole is then drilled and tapped through the center of one of the flats for securing the tool with a set screw.

To swage the terminal, a *setting tool*, such as one of those shown in Fig. 14.2c, is used. For single-ended terminals, a setting tool with a *solid*

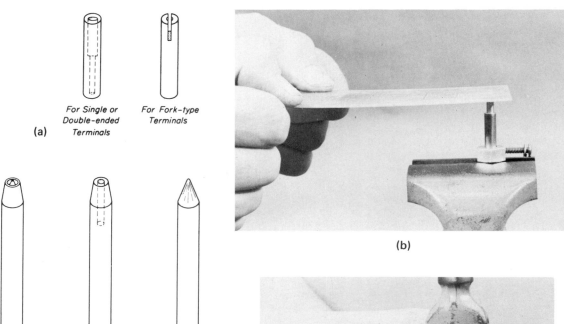

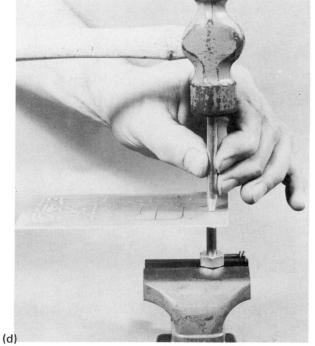

For Single or
Double-ended
Terminals

For Fork-type
Terminals

(a)

(b)

For
Single-ended
Terminals

For Double-ended
Terminals

For Flaring
Single-ended
Terminals

(c)

(d)

(e)

OUT

FIGURE 14.2. Hand staking operation for solder terminals to printed circuit boards. (a) Anvils for swaging. (b) Board positioning. (c) Setting tools. (d) Staking operation. (e) Properly swaged terminals.

face that will "cup" the *shank rim* down firmly against the terminal pad is required. The swaging of double-ended terminals requires that both the anvil and setting tool be hollow. The setting tool hole diameter should be slightly larger than that of the terminal post for clearance. With the correct setting tool positioned directly over the terminal shank, it is firmly tapped with an 8-ounce hammer (Fig. 14.2d). This will cup the shank rim over against the copper conductor, forming a sound mechanical joint. Figure 14.2e shows the result of hand staking on a terminal. A properly swaged terminal shank will have the appearance of a perfectly smooth cupped eyelet with no fractures around the shaped rim. In addition, no "play" should be detected when pressing down or pulling up on the terminal post.

Impact staking is performed in a similar manner to hand staking except that more consistent results are obtained when many swaging operations are required. The terminal may be staked by an *automatic impact staker tool*. This tool is similar to the automatic center punch discussed in Chapter 5 except that the point is replaced with a setting tool. Optimum striking force can be obtained by adjusting the upper portion of the impact tool. This striking force will be uniform for each terminal shank, resulting in identical cupping characteristics. Swaging is performed by positioning the impact staking tool over the terminal shank (Fig. 14.3). A steady downward pressure is applied to the handle until the setting tool strikes against the terminal shank. Inspection of the results will determine if the proper impact adjustment has been achieved.

FIGURE 14.3. Impact staking.

(a)

(b)

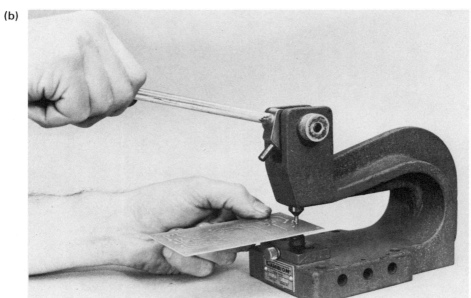

FIGURE 14.4. Pressure staking. (a) Arbor press. (b) Engaging setting tool with terminal.

Pressure staking is the preferred method of securing solder terminals to pc boards because there is least chance of fracturing the board by this method. Hand and impact staking impart mechanical shocks to the board material and unless extreme care is exercised, the base material could fracture. Pressure staking imparts a steady pressure to the terminal, eliminating the mechanical shocks and thus reducing the possibility of board fracture.

Pressure staking utilizes an *arbor press* such as that shown in Fig. 14.4a. The terminal is inserted into the anvil exposing the shank onto which the hole of the pc board is positioned. With the appropriate setting tool attached, rotating the operating handle forward will bring the tool in contact with the terminal shank rim. Additional rotation will apply pressure to the terminal to complete the swaging. This operation is shown in Fig. 14.4b. Because excessive pressure could damage the anvil, setting tool, terminal, and the pc board, most arbor presses are equipped with a threaded anvil. This anvil is threaded into the press table to a height that will provide adequate clearance for the thickness of the pc board. Adjusting for the correct travel of the setting tool is made by trial and error. Optimum anvil setting is achieved when full forward rotation of the operating handle properly swages the terminal without excessive pressure being applied to the insulating material or terminal shank.

When a terminal shank is formed by a V- or funnel-type swage instead of being fully cupped, a *solder ring* may be first slipped over the shank where it protrudes through the board prior to the swaging operation. For this type of swaged shank, solder rings are recommended to ensure a uniform, reliable solder joint, especially if a board is to be hand-soldered. This arrangement is shown in Fig. 14.5a and b.

Terminals with test leads attached and leads soldered are shown in Fig. 14.6a and b, respectively. Techniques for properly orienting leads and soldering them to terminals will be discussed in detail in Chapter 19.

FIGURE 14.5. Installation of solder ring with single-ended terminal. (a) Terminal assembly with solder ring. (b) Flaring operation.

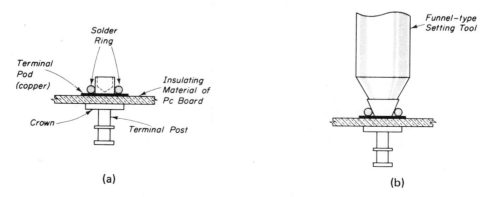

(a)

(b)

(a)

(b)

14.2 SOLDERLESS PRINTED CIRCUIT TERMINALS

An alternative method of providing randomly located metal studs to which external connections can be made *without* the necessity of soldering wires is through the use of *wire-wrapping* techniques. Wire-wrap terminals are square or rectangular pins having at least two sharp corners that are mechanically secured to the pc board by staking (press-fitted) or soldering. Lead connections are made to these pins with mechanical or pneumatic wire-wrapping tools using solid wire. The resulting connection is both mechanically secure and electrically sound.

Two types of wire-wrap pins are shown in Fig. 14.7, each having its own configuration and mounting arrangement. The most basic wire-wrap pin is shown in Fig. 14.7a. It consists simply of a 0.045-inch-square wire from ¾ to 2 inches long with both ends tapered for ease of initial insertion into the pc board. No special equipment is required for their installation since the pin is simply pressed into a predrilled hole. The recommended drill hole has a diameter of 0.050 inch with a 0.002-inch tolerance. Although commercially available vibrating tables and templates are employed to position the pins into the holes, the cost of such equipment is prohibitive in small volume or prototype work. These pins can be inserted easily by hand. The tapered end of the pin is first inserted into

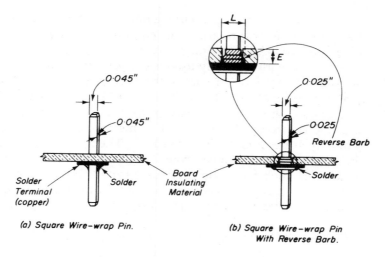

FIGURE 14.7. Styles of wire-wrap terminals. (a) Square wire-wrap pin. (b) Square wire-wrap pin with reverse barb.

the drill hole of the board using finger pressure. This procedure is sufficient to temporarily hold the pin in a staking position. Similar to the staking of solder-type terminals, an arbor press is used with an anvil and flat-faced setting tool. The guide hole in the anvil should be large enough to accept the square configuration of the pin with sufficient clearance. The arbor press then is used to force the pin through the pc board, allowing it to extend to any desired height on the opposite side of the board. Although only held in position by the contact between the corners of the pin and the inside edge of the hole, it is sufficiently gripped in place until it can be soldered to the terminal pad. Connecting wires are then wrapped along the length of the pin extending above or below either side of the board.

Another wire-wrap terminal that provides for uniform staking height is shown in Fig. 14.7b. It consists of double-ended square stock with either 0.025-inch or 0.045-inch flats made of brass and plated with gold alloy. Dimension E in Fig. 14.7b is generally selected to be equal to the thickness of the pc board insulating material. To insert this pin, a hole is drilled that is approximately 0.002 inch smaller than the maximum diameter of the *reverse barb*. For the 0.025-inch square pin, dimension L is 0.052 inch (hole size $\cong$ 0.050 inch) and for the 0.045-inch square pin it is 0.096 inch (hole size $\cong$ 0.094 inch). The pin is positioned in such a manner as to allow its crown to contact the foil side (terminal pad) of the board. The pin is then secured with an anvil and setting tool in an arbor press. The reverse barb prevents the pin from turning or backing out of the insulating material. Soldering between the crown and terminal pad will make a reliable electrical connection.

Several terminals properly wrapped with solid wire are shown in Fig. 14.8. (Specific information on equipment and methods for connecting solid wire to wire-wrap terminals is considered in detail in Chapter 19.) The type of wire normally used is copper-alloy solid wire with a tin, silver, or gold plating. Sizes of wire for this application range from AWG (American Wire Gauge) No. 18 to No. 26. (AWG is discussed in Chapter 19.) The size of the wire will determine the exact number of turns to make a properly wrapped connection. This information for two types of interconnections is given in Table 14.1. The type 1 interconnection requires that the wire be wrapped around the terminal with its insulation approximately $\frac{1}{16}$ inch from the first turn; type 2 interconnections require that at least one turn of insulation also be formed around the terminal. This second classification utilizes the flexibility of the wire insulation to reduce

FIGURE 14.8. Wire-wrap assembly. Courtesy of Augat, Inc.

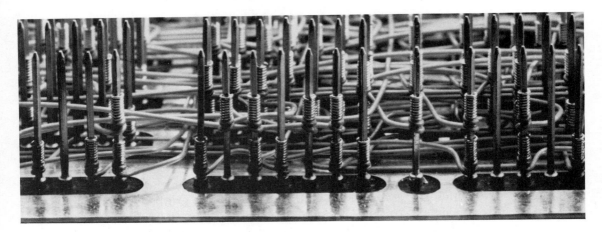

TABLE 14.1

Recommended Wiring Specification for 0.025-inch and 0.045-inch Square Wire-Wrap Terminals

AWG Number	Type 1 Connection Number of Turns	Type 2* Connection Number of Turns
18 20	4	5
22 24	5	6
26	6	7

*Type 2 includes one turn of insulated conductor contacting the terminal

stress encountered during vibrations of the wire at the point where it enters into the first wrap on the terminal. In addition, quality workmanship dictates that (1) no more than three independent lead wraps be made on a single terminal pin, (2) turns from one independent wrap do not overlap any portion of another wrap, and (3) adjacent turns on the same wrap touch but do not overlap.

14.3 PRINTED CIRCUIT BOARD CONNECTORS

The preceding two sections introduced two simple means of providing terminal points on pc boards for external connections to chassis- or panel-mounted components. This section introduces basic and commonly available hardware and methods for *interconnecting* boards. Selecting the most appropriate connector for a specific application will depend on such factors as (1) cost, (2) mounting limitations, (3) required number of contacts, (4) board thickness, (5) vibration, (6) current and voltage requirements, and (7) required insertion and removal operations. These considerations must be examined before an intelligent choice of connector can be made.

Connectors are available for *board-to-board, board-to-wire, board-to-flexible cable*, and *board-to-chassis* interconnections. These can be separated into two general categories: board edge *plugs* and *receptacles*. The edge-receptacle or *finger*-receptacle-type connector consists of a molded insulator containing a number of female-type contacts that accept the edge of the pc board and mate with foil fingers to make the electrical connections. A typical edge connector is shown in Fig. 14.9a. An edge plug-type connector is shown in Fig. 14.9b, which contains either in-line or staggered contact pins swaged into the pc board and soldered to the terminal pads at the edge of the board. Assembly time is increased with this connector but it is recommended when the unit is subjected to high vibration. The two-piece connector shown in Fig. 14.9c consists of a male plug and a female contact arrangement. One section is mounted to the pc board and the other section is attached to a chassis or another pc board. Electrical contact is made when these two connectors are mated.

Connectors are also specified in terms of female contact design. For the edge-receptacle type, which must contact the foil fingers extending to the edge of the board, two contact arrangements are necessary. The first is the type with contacts that mate with a row of fingers on a single-sided board, whereas the second mates with rows of fingers on *both* sides of a board. Several contact designs for single- and double-sided requirements are shown in Fig. 14.10. The *tuning fork* design shown in Fig. 14.10a has low contact resistance and is used for single-sided board applications because it has only one terminal point that extends through the rear of the insulator. The *ribbon*-type contact design shown in Fig. 14.10b provides for contact to double-sided boards. Each upper and lower pair of contacts have individual terminal pins that may be used independently or

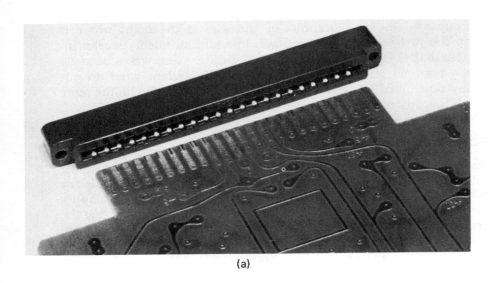

(a)

(b)

(c)

FIGURE 14.9. Printed circuit board connectors. (a) Edge (finger). (b) Plug style. (c) Two-piece.

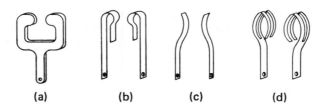

FIGURE 14.10. Female contact designs for finger receptacle connectors. (a) Tuning fork. (b) Ribbon. (c) Cantilever. (d) Bifurcated.

soldered together with a jumper wire. The *cantilever* contacts shown in Fig. 14.10c are preformed and preloaded so that a specific force is applied to the foil fingers when the pc board is inserted into the connector. Finally, the *bifurcated* contact arrangement shown in Fig. 14.10d furnishes two points of contact on each foil finger. Each section of the individual bifurcated contacts is different in width, imparting to each its specific *resonant frequency*. During a period of vibration, it would be expected that at least one leaf will always be in contact with the foil. These split contacts also ensure positive contact if there should be any nonuniformity in finger thickness.

The connectors just discussed are typically constructed of either *brass, phosphorous bronze,* or *beryllium–copper.* When cost is a major factor, brass is the least expensive. It does, however, degenerate spring tension, owing to aging and continual insertion and removal of the pc board. Phosphorous bronze contacts are superior in spring-retention characteristics, but they are slightly more expensive and have a higher contact resistance than brass. Beryllium–copper contacts, although more expensive, overcome all the disadvantages of the brass or phosphorous bronze type. Since the materials used in the construction of the contacts for these connectors are relatively soft, they tend to wear under continual insertion and removal of the pc board. For this reason, they are plated with a gold alloy to a thickness of 20 to 100 millionths of an inch.

Several contact terminations through the top or rear of the connector are shown in Fig. 14.11. The *eyelet* and *dip solder* types afford the most sound electrical connections and are recommended for use where vibrations are expected. *Wire-wrap* styles allow for more rapid assembly, with a resulting cost savings over the solder types. The *crimped tapered* pin style allows for easy and rapid modification or repair of assembled connectors.

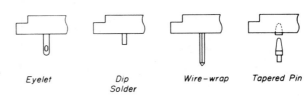

FIGURE 14.11. Typical connector terminations.

Eyelet   Dip Solder   Wire-wrap   Tapered Pin

Connectors have contact spacings that vary from 0.100 to 0.200 inch, the most common being an on-center spacing of 0.156 inch. The number of contacts per connector may vary from 6 to as many as 100 and will, naturally, depend on the number of contacts required on the pc board.

Most manufacturers design their connectors to function equally well with either $\frac{1}{16}$- or $\frac{1}{32}$-inch-thick pc boards. Insertion-type connectors for thicker boards are also available.

To eliminate the possibility of incorrectly inserting the pc board into the connector, especially in the case of double-sided boards, some method of connector polarization should be employed (Fig. 14.12). To insert the board into the connector in the correct position, a plastic or metal *polarizing key* is inserted either between pairs of contacts or beside them in slots provided for this purpose. For the board to be inserted into the connector when this type of keying is used, a notch in the leading edge of the board at the fingers must align with the key. The type of key that is positioned on the insulator beside contact pairs is sometimes preferable because it does not reduce the number of available contacts by one set. The leading end of the board that is inserted into the connector is chamfered at a 45-degree angle on both edges (Fig. 14.13). This configuration allows the board to be inserted easily into the connector and reduces the possibility of the foil fingers lifting on boards that are frequently removed and replaced.

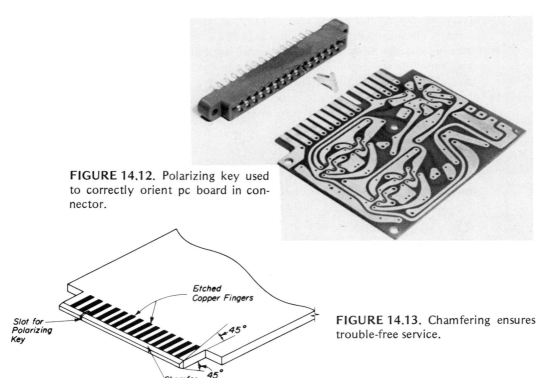

FIGURE 14.12. Polarizing key used to correctly orient pc board in connector.

FIGURE 14.13. Chamfering ensures trouble-free service.

Slot for Polarizing Key

Etched Copper Fingers

45°

Chamfer 45°

14.4 COMPONENT ASSEMBLY

The smaller components such as resistors and capacitors are first assembled. Under no circumstances should swaging be done on the board after any components have been assembled because of fragile components that might be damaged.

The components to be first considered are *axial lead* type (tubular shape having leads at each end). On single-sided boards, these components are normally mounted parallel to the pc board surface with the body of the component lying flush against the insulated side. The leads are bent at right angles to the component body and passed through the predrilled clearance holes to the foil for electrical connection. Once through the board, the leads are bent in the direction of the terminal pad entry. This operation keeps the components from slipping out of position as others are installed and ensures sound electrical connections after soldering. The proper methods of component assembly are shown in Fig. 14.14. Several considerations need to be taken into account before the components are assembled. First, only components that weigh ½ ounce or less and/or dissipate less than 1 watt should be considered for flush mounting against the board. Heavier components should be provided with some mechanical clamping arrangement, which will be discussed later in this section. Components dissipating more than 1 watt should be mounted above the surface of the board (also discussed later in this section), since the heat generated from these components tends to weaken the foil bond.

Second, to minimize stresses on the components and their leads, the following minimum specifications should be observed: (1) the minimum bend radius for all component leads should not be less than twice the lead diameter, and (2) this bend should begin no closer than $\frac{1}{16}$ inch from the body or end bead of the component. These dimensions are shown in Fig. 14.14a, labeled R and X, respectively.

Finally, the leads must be properly bent after passing through the insulating material. One of two types of bending procedures may be adopted; these are the *fully clinched* lead arrangement, as shown in Fig.

FIGURE 14.14. Two methods of assembling axial lead components to printed circuit boards. (a) Fully clinched leads. (b) Leads with service bends.

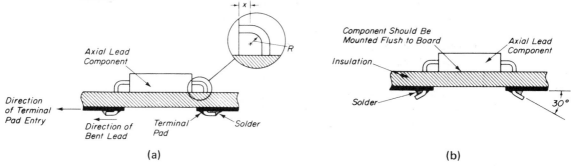

(a) (b)

14.14a, and a *service bend*, shown in Fig. 14.14b. The fully clinched lead is used when maximum rigidity of the component is desired. This technique also requires less solder to complete the electrical connection. Service bends, as the name implies, are adopted if components may have to be removed later on. This lead arrangement is made by bending the component lead to within approximately 30 degrees of the terminal pad. This technique makes it less difficult to remove a component after it has been soldered than one with fully clinched leads.

Although the specifications provided do not conform exactly to any standards, they are realistic values that have proven to be effective in pc boards not exposed to severe environmental or physical stresses.

The leads of tubular components must be accurately bent before being inserted into the pc board. Two simple bending tools are shown in Fig. 14.15. For nonuniform and nonstandard terminal pad center dimensions, the combination *caliper* and *lead bender*, shown in Fig. 14.15a, is first positioned with its pointed jaws aligned with the centers of the proper holes in the pc board. This automatically sets the correct spacing between the *bending tabs*, which are located between the jaws and the tool handle. A *knurled thumb screw* is provided to maintain this setting. With the component centered in the *bending jig*, the leads are sharply bent around the tabs until they are at right angles to the component body. This procedure is shown in Fig. 14.15b. When this bending operation is completed, the component leads are ready for insertion into the pc board.

For standard or uniform spacing between terminal pad centers, the *universal bending block*, shown in Fig. 14.15c, may be more conveniently used. This type of bending block is usually graduated in 0.05- to 0.1-inch increments, with increasing on-center dimensions from 0.5 to 1.5 inches. The component body is positioned in the *center channel* with the leads resting in the appropriate pair of lead *slots* that most closely agree with the terminal pad on-center dimensions on the pc board. The leads are then bent down over the edges of the block to properly form both the correct radius and on-center dimension.

If neither of the described bending tools are available, *tapered round-nose pliers*, such as those shown in Fig. 14.16, should be used to perform the lead-bending operation. Only pliers with smooth jaw surfaces should be used, to avoid scraping and nicking the component leads. A nicked lead could result in a complete fracture under operational stress or installation.

To form the bend with this tool, the lead is gripped at the proper position between the tapered jaws. Holding the pliers stationary, the lead is bent at right angles to the component body (Fig. 14.16b). Unlike bending tools, round-nose pliers do not allow for repeated accurate formation of on-center dimensions. With practice and care, however, the correct lead position in the tapered jaws can be determined quickly to achieve the desired bend radius and on-center dimensions.

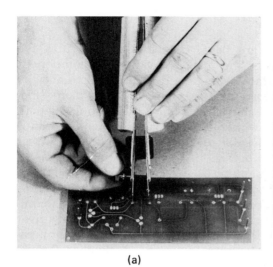

(a)

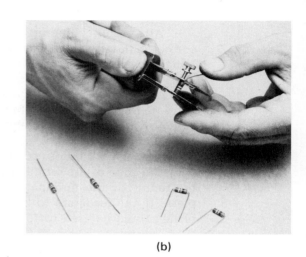

(b)

FIGURE 14.15. Methods and tools for accurate lead bending of tubular components. (a) Caliper setting for lead bending. (b) Once jaws are set to desired spacing, component leads can be uniformly bent. (c) Bending block.

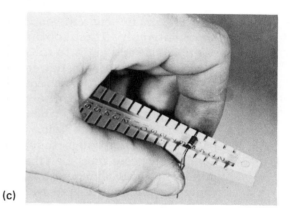

(c)

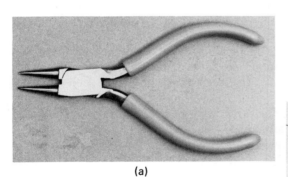

(a)

FIGURE 14.16. Pliers for component assembly. (a) Tapered round-nose pliers. (b) Component lead bending.

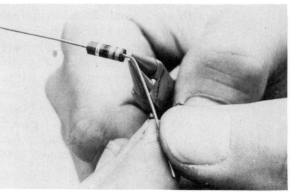

(b)

After the component leads are bent, they are inserted into the correct holes in the pc board from the insulated side. The component body is held firmly against the surface of the board with one hand and the leads are initially bent back against the foil with the other hand (Fig. 14.17a). The direction of the bend should be toward the conducting path entry at the terminal pad through which the lead passes. When the lead is bent to within approximately 30 degrees of the foil, it is cut with a pair of small *diagonal cutters*, as shown in Fig. 14.17b. (More information on these cutters will be provided in Chapter 19.) The lead should be cut even with the edge of the terminal pad so that the length of the lead is equal to the radius of the terminal pad. For service bends, the assembly is complete. If the leads are to be fully clinched, the flat side of a screwdriver blade should be used to force the lead firmly against the foil. This operation is shown in Fig. 14.17c. Leads should never be clinched before cutting because the jaws of the diagonal cutter could gouge the copper conducting path.

Pneumatic cut-and-bend tools are commercially available for pc board assembly. These tools will, in one operation, bend the lead to the desired angle and cut it to the correct length.

Both sides of the power amplifier board assembled with wire-wrap terminals and tubular components are shown in Fig. 14.18.

The assembling of disc capacitors presents mounting problems due to their physical design. Whenever possible, they should be mounted with the edge between the leads flush against the insulation and the leads passing straight through the board before bending. Since disc capacitors are circular, care must be taken to prevent the component from rotating during the lead clinching operation, which could upset the vertical and center positioning. Figure 14.19a shows the proper method of mounting a disc capacitor and the incorrect mounting method. It is advisable to fully clinch the leads of disc capacitors to secure these components firmly to the pc board prior to soldering.

Vertical-mounted capacitors, such as those shown in Fig. 14.19b, are specifically designed for more efficient use of available area on pc boards. In mounting these capacitors, the access holes must be in alignment with the base layout of the component leads so that they may feed straight through the board. Bending the leads on this type of component from their original direction defeats the advantage of mechanical securing obtained when the base is flush against the board insulation.

When space is extremely critical on a pc board, axial lead components may be mounted vertically. This is accomplished by inserting one lead of the component, without bending, directly into the pc board. The other lead is bent 180 degrees around the body of the component and then passed through the board. This technique is shown in Fig. 14.20a. As can be seen from this figure, the mechanical security of even a small component mounted in this fashion is extremely poor. Under shock and vibration, stresses encountered between the component and the pc board could

(a)

(b)

(c)

FIGURE 14.17. Lead-forming technique. (a) Lead first bent to within 30 degrees of conductor pad. (b) Excess lead cut with diagonal cutters. (c) Full-clinched lead is formed with screwdriver blade.

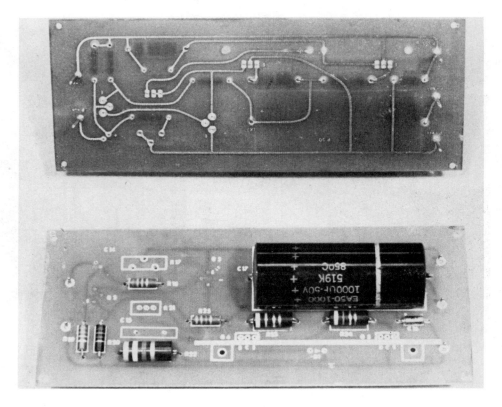

FIGURE 14.18. Assembly of all tubular components of power amplifier.

FIGURE 14.19. Capacitor assembly to pc board. (a) Disc capacitor assembly to pc board. (b) Vertical-mounting-style component.

Right   Wrong   (b)

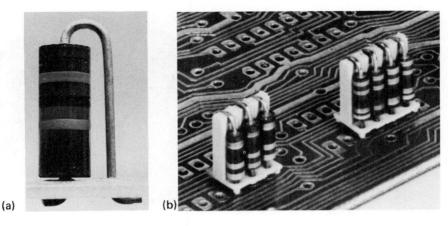

(a)　　　(b)

FIGURE 14.20. Vertical-mounting techniques of axial lead components. (a) Unsupported mounting method. (b) Plastic supports for vertical mounting of axial lead components. Courtesy of Robison Electronics, Inc.

result in mechanical failure. To minimize this possibility, plastic supports, such as those shown in Fig. 14.20b are used. They significantly improve mechanical security.

Physically large components (greater than $\frac{1}{2}$ ounce) may present undue strain on their leads if not secured. A common type of support used to secure large components to pc boards is shown in Fig. 14.21. This clip-style holder is provided with an expansion lug that secures the clip to the board by a simple press fit. This type of component clamp is desirable since it does not require any disassembly to remove the component.

The mounting of small trim-pots (screw-adjusted potentiometers) and thumb-pots (thumb-adjusted potentiometers) shown in Fig. 14.22 requires special consideration since their leads are in the form of rigid pins or tabs that may not be crimped or bent. These leads must therefore be fed straight through when mounted. For mechanical security prior to soldering, the access holes for the pins are drilled for a press fit.

FIGURE 14.21. Plastic clamps for rigidly mounting large components. Courtesy of Lorain Tool and Manufacturing Company.

FIGURE 14.22. Trim and thumb potentiometers.

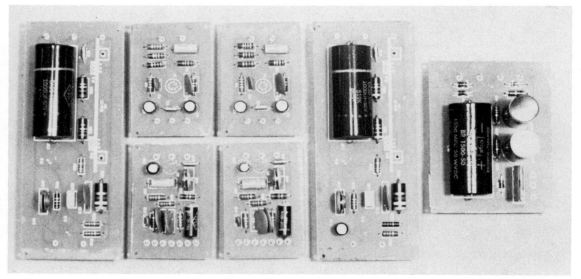

FIGURE 14.23. All amplifier boards assembled minus devices.

All the pc boards for the amplifier with their basic hardware and components assembled using the techniques discussed in the preceding sections are shown in Fig. 14.23.

14.5 DEVICE ASSEMBLY

Assembling devices to pc boards presents special problems in terms of lead bending, aligning, inserting, crimping, and protecting the device before and during soldering. All of these points need attention if the finished pc board is to be reliable. Devices that are considered in this section for direct mounting onto the pc board are *transistors, stud-style zener diodes* and *SCRs*, *FETs*, *TO-5-style integrated circuits* (ICs), *dual-in-line plastic pack ICs* (DIPs), and *flat-pack ICs*. Unique assembly problems for each of these devices will be considered separately. For a complete understanding of the various assembly techniques, continual reference should be made to Appendix III, which provides a listing of device outlines, dimensions, and lead orientations for some of the more common case styles used to package solid-state devices. A large number of solid-state device sizes and case configurations have been standardized by the *Joint Electronic Devices Engineering Council* (JEDEC). Several diode and transistor case outlines are identified by numerical designations prefixed by *DO* or *TO* to indicate a specific outline. Although many power transistors and rectifiers do not follow any established standard for style and size, such as the JEDEC package identification system, the manufacturer's case numbers and specifications will readily provide the necessary device outline and spacing dimensions for assembly.

Before the devices are assembled, an important consideration regarding their protection will first be discussed. Solid-state materials are extremely temperature sensitive and, as such, precautions must be taken during assembly that will protect them from excessive heat while soldering. This heat is transferred from the soldering iron tip through the lead to the device and effects the lead bond. This thermal problem often dictates the choice of specific mounting arrangements for the device. Manufacturers often provide assembly information together with maximum allowable time at specific temperatures that the device can tolerate without ruining the lead bonds. For the 2N3053 transistor used in the amplifier, the manufacturer specifies that the maximum allowable thermal conditions while soldering is 10 seconds at 491°F (255°C). If these conditions are expected to be exceeded, some method of heat transfer is necessary. The most common device used is a *lead heat sink*. This aid is mechanically connected to the lead between the device case and the soldering iron tip. The heat sink absorbs and draws off excessive heat from the device during soldering. Two common methods of heat sinking are shown in Fig. 14.24.

FIGURE 14.24. Lead heat sinking to prevent damage to device. (a) Smooth-jaw pliers. (b) Spring-loaded lead heat sink.

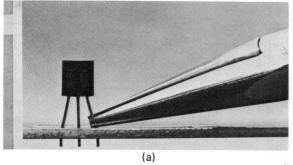

(a)

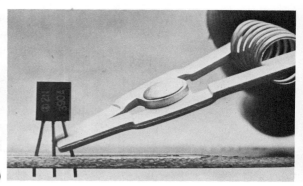

(b)

In Fig. 14.24a, the flat jaws of a pair of long-nose pliers are used. Figure 14.24b shows a commercially available heat sink that is made of aluminum and is spring-loaded. Aluminum, copper, or brass heat sinks are more efficient than steel pliers because these materials are better conductors of heat. As can be seen from Fig. 14.24, some space must be provided between the device case and the insulation side of the pc board, where sinking is necessary, to attach the heat sink. This obviously is a disadvantage in terms of mechanical security because it prevents the device from being flush-mounted to the pc board. Heat sinking of leads is not normally necessary when dip or wave soldering techniques (discussed in Chapter 15) are employed but should be considered for hand soldering. Therefore, when assembling devices the technician must consider the necessity of lead heat sinking and provide for it where necessary.

A popular case configuration used for low-power applications below 1 watt is the *TO-5* style shown in Fig. 14.25a. As will be shown, this case configuration is relatively simple to assemble to pc boards. The wide *sealing plane* provides excellent mechanical security when mounted flush against the insulated side of the board. This type of mounting requires that lead holes in the board have the exact spacing and orientation as the transistor leads. Straight feed-through connections to the pc board are made, allowing for either service bends or fully clinched leads onto the foil side of the board. This mounting arrangement is shown in Fig. 14.25b. One possible disadvantage could result from this type of mounting. The coefficients of expansion of the transistor leads and the insulating material of the pc board are not the same. This being the case, sufficient force may be generated to crack the solder bonds. An alternative method for mounting devices that provides for stress relief is to employ an insulated *spacer.* Several styles of spacers are shown in Fig. 14.25c. The transistor leads are first provided with a "C" bend close to the sealing plane to allow for movement. Special pliers of the type shown in Fig. 14.25d are used to form this bend. The leads are then inserted through the spacer before assembly to the board. A completely assembled device is shown in Fig. 14.25e. The leads of the TO-5-style case initially are inserted most easily into both the spacer and the pc board by positioning the center lead over its respective lead hole. The case is then tilted at an angle to *raise* the other two leads above the surface of the board or spacer. This procedure is shown in Fig. 14.26. Once this first lead is properly positioned, the case is rocked to align and insert another lead. The last lead is inserted in a similar manner. Once the leads have been inserted through the spacer and the pc board, the assembly can be firmly seated. The leads are bent over toward the conductor pad, cut to the desired length, and clinched similar to component leads, as discussed in Sec. 14.4. The transistor is now ready to be soldered.

Another popular case style is the *TO-92* package shown in Fig. 14.27a. These in-line leads are best assembled by first orienting an end lead over its lead hole on the pc board (Fig. 14.27b). The middle lead is

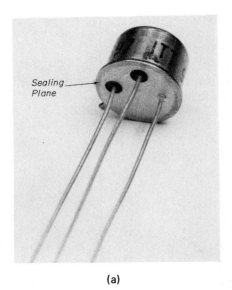

Sealing
Plane

(a)

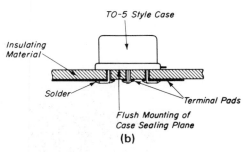

TO-5 Style Case

Insulating
Material

Solder

Flush Mounting of
Case Sealing Plane

Terminal Pads

(b)

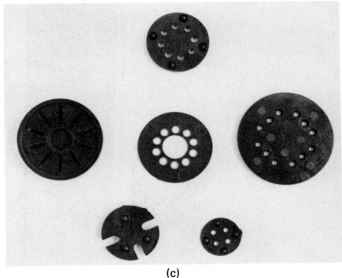

(c)

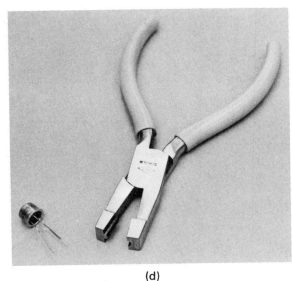

(d)

FIGURE 14.25. Methods of assembling the TO–5 case style to pc boards. (a) TO–5-style case configuration. (b) Flush mounting of TO–5-style case. (c) Insulated device spacers. (d) Pliers to form "C" bend in device leads for strain relief. (e) Strain-relief mounting technique.

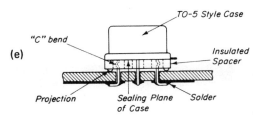

TO-5 Style Case

"C" bend

Insulated
Spacer

Projection

Sealing Plane
of Case

Solder

(e)

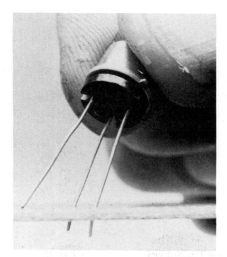

FIGURE 14.26. Case is tilted for insertion of second lead.

(a)

(b)

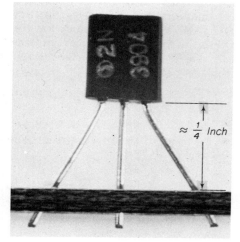

$\approx \frac{1}{4}$ *Inch*

(c)

FIGURE 14.27. Typical mounting technique for small plastic transistors. (a) TO-92 case style. (b) Initial lead insertion. (c) Final installed position.

inserted next and finally the remaining lead is fed into its hole. The hole centers on the board are usually drilled farther apart than the lead orientation on the transistor to provide sufficient spacing among terminal pads because the leads are so closely arranged on this type of device. The transistor is mounted so that the height of the sealing plane is approximately $\frac{1}{4}$ inch from the board surface. Lead heat sinking can then be easily employed. The completed assembly of a TO-92 transistor is shown in Fig. 14.27c.

The *stud-mounted zener diode* or similar-style SCR is provided with a threaded stud for mounting. Since the electrode electrically connected to the stud is usually the anode, the use of an insulating washer may be necessary when mounted to a metal frame. For pc board mounting, a hole large enough to clear the threads of the stud is formed. Sufficient copper foil at the terminal pad must remain after drilling this clearance hole to ensure a sound electrical connection between the stud and a flat metal washer used for mounting. The arrangement of this device and hardware is shown in Fig. 14.28a. The mounting nut should be tightened onto the lockwasher with a *torque wrench* such as those shown in fig. 14.28b, c, and d to prevent fracturing the insulation around the edges of the access hole. A torque wrench allows a specific pressure to be applied to the nut. Calibrated in inch-pounds, the torque wrench is rotated with an even force until the desired pressure is reached. The *beam-type* torque wrench shown in Fig. 14.28b is rotated in a clockwise direction until the pointer has moved to the scale marking that indicates the desired torque value. Zero torque position is at center scale, allowing torque measurements in both left- and right-hand directions.

Another style of torque wrench is the dial-indicating type shown in Fig. 14.28c. This wrench allows the operator to determine the applied torque in either direction on a finely calibrated dial indicator.

Torque screwdrivers, as shown in Fig. 14.28d, are also available. These tools have bit holders that allow screwdriver blades, hex keys, or hex sockets to be interchanged quickly and have micrometer style scales that are set to the desired torque. The barrel scale mark is aligned with the desired micrometer setting on the handle. This tool is extremely useful where assembly accessibility is limited. When the set torque value is reached, the tool disengages with a clearly audible *"click."*

Manufacturers' torque specifications for mounting devices must be consulted to ensure that the device or pc board is not damaged. On stud-mounted devices, excessive torque can deform the sealing plane enough to form a gap in the thermal circuit that exists between the device material and the case. This would cause an increase in the internal thermal resistance that could damage the device.

The cathode of the zener diode, normally connected to a lug extending from the top of the device, can be electrically joined to its respective terminal pad with a piece of insulated wire. This connection is also shown in Fig. 14.28a.

(a)

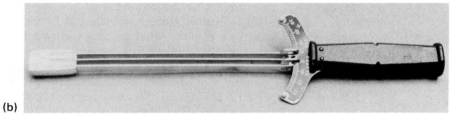
(b)

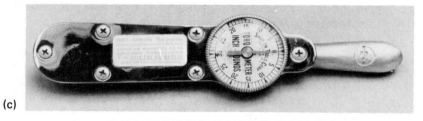
(c)

(d)

FIGURE 14.28. Stud-style devices should be assembled with torque wrenches. (a) Hardware arrangement for mounting stud-type devices to pc boards. (b) Beam-type torque wrench. (c) Dial-indicator torque wrench. (d) Torque screwdriver.

Some insulated-gate field-effect transistors (IGFETs), or metal oxide semiconductor field-effect transistors (MOSFETs), require utmost care during assembly. This device is extremely sensitive to the most minute static voltage at its gate terminal. For this reason, the manufacturers of these devices have provided a metal ring assembly near the sealing plane of the case that electrically connects all the leads together. This assembly is shown in Fig. 14.29a. This arrangement prevents any static charge from building up at the gate terminal, which could easily destroy the device. This ring must not be damaged or cut prior to soldering. The leads of the MOSFET are first inserted into the pc board to a depth that allows access to the ring between the sealing plane and the board surface with diagonal cutters. The leads are then bent and cut. Only after they are soldered to the foil side may the ring be cut, using diagonal cutters very carefully so that no leads are damaged. Figure 14.29b shows the ring being removed.

Some manufacturers of MOSFETS, instead of providing a binding ring, simply twist the device leads together for static charge protection. For the assembly of this type of arrangement, each lead must first be identified without untwisting them. These leads are then carefully inserted into the proper holes in the pc board up to the twisted lead section just below the sealing plane. Finally, the leads are bent and cut but must not be untwisted prior to soldering.

FIGURE 14.29. Field-effect transistor assembly. (a) FET with shorting ring. (b) Ring is removed with diagonal cutters.

(a)

(b)

The *TO-99* style of case and lead configuration, shown in Fig. 14.30a, presents uniquely difficult assembly problems. Integrated circuits of this style have at least eight leads attached to a case with a diameter of 0.200 inch. The mounting problems arise when attempts are made to align and insert all eight leads simultaneously, since the manufacturer normally cuts all device leads to an equal length. Assembling these devices can therefore be simplified by *staircasing* — that is, by cutting each lead, starting at the *key lead*, succeedingly shorter than the next in a descending direction toward the sealing plane. The key lead is then used as the reference for lead alignment and placement into the pc board. The *shortest* lead must be cut to a length that will allow it to pass through the pc board and be easily soldered to the terminal pad. With the leads cut in staircase fashion, the longest lead is inserted first into the proper hole without aligning the others. Each succeeding shorter lead is then inserted, *one at a time*, until all the leads are inserted and the component is properly mounted. The assembly of an IC with staircased leads is shown in Fig. 14.30b. An IC assembled with fully clinched leads and soldered is virtually impossible to remove without damaging the device. For this reason, service bends are recommended if there is any chance that the device will have to be removed at a later time, which is sometimes the case in prototype work.

Assembling 14-pin *dual-in-line packaged (DIP)* ICs (TO-116 package) presents equally difficult problems in terms of lead insertion. Since the leads on these devices are too short to staircase, an alternative approach is necessary. The DIP is first tilted back with its leads positioned above the pc board and approximately over the lead holes. This step is shown in Fig. 14.31. Finger pressure is applied evenly to the outside edges of the first two opposite leads closest to the board until they align with their respective holes. These leads are partially inserted into the board. The device is then tilted down closer to the board surface until the next two adjacent pins are against the board. Again, finger pressure is applied to align and insert these leads. This technique is continued until all the leads have been inserted into the board. Because the leads provided with

(a)

(b)

FIGURE 14.30. Method of asembly for TO-99 case style. (a) TO-99-style IC package. (b) "Staircased" leads simplify their insertion into pc board.

FIGURE 14.31. Assembly of
dual-in-line (DIP) package.

the DIP are usually too short for full clinching, they are normally soldered
while extending straight through the board. The DIP remains positioned
during assembly as a result of the slight pressure exerted outward by
the leads against the edges of their access holes. However, the package
position should be checked just prior to soldering the board to ensure
proper mounting.

The final device package to be considered for assembly is the flat-
pack configuration shown in Fig. 14.32a. This device may be mounted
using one of two methods of assembly. These are (1) mounted onto the
insulated side of the pc board with leads bent at 90-degree angles to the
base of the package for insertion into the access holes, or (2) mounted
directly to the foil side of the pc board and directly soldered to foil
conductor fingers.

The first method of mounting is similar to those already discussed.
However, the leads in this device must be bent as shown in Fig. 14.32b.
This configuration is necessary when the leads are to be soldered to
terminal pads. Since the lead spacing on the device is so small, the terminal pads in the conductor pattern must be staggered, to avoid short
circuits. The second assembly method makes use of a lead configuration
such as that shown in Fig. 14.32c. However, if the leads extend straight
out from the side of the package and the *double* 90-degree bend is not
provided by the manufacturer, they can be pressed down against and
aligned with the appropriate conductor fingers and soldered (Fig. 14.32d).
To maintain the proper position as well as provide lead heat sinking
(necessary for low-power flatpacks) for either lead configuration, a
special U-channel device *clamp-sink* is employed. This can easily be
constructed with a short length of aluminum channel stock. The clamp-
sink is positioned with the inside surface of the channel base over the
IC and the channel legs against the device leads. The clamp-sink and
the device are then held in place with the weight of an arbor press striker
brought to rest on top of the channel stock. This arrangement is shown
in Fig. 14.32e. After securing the package, the leads are soldered by hand.
Soldering at this time is necessary to mechanically secure the flatpack to
the pc board after the clamping arrangement is removed. This technique
of mounting flatpacks is often used with double-sided boards wherein it is
not possible to employ wave or dip solder methods.

(a)

FIGURE 14.32. Methods of assembling flat-pack style cases. (a) Flat-pack configuration. (b) Lead-bending arrangement for installation into pc board through insulation. (c) Lead-bending arrangement for soldering directly to conductor pattern. (d) Flat-pack soldered to foil without double 90-degree bend. (e) Clamp-sinks are employed for flat-pack soldering.

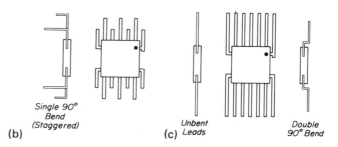

Single 90°
Bend
(Staggered)

(b)

Unbent
Leads

(c)

Double
90° Bend

(d)

(e)

14.6 DEVICE SOCKETS

An alternative method of assembling devices to pc boards is with the use of commercially available device *sockets*. These sockets are obtainable in all the standard case configurations. Device sockets have two decided advantages over direct device soldering to the foil. First, the use of sockets allows devices to be removed and replaced easily. Second, lead heat sinking is unnecessary. The only disadvantage of sockets is the additional cost, a factor that must be emphasized throughout the fabrication of a prototype. Where cost is of prime importance, sockets should only be used for devices that may need occasional replacement because of aging or other anticipated design changes.

Some commercially available sockets for transistors and ICs are shown in Fig. 14.33. These sockets are provided with round pins for direct soldering into pc boards or with long square pins that serve the dual function of pc board mounting and wire-wrap lead assembly.

FIGURE 14.33. Transistor and integrated circuit sockets. (a) 3-, 4-, 8-, 10-pin sockets for TO-5 case style. (b) 14-pin DIP socket with wire-wrap terminals. (c) 14-pin DIP socket with solder terminals. (d) 24-pin socket for large-scale integration (LSI). Courtesy of Augat, Inc.

(a)

(b)

(c)

(d)

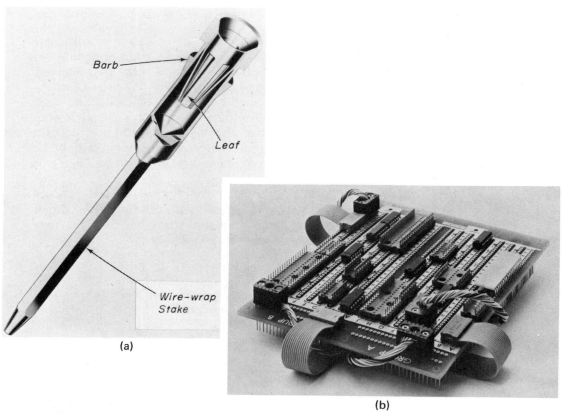

(a)

(b)

FIGURE 14.34. Socket terminal nomenclature and DIP test panels. (a) Wire-wrap socket terminal. (b) High-density test panels using wire-wrap socket terminals. Courtesy of Augat, Inc.

Another method of assembling devices that has all the advantages of sockets plus the additional benefit of versatility in positioning involves the use of *socket terminals* such as those shown in Fig. 14.34a. These individual terminals consist of a *wire-wrap stake* for interconnections at the base, a *four-leaf contact socket* arrangement for device lead insertion, and a *barb* on the side of the socket for positive positioning and securing the terminal when it is inserted into the pc board. To mount this type of socket, the appropriate diameter mounting holes are first drilled into the pc board in accordance with the socket lead orientation. The socket terminals are then inserted into the board and swaged into place with an arbor press. If desired, the pins can later be soldered to the copper foil of the pc board. High-density *DIP test panels* using the socket terminals are shown in Fig. 14.34b.

14.7 DEVICE HEAT SINKING FOR PRINTED CIRCUIT BOARD APPLICATIONS

Heat sinking devices on pc boards is usually less difficult than heat sinking chassis-mounted devices. When devices and heat sinks are assembled on the pc board, insulators are not necessary because the board itself is an insulator. This is not the case with chassis-mounted heat sinks (as was discussed in Chapter 1).

Several commercially available heat sinks used in low- and medium-power applications and the case styles to which they are adaptable are shown in Fig. 14.35. These heat sinks are fabricated from aluminum or beryllium–copper alloys. They are secured to the device case by either a press fit resulting from the spring tension of the sink material, or by mechanical clamping techniques. *Silicon grease* (thermal joint compound) used where the device case contacts the heat sink can reduce the thermal resistance between the two by as much as 100%. These sinks are finished in either a black anodized or natural unaltered metal surface. The black finish is more efficient in radiating heat away from the device case.

Heat sinks of the type used to mount two or more identical devices to the same thermal environment, such as push-pull drivers or complementary-symmetry power applications, are shown in Fig. 14.36. In these configurations, pairs of devices are mounted on the same heat sink, which may be either assembled onto a pc board directly or may be in turn mounted onto another sink to maintain the same thermal environment.

Medium-power plastic devices, such as those styles shown in Fig. 14.37a, may require heat sinking. There are several methods of heat sinking these devices, one of which is with a commercial staggered *fin-type* heat sink also shown in Fig 14.37a. This sink is first mounted on the pc board. The transistor is then set onto the sink with the leads bent downward and at right angles to the base of the device before they are inserted into their appropriate holes in the board.

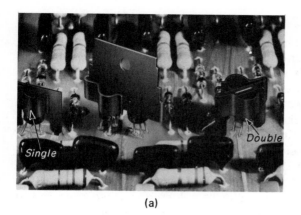

(a)

(b)

(c)

FIGURE 14.35. Printed circuit heat sinking for low- and medium-power devices. (a) Sink for TO-92 style "plastic" transistors. (b) Sink for TO-5 and TO-18 case styles. (c) Dual TO-5 sink. Courtesy of International Electronic Research Corporation.

FIGURE 14.36. Dual and quad device sinking arrangements. Courtesy of International Electronic Research Corporation.

(a)

(b)

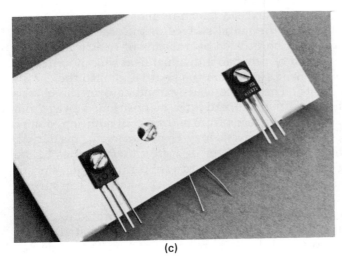

(c)

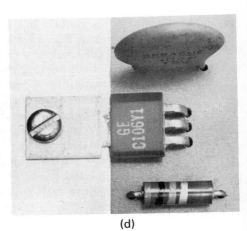

(d)

FIGURE 14.37. Several methods of heat sinking medium-power plastic devices. (a) Horizontal mounting of staggered fin-type heat sink. Courtesy of International Electronic Research Corporation. (b) Vertical mounting of staggered fin-type heat sink. Courtesy of Thermalloy. (c) Economical method of dual-mount heat sinking. (d) Printed circuit board to chassis heat sinking.

Another fin-type heat sink, shown in Fig. 14.37b, is specifically designed for use with metal tab-type plastic devices and may be rested horizontally on the pc board or left free standing in a vertical position. When mounted in the free-standing position, the device leads may be fed directly through the board without bending them at a 90-degree angle. The free-standing position, however, is not recommended if shock and vibration is expected.

A less expensive method for heat sinking plastic devices is to construct a sink from sheet aluminum to the dimensions required to provide adequate heat sinking. (These dimensions are determined by the designer of the circuit.) One such arrangement is shown in Fig. 14.37c for the power amplifier pc board. The devices are mounted to the heat sink with machine screws and nuts. The optimum position for the devices is as close to the bottom edge of the sink as possible but allowing some heat-sink surface below the base of the device for more efficient cooling. Since the sink is vertically mounted, it allows for excellent heat dissipation with normal air convection. After the devices have been secured, the sink is mounted to the pc board through its mounting tabs. Since these plastic-type devices have their collectors connected to a metal surface on the case for heat-dissipating purposes, an insulating mica washer must be used to electrically separate them when mounted on a common heat sink. When securing these plastic cases to a heat sink, a torque screwdriver should be used. Care must be taken not to exceed 5 inch-pounds of torque to maximize thermal transfer between the metal surface of the case and the sink.

A metal chassis can also be utilized as a means of heat transfer. As shown in Fig. 14.37d, instead of using an individual heat sink mounted on the pc board, the device is bent away from the board and onto the chassis that is supporting the board. The device leads extending over the chassis and bent into access holes on the pc board must be positioned away from the chassis in order to prevent the leads from shorting. In addition, a mica washer between the device and the chassis may be necessary for insulation.

Techniques for heat sinking DIP-type ICs are shown in Fig. 14.38. The sink shown in Fig. 14.38a requires no hardware for mounting, since it is held by a friction fit between the base of the device and the top surface of the socket.

Some DIP types are equipped with heat-sink tabs protruding from the case (Fig. 14.38b). The tabs are inserted through the insulated side of the pc board and soldered to a large foil area provided on the conductor side. This technique of using a large foil area that would normally be etched away is quite economical since it eliminates the cost and assembly of a separate sink.

For those DIP-type ICs that are to be mounted directly to the pc board without the use of a socket, the heat sink with its method of assembly as shown in Fig. 14.38c may be used. This arrangement allows heat to be removed from both the top and bottom surfaces of the device. The mounting used to secure the heat sink to the device is also used to secure the assembly to the pc board.

Thyristors packaged in a modified form of the TO-5 case configuration present special thermal conduction problems. Because the average forward power dissipation for these devices ranges from 2 to 12 watts, an extremely efficient method of heat sinking must be employed in order to operate them safely. The simple heat-sinking methods previously discussed for this case style are suitable only if the power dissipation expected is

(a)

(b)

(c)

FIGURE 14.38. Methods of heat sinking DIP integrated circuit packages. (a) Sinking for socket assembly. Courtesy of Astrodyne, Inc., Wilmington, Mass., subsidiary of Roanwell Corp. (b) Tab for direct soldering to large copper-clad area of pc board for heat sinking. (c) Assembly technique for dual surface heat sinking. Courtesy of Thermalloy.

below a few watts. For the larger power dissipations that are typical of thyristors, a more efficient method of reducing the thermal opposition to heat flow between the case and heat sink must be employed. The most effective means of accomplishing this and effecting a larger output power capability is to solder the device case *directly* to a *heat spreader*, shown in Fig. 14.39a, prior to assembly to the heat sink. Manufacturers provide tin-plated cases for this purpose to improve solderability. When soldering to the case of the device, the spreader allows heat to be transferred over a larger surface area than would be available with the case alone. The spreader can then be mechanically mounted to a chassis to further extend the power capability of the device. Heat spreaders are available in copper, brass, or aluminum. The aluminum type must first be plated to ensure solderability.

The thyristor case is first assembled onto the spreader between the mounting tab and the self-jigging tab with a solder ring sandwiched between the sealing plane of the case and the mounting tab. Sufficient heat

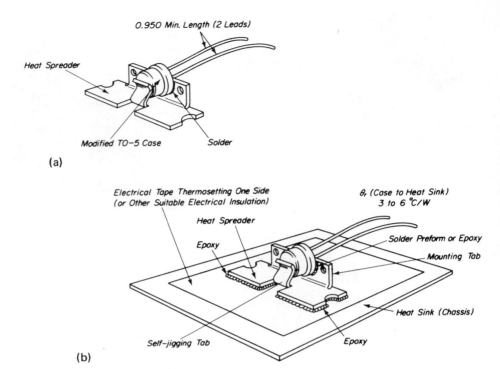

0.950 Min. Length (2 Leads)

Heat Spreader

Modified TO-5 Case

Solder

(a)

Electrical Tape Thermosetting One Side
(or Other Suitable Electrical Insulation)

Heat Spreader

Epoxy

θ_r (Case to Heat Sink)
3 to 6 °C/W

Solder Preform or Epoxy

Mounting Tab

Heat Sink (Chassis)

Self-jigging Tab

Epoxy

(b)

FIGURE 14.39. Effective method of heat sinking thyristors. (a) Heat spreader. (b) Mounting technique for greater heat dissipation.

to liquefy the solder ring is applied to the heat spreader only long enough to allow the solder to flow and form an alloy between the case and the mounting tab. Heat can be applied by a soldering iron or a hot plate. The use of the hot plate is recommended since it distributes the heat more evenly over the metal surface of the heat spreader. Once secured (soldered) to the heat spreader, the thyristor can be mounted to, but electrically insulated from, a chassis for additional surface area to improve heat-transfer capabilities. Epoxy or screws and nuts with mica washers may be used for mounting. One arrangement is shown in Fig. 14.39b.

The selection of the proper size heat sink is dictated by design factors discussed in Chapter 1. However, it is the responsibility of the technician to select the most appropriate type and style to meet the packaging requirements.

All the assembled pc boards for the amplifier's circuits, using all the assembly techniques discussed in this chapter, are shown in Fig. 14.40.

The assembled boards are now ready to be soldered. The following chapter discusses types of solder and fluxes, and methods and techniques of soldering pc boards.

FIGURE 14.40. Completely assembled amplifier boards.

PROBLEM SET

14.1 Cut two 1½-inch lengths of ⅛-inch diameter brass or copper rod. File a taper on one end and a flat side over the entire length of each piece. Remove the teeth from only the ends of the jaws of a small alligator clip and construct a lead heat sink similar to that shown in Fig. 14.41.

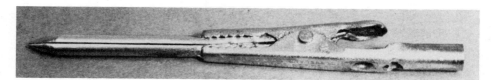

FIGURE 14.41.

14.2 Install turret-type terminals and socket pins into the printed circuit device test jigs designed in Problem 11.2 and fabricated in Problem 13.3. Soldering is performed in Problem 15.2.

14.3 Assemble the components onto the pc board designed for the seven-fixed-interval UJT timer of Problem 11.3 and fabricated in Problem 13.5. The assembled pc board is soldered in Problem 15.3. The indicating lamp (L1), on–off switch (S8), and battery clip are mounted to a chassis in Problem 18.5.

14.4 Mount the components on the pc board designed for the series feedback regulator in Problem 11.4 and constructed in Problem 13.6. The assembled board is soldered in Problem 15.4.

14.5 Install the components on the pc board designed for the tachometer in Problem 11.5 and fabricated and labeled in Problem 13.7. The assembled board is soldered in Problem 15.5. The tachometer pc board and meter are chassis- or panel-mounted in Problem 18.6.

15

Printed Circuit Soldering

15.0 INTRODUCTION

With the components properly assembled, the final process necessary to complete the construction of the pc board is to produce sound electrical connections between the leads and the foil pattern. These connections can be produced by several methods, the most common of which are *welding* and *soft soldering*. Welding requires the use of complex and expensive equipment and, as such, is rarely used in fabricating a prototype pc board. Soldering, on the other hand, can be performed quickly with less expensive equipment and results in excellent mechanical and electrical connections and protects the joint from oxidation. For these reasons, soldering will be discussed exclusively in this chapter.

Soldering is a metal solvent or chemical alloying action of the solder with the surfaces of the metal parts between which an electrical connection is formed. This completely metallic contact is produced by the application of soft solder with the heat of a soldering iron to the joint between the component lead and the terminal pad. The resulting connection is electrically sound, with the new alloy formed having different electrical and mechanical characteristics than either the solder or the metals joined.

The soldering of pc boards requires the development of proper techniques if quality results are to be obtained.

The three methods of soldering discussed in this chapter are (1) *hand*, (2) *dip*, and (3) *wave* soldering. These three methods are used extensively in pc board applications.

Specific information is also provided on the following topics: soft solder and its characteristics; the types of *flux* available and its application; the soldering iron, including the *tip*, *power rating*, and *tinning* processes; and hand-soldering techniques, with emphasis on the proper soldering of leads, swaged terminals, and conductor paths (tinning). A discussion of the characteristics of correct solder connections is presented to aid in visual inspection. *De-soldering* and the solvents used to remove the flux residues formed on soldered connections are also included. Finally, information on production-line methods of soldering, specifically dip and wave soldering, is presented.

15.1 SOFT SOLDER AND ITS CHARACTERISTICS

Soft solder used extensively in electronic equipment construction is an alloy principally of *tin* and *lead*. Soft solder is differentiated from hard solder by its tin content and lower melting point. The amount of tin contained in soft solder ranges from 50 to 70%. The tin–lead ratio determines the strength, hardness, and melting point of the solder.

Solder liquefies at temperatures between 361 and 621°F (183 to 327°C), the exact temperature depending on the tin–lead ratio. A metal such as copper, which has a melting point of 1981°F (1083°C), can be successfully alloyed with solder at temperatures well below this value because of the *solvent action* of solder when it is liquefied. At the melting point of solder, a thin film of metal is dissolved from the copper surface, forming an alloy and establishing an electrically continuous joint. The formation of this alloy between the metal and the solder has its unique physical properties, such as torsional, shear, and tensile strength, which are different from those of either the solder or the metal. These properties will vary widely and will depend on the depth of alloying into the metal surface.

Pure tin melts at 450°F (232°C) (point *B*, Fig. 15.1) and the melting point of pure lead is 621°F (327°C) (point *A*). When these two elements are combined, the melting point of the solder formed can be below that of either pure metal. A composition of 63% tin and 37% lead melts at 361°F (183°C), which represents the lowest melting point and most rapid transition from the solid to the liquid state of any other tin–lead ratio. This point, *C*, is also shown in Fig. 15.1. The 63/37 solder is termed the *eutectic* composition. All solder, with the exception of the eutectic composition, which melts sharply from solid to liquid, passes through a stage of softening between its solid and liquid stages known as the *plastic*

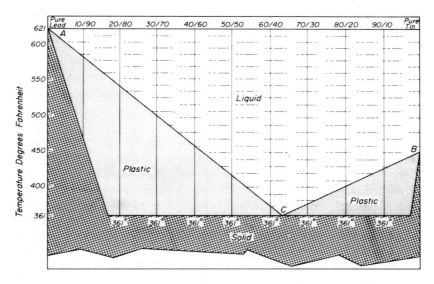

FIGURE 15.1. Tin-lead fusion diagram. Courtesy of Kester Solder, Division of Litton Systems, Inc.

range. A fusion diagram showing the temperature ranges for the various states of solder for all tin–lead combinations is shown in Fig. 15.1.

In order to determine the proper tin–lead ratio for a specific application, the function of the connection must be examined. The necessary properties that will dictate the composition of solder are mechanical resistance to fractures due to stress, ability to form a continuous metallic connection at low temperatures, and cost. These factors are discussed here.

Since tin is more expensive than lead, solder with a higher tin content is more costly. It has been shown empirically that the highest joint resistance to stress exists with a 63/37 tin–lead ratio (eutectic solder). This concentration, therefore, affords the best alloying qualities in addition to the lowest melting point.

For most hand-wiring and printed circuit applications, solder with a tin–lead ratio of 60/40 is commonly used, because of its excellent wetting action. *Wetting* is the term used to describe the ability of the solder to readily spread and alloy uniformly over the entire metal surfaces to be joined. Eutectic solder is sometimes selected to take advantage of the lowest and sharpest melting-point characteristics wherein maximum precautions are necessary to avoid component heat damage and upsetting the joint as it cools. If these two considerations are not critical, 60/40 solder is an excellent compromise.

Solder for electronic applications is available in bars, sheets, wire spools, and special forms, such as pellets, rings, and washers. For hand-wiring purposes, solder wire ranges from 0.030 to 0.090 inch in diameter.

The larger sizes are used for general-purpose work and the smaller for delicate soldering applications such as pc boards and solder cup-type pins (see Chapter 19) found on certain connectors.

Solder wire is also available with a core containing flux (see Sec. 15.2) in specific amounts to promote sound solder connections. For this reason, flux-core solder wire is used almost exclusively for electronic applications.

15.2 FLUX

The interaction of metal parts with the atmosphere forms a thin layer of oxide on their surfaces. This oxidation increases as the metal is heated and will severely interfere with the solvent action of solder, thus preventing alloying and the formation of an electrically continuous joint. Consequently, the oxide must be removed. *Fluxes* are used for this purpose. They are chemical agents that aid in soldering by removing thin films of oxide present on the metal surfaces to be soldered. When applied to the joint, the flux attacks the oxides and suspends them in solution, where they float to the surface during the soldering process. When the joint is heated, the presence of flux also prevents further oxidation in addition to lowering the surface tension of the metals, thereby increasing the wetting action. It is important to remember that flux is not a cleaning agent for removing grease or other contaminants. Its sole function is to remove the oxide film. For optimum soldering results, the parts must be thoroughly cleaned before the flux is applied.

The flux in no way becomes a part of the soldered connection but aids in the process. Upon completion of the soldered joint, a flux residue appears on its surface, which contains the captured oxides. This residue should be removed with an appropriate solvent.

The ability to rapidly remove oxide films from metal surfaces constitutes *activity* of the flux. It would appear that a highly active flux is ideally suited for electronic construction since it would afford rapid alloying and thereby reduce the possibility of heat damage to the components. This, however, is not the case. Highly active fluxes may be corrosive at room temperature and if allowed to remain as a residue will deteriorate the conductor surfaces or reduce the resistance of the insulation between soldered connections. Corrosive damage to components may also occur, since some of the active residues will gradually spread as they absorb moisture from the atmosphere. Even with suitable solvents, there is no assurance of complete flux residue removal, especially around closely spaced terminals, connectors, or conductors on pc boards where complete solvent flushing is not possible.

There are three major classifications of flux: (1) *chloride* (inorganic salts), (2) *organic* (acids and bases), and (3) *rosin* fluxes. The chloride types are the most active (highly corrosive) fluxes. They absorb moisture from the atmosphere and strongly react with acid even at room temperature. Organic fluxes are slightly less active than the chlorides and are used mainly for confined areas in which fast soldering time is important and corrosion problems are not critical. Many of the organic fluxes are converted to an inert residue after thermal decomposition. They do not absorb moisture and are difficult to remove. For the reasons given above, chloride- and organic-type fluxes are not recommended for use in electronic construction. The rosin-type fluxes are used almost exclusively because of their noncorrosive characteristics at room temperature. They are corrosive at temperatures near the melting point of solder. Consequently, they attack the oxide film during the heating cycle but are inactive when room temperature recurs. Rosin fluxes are available with activating agents that greatly improve their activity. These activated rosin fluxes are much more corrosive than pure rosin when heated and present the appearance of an instantaneous melting, wetting, and flowing action of the solder. They are essentially as noncorrosive at room temperature as the pure rosin types and are often preferable if a higher degree of flux activity is dictated such as in dip or wave soldering.

Liquid flux may be applied by *wiping, dipping, spraying,* or *sponging.* Wiping or dipping methods are not extensively used. Uniform flux coatings are difficult to realize when wiping with a brush and thorough wetting of all surfaces may not result when dipping because of air pockets or cavities created during this process. Another disadvantage of these two methods is that application of an excessive amount of flux requires extensive removal of residue after the soldering has been completed.

Spraying flux involves applying a fine mist to the joints. The use of this method ensures that leads and terminals are more uniformly and completely coated with flux. When spray methods of flux application are to be used, the manufacturer's literature should be consulted regarding the recommended spray gun, nozzle, pressure, and solvent to use.

The application of flux with a sponge is perhaps the most effective and least messy method. To apply the flux, the board is firmly pressed against a sponge that has been saturated with liquid flux.

When hand soldering, the proper amount of flux can best be applied with the use of *flux-core wire solder.* This form of solder contains a core of solid rosin flux in a single or multiple core. However, there is no significant advantage in using multiple-core solder since it is essentially the volume ratio (amount of flux to solder) that determines optimum soldering conditions. Core sizes are available that provide a ratio of rosin flux per unit volume of solder of 0.6 to 4.4%. These ratios can be obtained for any size of wire solder. Indications are that 60/40 rosin flux core solder with a diameter of 0.040 inch and 3.6% flux is ideally suited for hand soldering pc boards and other electronic precision work.

15.3 THE SOLDERING IRON

The soldering iron, shown in Fig. 15.2a, consists of four basic parts. These are the (1) *tip*, (2) *heating element*, (3) *handle*, and (4) *power cord*. Some soldering irons are equipped with either a cork finger grip or a heat deflector. These are shown in Fig. 15.2a and b, respectively, and are designed to improve thermal insulation in the handle, which tends to become hot after prolonged use. In addition, the cork finger grip allows for more comfortable use of the iron.

When power is applied to the soldering iron, the tip is heated by direct thermal contact with the heating element into which it is set. The type of tip shown in Fig. 15.2b is inserted into the element and secured with a set screw in the side of the element jacket. The tip must be fully seated before the set screw is tightened so that maximum heat is transferred to the tip. Some tips are an integral part of a porcelain-type heating element (Fig. 15.2a). Others (tiplets) thread into the metal end of an integrated-type porcelain heating unit (Fig. 15.2c). The solder is liquefied when it comes in contact with a tip that has reached its operating temperature.

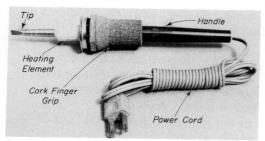

(a)

FIGURE 15.2. Types of soldering irons and tips used in electronics. (a) Soldering iron nomenclature. (b) Interchangeable soldering tip. (c) Integrated-type procelain heating unit with tiplet.

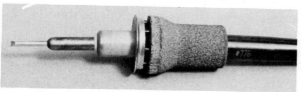

(b)

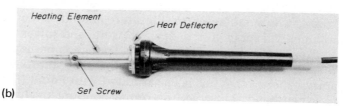

(c)

A soldering iron is selected for a specific application by considering the following factors: (1) size and style of the tip, (2) tip material, (3) required tip temperature, and (4) tip-temperature recovery time.

Selecting a tip style is somewhat a matter of personal preference. However, the shape of the particular tip used must provide the largest contact area to the specific connection for maximum heat transfer while minimizing the possibility of heat damage to surrounding leads or components. Some of the widely used tip styles are shown in Fig. 15.3. Each of the available tip configurations is designed for a specific soldering application. A brief description and application of each tip shown in Fig. 15.3 follows:

Chisel and *pyramid* style tips are commonly used for hand wiring and general repair work. The large flats of these tips allow large areas to be heated rapidly. The *turned chisel* and *conical chisel tips* lend themselves very well to soldering in confined areas, such as hand soldering components to double-sided pc boards.

Bevel designs are suitable for soldering terminal pad connections on single-sided pc boards if extreme conductor pad density does not exist. This style allows rapid heat transfer owing to the large tip surface area.

Conical tips are preferable for soldering high-density wiring, eyelets, and small heat-sensitive parts.

Radius groove tips also work well in high-density soldering applications and on round configurations such as pin connectors, pot-type terminals, and turret terminals.

FIGURE 15.3. Solder tip configurations.

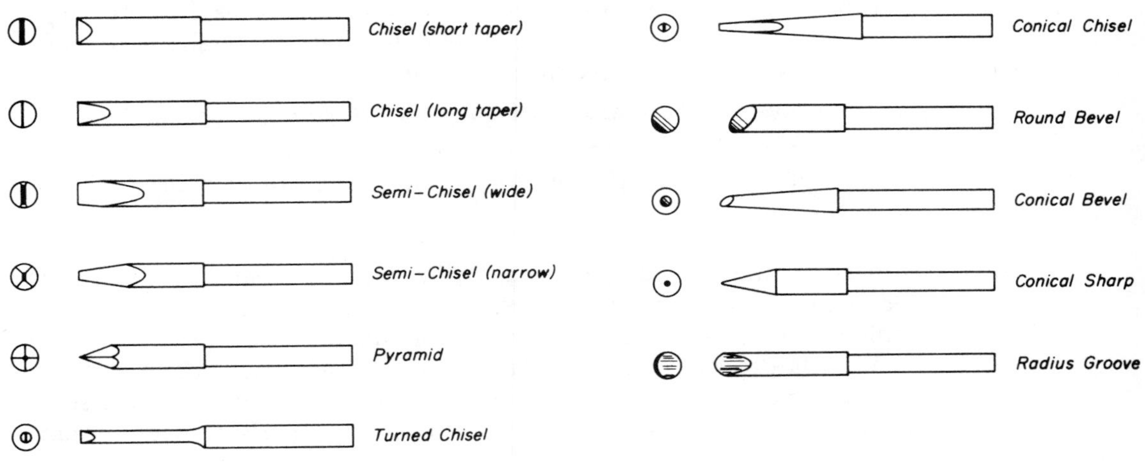

347

Soldering iron tips are manufactured chiefly from copper and are available both in plated and unplated finishes. Typical platings for tips are *iron*, *gold*, and *silver*. These platings protect the copper tip from corrosion and pitting, which become pronounced over extensive periods of use. The plated tips should never be filed or cleaned with harsh abrasives that would remove the plating and expose the copper. The tips should instead be periodically cleaned by first dipping them cold into liquid rosin flux and then bringing them up to operating temperature to loosen any surface oxidation or contaminants (burned solder and flux residues) that may be present. Solder is then applied to the hot tip. Contaminated areas are easily detected by observing those areas that resist wetting by the solder. Sufficient solder has been applied when it begins to puddle. The tip is then wiped on a moist fine-pore cellulose sponge to remove the contaminants.

The plain copper tips oxidize quickly under normal use. Since this copper oxide acts as an insulator, it will severely impair heat transfer from the tip to the work unless it is removed. In order to remove the oxide and to maintain these tips in optimum working condition, they must be prepared prior to use by a process termed *tinning*, which simply involves applying a thin layer of solder to the tip. The procedure for tinning follows:

1. The surface contaminants and oxides are removed by buffing the tip thoroughly with No. 00 steel wool or fine emery cloth.
2. The tip is dipped into liquid rosin flux.
3. The iron is plugged in and allowed to heat. Depending on the power rating and the size of the tip, it will take from approximately 30 seconds to 3 minutes to reach the melting point of solder.
4. During the heating period, the tip should be checked for temperature with a piece of rosin-core wire solder. As soon as the core flux starts to flow and the solder begins to melt, a liberal amount of solder is applied directly to the working end of the tip.
5. The excess solder is wiped off onto a moist sponge.
6. This process of applying solder and wiping continues until a thin film of solder completely forms on the entire working surface of the tip.

The tip must be dipped into a flux prior to heating since a dense layer of oxide would otherwise form, which the core flux alone could not completely remove, making proper tinning impossible.

When a properly tinned soldering tip has been used for an extensive period of time, it will become pitted and contaminated with burned solder and flux residue. To repair and refinish an improperly tinned, contaminated, or pitted plain copper tip, the following procedure must be used before the iron can be tinned:

1. The tip is firmly secured in a bench vise having smooth clamping surfaces (Fig. 15.4). *Caution:* Only the tip is secured in the vise, to avoid damaging the heating element.
2. To reshape the tip flats, initial filing is performed with a second-cut or bastard file.
3. Since the surfaces of the flats must be as smooth as possible for optimum heat transfer, the initial shaping is followed by fine filing with a smooth-cut file.
4. Final surface smoothing can be achieved with fine-grade *crocus cloth* and a block of wood.

Care must be exercised when filing tips to maintain the original contour. Round files are required for shaping radius groove tips. Conical tips are more easily shaped if the file rather than the tip is secured in the vise. The tip is held at the correct angle to the face of the file and rotated as it is drawn along the file length. This procedure is repeated on the crocus cloth. With the reshaping and smoothing completed, the tip is re-tinned as previously outlined.

Tips, especially of plain copper, should periodically be removed from the soldering iron because oxide will build up on the shank or threaded portion that fits into the heating element. If this oxide formation is not occasionally removed, the tip will seize in the element, making it difficult to remove and possibly damaging the element.

The required tip temperature is based on its application. For general-purpose soldering, such as to terminal strips and solder lugs, a tip temperature of 600 to 900°F (316 to 482°C) is sufficient. Printed circuit soldering requires fast heat transfer to the foil, yet excessive heat that could upset

FIGURE 15.4. Plain copper tips can be repaired by light filing.

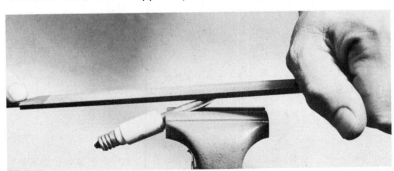

the foil bond must be avoided. For soldering to conductor patterns having widths of $\frac{1}{32}$-inch or larger and for terminal pads having $\frac{3}{32}$-inch diameters or larger, irons with tip temperatures of 800 to 850° F (427 to 454° C) are recommended. When soldering to more delicate pc boards or extremely fine wire (28 gauge or smaller), tip temperatures of 550 to 700° F (288 to 371° C) are suggested. However, when fine-component wire is to be soldered, especially in the case of heat-sensitive devices, a tip temperature of at least 900° F (482° C) is recommended (if lead heat sinks are not used) so that rapid heat transfer occurs with a minimum amount of tip contact time.

Manufacturers normally rate their irons in *watts*, typically available from 20 to 60 watts for electronic application. It is difficult to compare exact tip temperature with wattage rating since the length and the diameter of the tip largely influence tip temperature. For example, a 25-watt iron with a $\frac{3}{16}$-inch tip diameter and $\frac{3}{4}$-inch tip length will produce a temperature of approximately 680° F (360° C). Another 25-watt iron with a tip having the same $\frac{3}{16}$-inch diameter but a $1\frac{1}{2}$-inch length will reach a temperature of approximately 605° F (318° C). As a rule, tip temperature decreases linearly, approximately 10° F for every 0.10-inch increase in tip length. Tip temperature is inversely related to tip diameter. However, this relationship is not linear, as is the case of length versus temperature. Therefore, empirical data, such as that provided in Table 15.1, must be consulted. This table shows comparative values of tip temperatures to specific wattage ratings for three common tip diameters. The values given in this table are for iron-clad chisel-style tips having a length of $\frac{3}{4}$ inch.

A more general relationship of wattage to tip temperature is given in Table 15.2. This table provides a comparison of the common wattage ratings to a range of tip temperatures that can be expected from any style or size tip. In general, approximately a 50° F change in tip temperature can be expected between the successive wattage ratings listed.

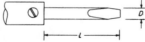

TABLE 15.1

Comparison of Wattage Rating and Tip Temperatures for Various Sizes of Solder Tips

Diameter (D) in Inches	Length (L) in Inches	20W	25W	30W	35W	40W	50W	60W
3/64	3/4	675°F	725°F	775°F	825°F	875°F	925°F	975°F
1/8	3/4	640°F	690°F	750°F	800°F	860°F	910°F	960°F
3/16	3/4	630°F	680°F	730°F	790°F	850°F	900°F	950°F

TABLE 15.2

Soldering Iron Wattage Rating vs. Range of Tip Temperatures

Wattage Rating	Range of Tip Temperatures
20 Watts*	550 to 750°F
25 Watts*	600 to 800°F
30 Watts*	650 to 850°F
35 Watts*	700 to 900°F
40 Watts	750 to 950°F
50 Watts	800 to 1000°F
60 Watts	850 to 1150°F

*Commonly selected wattage ratings for electronic applications.

There is a useful technique for determining the approximate tip temperature of iron-clad tips by visual inspection. When the iron is heated to its *idling* (maximum operating) temperature, the flats of the tip are alternately wiped several times across the surface of a moist sponge. The color of the tip is observed 2 to 3 seconds after wiping. There is a relationship between the color obtained and the approximate tip temperature. This relationship is given in Table 15.3. Tip temperature can be periodically monitored using this technique.

Recovery time is the rate at which a tip will return to its idling temperature after transferring heat (tip cooling) to the work during soldering. In mass production applications wherein a rapid succession of soldered connections are to be made, fast recovery time is absolutely essential. It is not of major concern in prototype work, however, since the recovery rate is normally much greater than the rate at which soldered connections are made.

TABLE 15.3

Reference for Iron-clad Tip Temperature by Visual Inspection

Tip Temperature	Tip Color (2 to 3 seconds after sponge wipe)
700°F	Silver
800°F	Gold
850°F	Gold with streaks of blue at tip
900°F	Blue to Purple
1000°F	Ash Black

The following example will serve to illustrate how to evaluate the many available irons and tips to select the most appropriate one for a specific application. An iron will be selected for soldering a *double-sided pc board having typical conductor widths of* ¹/₃₂ *inch with terminal pads of approximately* ¹/₁₆ *inch in diameter.* Ideal tip temperature for pc board soldering is 800 to 850°F (427 to 454°C). As can be seen in Tables 15.1 and 15.2, a wattage rating of 35 watts will provide this necessary temperature with a ³/₆₄ - by ³/₄ -inch tip. The style of tip should permit ready access between components. A *conical chisel* design is selected since this configuration provides the desired contact area for component lead soldering to terminal pads. Finally, the iron-clad finish is chosen for durability. The tip selected for this example is quite versatile because of its thin chisel shape, which works well with both double- and single-sided boards.

In general, the soldering iron that produces the best performance over an extended period of time is one that has high heat conductivity and low heat loss when in contact with connections to bring them up to soldering temperature. It should also have a cool, lightweight handle, a no-burn power cord with strain relief and a simple method for removing the tip and disassembling the iron to replace the heating element.

Where precise control of tip tempertature is required, *temperature-controlled* soldering irons having tips with built-in temperature sensors are available. This type of iron is shown in Fig. 15.5. Temperature-controlled units allow for tip temperature settings from 500 to 800°F (260 to 427°C). They can maintain the set idling temperature to within seconds after each soldering operation. This type of temperature control ensures consistent soldering temperature condition.

FIGURE 15.5. Temperature-controlled soldering iron. Courtesy of Hexacon Electric Company.

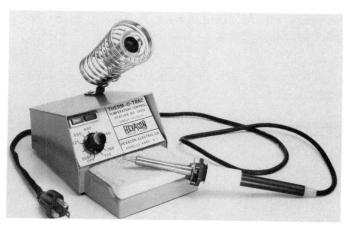

15.4 HAND SOLDERING PRINTED CIRCUIT BOARDS

Of prime importance in soldering is that the surfaces to be joined are clean and void of oxides. For pc board soldering, this is especially true, for a poorly cleaned terminal area would prolong heating time excessively when soldering, which could upset the foil bond. The terminals should not only be cleaned but it is imperative that liquid flux be applied to all joints prior to soldering even if terminal areas have been tinned or protected with Sealbrite (discussed in Chapter 13). The flux may be applied to the pc board by any of the methods discussed in Sec. 15.2. However, wiping is preferable in prototype work when a solder mask is to be employed, such as in the case of the amplifier's pc boards. This method allows the flux to be selectively applied only to those terminal areas that are to be soldered. The pc board is first secured in a *circuit board holder* and tilted to a horizontal position. Liquid flux is then applied with a fine artist's brush or toothpick onto the areas to be soldered. This arrangement is shown in Fig. 15.6 for applying flux to the power supply board. When all the surfaces to be soldered have been coated with flux, soldering may begin.

To solder leads to pc boards, it is important that heat and solder be applied quickly and accurately. A small amount of solder is initially applied to the flat of the soldering iron tip (solder bridge) to promote rapid heat transfer to the joint (Fig. 15.7a). The entire flat of the tip surface is then placed in contact with the terminal area and lead simultaneously. The more surface area of the tip contacting the terminal pad and lead, the more efficient the transfer of heat (Fig. 15.7b). The joint is allowed to heat to the melting point of solder, usually taking from 1 to 2 seconds for most connections. Solder is then applied to the terminal pad

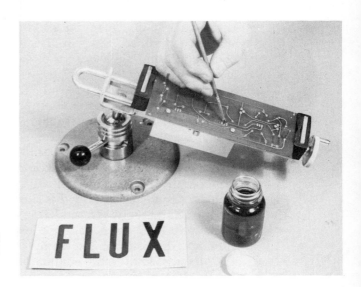

FIGURE 15.6. Application of liquid flux to selected areas.

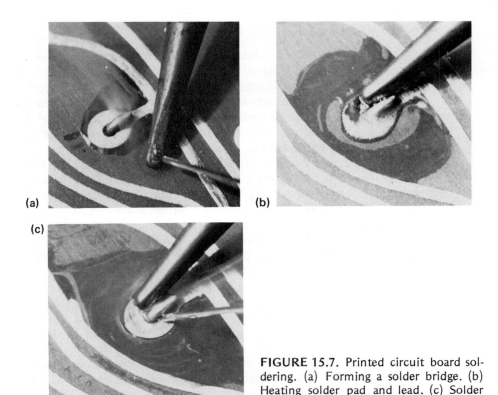

(a)

(b)

(c)

FIGURE 15.7. Printed circuit board soldering. (a) Forming a solder bridge. (b) Heating solder pad and lead. (c) Solder application.

and lead simultaneously and at a point on the opposite side of the lead from where the tip is contacting. The correct orientation of tip and solder is shown in Fig. 15.7c. As soon as the solder begins to flow, no more should be applied. Since solder follows heat flow, it will wet the entire connection and flow over the lead, forming a smooth contour of solder around the lead and terminal pad. The soldering iron is then removed and the joint allowed to cool. The joint must never be moved until the solder has completely solidified because any movement of the lead as it is cooling will prevent positive alloying from taking place. *It is important to apply only a minimum amount of solder sufficient to ensure proper alloying.* A properly formed solder joint is smooth and shiny in appearance. All surfaces must be completely wetted and the contour of the lead on the terminal pad clearly visible. A correctly soldered lead to a printed circuit terminal pad is shown in Fig. 15.8. A precaution should be mentioned here. When soldering the individual terminal pads, it is important to keep track of those leads that must be heat sinked. Lead heat sinking is discussed in Sec. 14.5.

Excess solder and contaminated flux residue on a tip must be removed before another solder joint is formed. Cleaning is accomplished quickly and easily by wiping the tip on a moist sponge. Soldering should

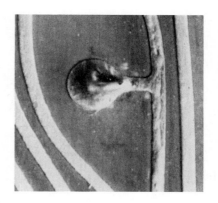

FIGURE 15.8. Correctly soldered terminal.

never be attempted with a tip covered with excess solder or contaminants if quality results are to be expected.

Swaged terminals are soldered to pc boards in a similar manner to lead soldering, with one important exception. Since the terminal pad area for swaged terminals is larger than those used for lead connections, it is necessary to apply heat for a longer period of time. The flat area of the soldering tip is placed in contact with the swaged portion of the terminal and solder is applied to the terminal pad. When heated sufficiently the solder will completely wet the terminal pad and flow uniformly around the swaged terminal in the area where it contacts the copper foil.

Since many pc boards do not employ solder masks, and if conductor pattern plating is not economically feasible, the technician may wish to tin the entire conductor area of the board to retard oxidation. This can be done during the soldering operation. Flux is first applied to the entire board surface. A small amount of solder is applied to the flat of the tip and the tip is slowly drawn along the conductor path. A thin layer of solder will be deposited on the surface of the copper. Several passes may be necessary to completely tin a conductor path. This will prevent excessive solder from forming in addition to minimizing foil bond damage. For this technique, a soldering iron with a tip temperature of between 500 to 600°F (260 to 316°C) should be used. The application of solder via the soldering tip is restricted to tinning conductor patterns. This technique is never employed for making soldered connections.

Properly soldered connections are uniform in appearance and their quality may be judged by visual inspection. Each solder joint may be inspected immediately upon soldering or after completion of the entire board. To aid in this inspection, a ×5 to ×10 magnifying glass may be used. Following is a list of deficiencies and their causes common to improperly soldered connections. These are shown in Fig. 15.9.

1. *Solder peaking* — characterized by a sharp point of solder protruding from a connection. Peaking is caused by the rapid removal of heat before the entire joint has had an opportunity to completely

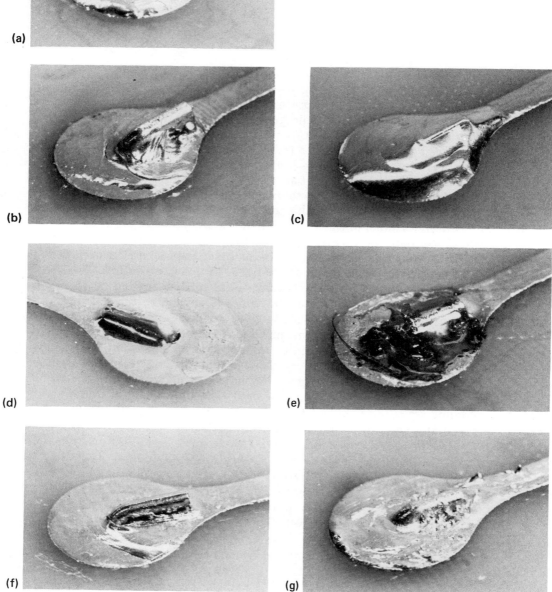

FIGURE 15.9. Typical inferior solder connections. (a) Solder peaking. (b) Incomplete wetting. (c) Excessive solder. (d) Cold solder joint. (e) Rosin joint. (f) Fractured joint. (g) Porous joint.

(a)

(b)

(c)

(d)

(e)

(f)

(g)

reach its soldering temperature. The solder follows the hot tip as it is removed from the connection, resulting in this peaking condition. Reheating the joint will correct this deficiency.

2. *Incomplete wetting* — occurs when portions of the soldered connections have not been alloyed with solder and are completely visible. This may be the result of both insufficient heat and solder. It may also be the result of contaminants on the soldering tip or terminal pad. Reheating the joint and applying additional solder will correct this fault if the condition is not caused by contaminants. In that case, de-soldering and cleaning are necessary before a new joint is attempted. The terminal pad can be cleaned with a sharpened pencil-style typewriter eraser. (De-soldering techniques are discussed in Sec. 15.5.)

3. *Excessive solder* — evident when the lead contour is not plainly visible. This is extremely undesirable since it prevents complete inspection of the alloying action of the solder, thus obscuring potential troubles. This condition can be rectified by removing some of the solder with the use of de-soldering aids.

4. *Cold solder joint* — an inferior connection easily detected by its dull-gray, grainy appearance, or as a cluster of solder that has not properly wetted all the surfaces. This is the result of applying insufficient heat, which prevents solder from alloying with the metal parts. Cold solder joints can generally be corrected by reheating the connections. If, however, the puddling of solder is caused by a connection that has not been thoroughly cleaned, it must be de-soldered and cleaned before it can be properly soldered.

5. *Rosin joint* — a joint in which a layer of flux residue is formed between the terminal pad and the solder. Since flux is an insulator, this condition could result in electrical discontinuity between the lead and the terminal pad. Rosin joints are the result of too short a heating time. The soldering iron tip must remain on the connection for a short time after the applied solder is withdrawn. This allows proper wetting action and causes all the flux to float to the surface with the captured oxides. This defect can be corrected by reapplying heat.

6. *Fractured joint* — characterized by the appearance of minute cracks in the solder. These cracks appear if the lead is moved before the solder has had an opportunity to completely solidify. This may be corrected by the reapplying of heat, if the lead will not again be moved during the cooling period. This problem of component and lead movement arises when a pc board is soldered with the components resting on a work surface. It is preferable to clamp the board by its edges in a circuit board holder, such as that shown in Fig. 15.6.

7. *Porous joint* — characterized by pinhole imperfections visible in the surface of the solder which is due to trapped gases caused by insufficient heating time, a period required to allow volatizing flux to escape. This fault can be remedied by reheating.

Flux residue remaining on the connection after soldering is undesirable. This residue presents a poor board appearance and contaminants suspended in this residue could cause troublesome electrical leakage paths. If more active fluxes must be used, their removal is doubly important to prevent corrosion. Flux removal can be accomplished by *hand brushing*, *dipping*, or with *ultrasonic equipment.*

Brushing with a stiff-bristle brush dipped in a flux solvent is often acceptable but may not remove flux completely from inaccessible cavities characteristic of pc boards in which swaged eyelets and terminals are used.

Dipping and mildly agitating the board in a solvent bath is preferable since it will improve solvent action in some areas that are not possible to reach by hand brushing (Fig. 15.10).

A more efficient means of flux residue removal, especially when activated fluxes are used, is to use an ultrasonic tank containing the flux solvent. Several minutes in this system ensures the most thorough cleaning.

FIGURE 15.10. Mild agitation for flux removal is accomplished by use of a rocker table.

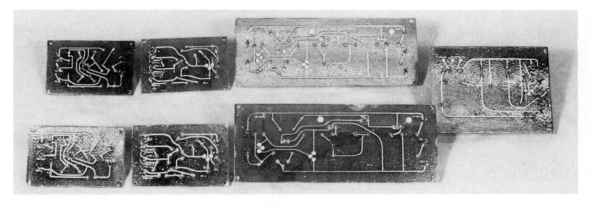

FIGURE 15.11. Hand-soldered amplifier boards.

For most electronic soldering applications, rosin flux residue is the type most generally encountered. Alcohol and trichlorethylene are excellent solvents for this residue. As in the case of most solvents, certain hazards present themselves. Alcohol is flammable and trichlorethylene should be used in a well-ventilated area and kept away from the skin. Finally, the remaining solvents are quickly evaporated by the use of an oil- and moisture-free air hose.

The seven pc boards for the amplifier, hand-soldered and using the techniques discussed in this section, are shown complete and ready for assembly to the main chassis in Fig. 15.11.

15.5 DE-SOLDERING PRINTED CIRCUIT BOARDS

When a lead is soldered to a terminal pad, it is difficult to remove without damaging the component or the terminal area. Several commercially available de-soldering aids, however, simplify this task. These aids are the *solder wick, de-soldering bulb, solder sucker, de-soldering tips,* and *extraction tools*.

A solder wick is made of finely woven strands of tinned copper wire such as that used for shielding coaxial cable (see Chapter 19). This flattened wick is first dipped into a liquid flux to promote solderability. It is then placed over the terminal area and lead to be de-soldered. This arrangement is shown in Fig. 15.12a. The soldering iron tip is placed in contact with the solder wick and pressed down against the connection. As the heat from the iron is transferred to the wick and connection, the solder will melt and flow in the direction of the heat transfer. The solder is thus trapped by the solder wick as it flows up through the weave. The result is shown in Fig. 15.12b. The solder can be completely removed from the joint with this method. The used portion of the wick is discarded. Solder wick is available in rolls for many de-soldering applications. With the solder removed from the connection, the lead can be bent away from the terminal pad and the component easily withdrawn.

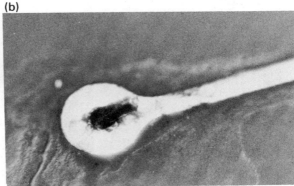

FIGURE 15.12. De-soldering technique employing solder wick. (a) Position of solder wick and soldering tip. (b) De-soldered terminal pad.

The de-soldering bulb and solder sucker, shown in Fig. 15.13, can also be used for removing excess solder or for de-soldering component leads. Both tools have hollow Teflon tips with high heat resistance and will not scratch or mar delicate pc board conductors. The de-soldering bulb, shown in Fig. 15.13a, is employed by depressing the bulb and then placing the hollow tip alongside the soldering iron tip on the joint. As the solder begins to melt, the bulb pressure is released. The liquid solder is drawn up into the bulb by the suction. Both tools are then removed from the connection for inspection. If solder remains, the process must be repeated. The solder-sucker shown in Fig. 15.13b is employed in basically the same manner except that the suction is produced by a spring-loaded piston. The piston handle is first pushed downward. The handle is then rotated to engage the release pin. As the solder begins to melt, the pin is disengaged and the solder is drawn up through the hollow tip as the piston snaps upward inside the tubular handle. Both tools are easily disassembled to remove the accumulated solder. Some manufacturers provide soldering irons and de-soldering bulbs as an integral unit so that one hand may be used to de-solder while the other is used to simultaneously extract the component. This tool is shown in Fig. 15.13c. These de-soldering tools do not generally remove sufficient solder on the first attempt to allow components to be removed, thus rendering the solder wick the preferable method.

For removing ICs, the previously discussed de-soldering aids would have to be used on each lead individually, making it a time-consuming process. De-soldering *tips* specifically designed to simultaneously heat all the leads for removing devices quickly and easily are available for this

FIGURE 15.13. De-soldering equipment. (a) De-soldering bulb. Courtesy of Ungar, Division of Eldon Industries, Inc. (b) Solder sucker. (c) Soldering iron with de-soldering bulb attachment. Courtesy of Enterprise Development Corporation.

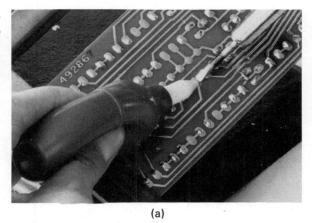

(a)

(b)

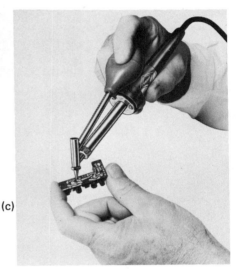

(c)

purpose. De-soldering iron tips for TO-5-style cases and dual-in-line configuration ICs are shown in Fig. 15.14.

The de-soldering tip is placed in contact with all the device leads simultaneously. As the leads are de-soldered from the foil side of the board, the device is removed from the component side with an extraction tool such as that shown in Fig. 15.14b and c. With this tool, the device is lifted away as the solder is melted. The remaining solder present on the terminal pad may obstruct the access holes and can be removed with any of the de-soldering aids discussed. The recommended type of extraction tool is one whose metal clamps grip the leads, thus providing some degree of heat sinking. It must also be mentioned that lead heat sinking is just as important when de-soldering as it is during soldering to avoid damage.

(a)

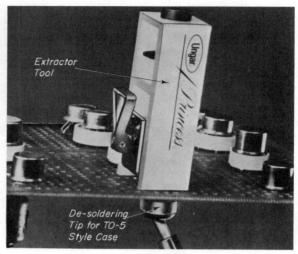

(b)

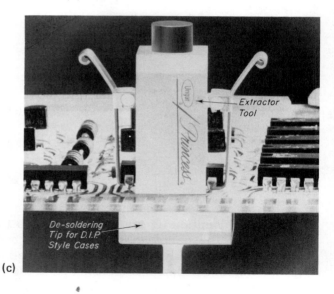

(c)

FIGURE 15.14. Device de-soldering equipment. (a) De-soldering tips for TO-5 and DIP style cases. (b) De-soldering TO-5 style devices. (c) De-soldering DIP style devices. Courtesy of Ungar, Division of Eldon Industries, Inc.

The common methods of soldering pc boards in mass production are *dip* and *wave* soldering. Each of these methods will be discussed here.

Dip soldering completely solders all connections and exposed copper conductors of the pc board in one operation. This process requires (1) *solder pot*, (2) *temperature control unit*, (3) *bar solder*, (4) *dross removal tool*, (5) *board holder*, and (6) *flux applicator*. After the flux is applied to the conductor side, the board is floated on the surface of molten solder. An even coating of solder will be deposited onto the entire surface of the conductor pattern. Although this process appears basically simple, the preparation and implementation are somewhat involved.

The preparation of the surfaces to be soldered is critical if quality results are to be realized. Absolute cleanliness is essential. The use of solder masks presents the fewest problems in dip soldering since the exposed areas to be soldered are small and the possibility of *solder bridging* is eliminated. (Solder bridging is the undesired formation of solder *between* conductor paths.) Flux is applied to the surface just prior to soldering even if the conductor pattern has been tinned or protected with chemicals such as Sealbrite.

Flux may be applied, as previously mentioned, by either *dipping*, *brushing*, *sponging*, or *spraying*. The object is to apply a uniform layer of flux over the entire surface of the board. This layer should be kept to a minimum thickness to promote improved solderability. If excessive flux is applied, it will not completely decompose at temperatures below those of molten solder, resulting in the formation of gas pockets between the board and the surface of the solder. These gas pockets will cause an uneven wetting of solder. Because of these limitations, *sponging* is the most effective way of applying flux onto the board (Fig. 15.15). When the flux has been applied, the board must be soldered immediately.

FIGURE 15.15. Application of flux onto conductor pattern.

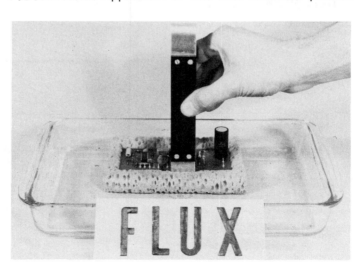

The specified tin–lead ratio for dip soldering is normally 60/40, which melts at approximately 370°F (188°C). Placing a pc board onto the surface of the molten solder will lower the solder's temperature. Consequently, a *guard temperature* must be included to allow for this cooling. A guard temperature of between 50 and 70°F (10 to 21°C) is normally sufficient. This means that the solder pot temperature should be maintained between 420 and 440°F (216 to 227°C) and accurately controlled. Although higher temperatures will hasten soldering, problems such as board warping and strain effects on the foil bond will be encountered. The exact temperature for optimum results for any particular system is best determined by trial and inspection.

As the solder pot is operated at elevated temperatures (above 370°F or 188°C), an oxidizing scum, referred to as *dross*, will form on the surface of the molten solder. If not removed, this dross will adhere to the areas to be soldered, resulting in cold solder joints. Just prior to floating the board onto the solder, the dross must be removed. This can be done by skimming the surface with a piece of aluminum mounted to an insulated handle (Fig. 15.16a).

FIGURE 15.16. Sequential method for dip soldering. (a) Dross removal. (b) Initial board entry angle. (c) Board is moved about solder surface. (d) Soldered conductor pattern.

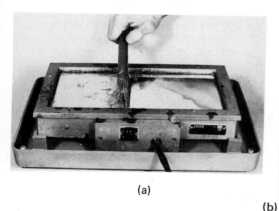

(a)

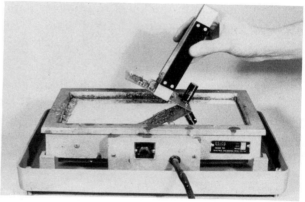

(b)

(c)

(d)

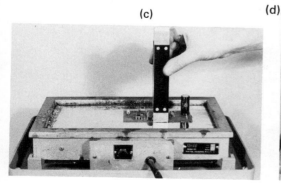

The board is placed in the pot and held firmly along its edges with tongs equipped with insulated handles. First one end of the board is lowered in direct contact with the solder at an angle of approximately 45 degrees with the surface. This procedure is shown in Fig. 15.16b. The board is then slowly tilted downward until the entire conductor pattern surface is contacting the molten solder. This technique of slowly tilting the board will force any gases generated during the flux decomposition period to escape from between the solder and board surface. With the board in complete contact with the surface of the solder, it is moved about the pot (Fig. 15.16c) to allow uncontaminated solder to continually contact the conductor pattern. The soldering operation should take from 4 to 7 seconds, depending on the solder temperature and type and amount of flux used.

After the board has been allowed to remain in contact with the solder for the prescribed amount of time, it is slowly raised to a 45-degree angle before it is completely removed from the pot. This technique is intended to reduce the possibility of the formation of *solder peaks* (points of solder protruding from the conductor pattern). Visual inspection should be made to ensure that no solder bridges between closely spaced conductors have formed. Although solder cannot alloy with the insulating material, bridges may still form between closely spaced conductors. The resulting appearance should be a completely wetted conductor surface with a smooth, uniform coating of solder. Once the flux residues are removed, the board is completed. A properly soldered single-sided pc board is shown in Fig. 15.16d.

Dip soldering, owing to its reasonable setup costs, can be adapted equally well to prototype applications. However, the results are not uniformly consistent or predictable. Often, hand-soldering touch-up work may be necessary to achieve quality results.

Extreme caution must be exercised when dip soldering. This process requires continual safety precautions because of the ever-present danger inherent when working with molten metal. Protective glasses, gloves, and an apron should be worn to guard against solder splatter. Although a pronounced "sizzling" sound will be heard when the cool surface of the board first contacts the solder surface, this is normal and should not cause concern. This effect is the result of the rapid volatilization of the flux. To prevent serious personal injury from a violent solder eruption, all liquids should be kept away from the solder pot.

A much more elaborate system of production line pc board soldering is the process of wave soldering. This method eliminates many of the inconsistencies resulting from manual dip soldering. In wave soldering, a standing wave of solder is mechanically generated, which rises $\frac{1}{2}$ to 1 inch above the surface of the solder. The boards to be soldered are secured to holders on a conveyor belt that moves them through a series of operations at the rate of approximately 1 inch per second. In the first stage of this process, flux is applied with a foam flux applicator. Air pressure passing through a sponge soaked in liquid flux generates flux bubbles, which contact the surface of the boards as they move along the conveyor belt. Some systems next incorporate an infrared preheater to gradually increase the surface temperature of the boards so that the flux solvents will vaporize just prior to soldering. Soldering is accomplished with the boards moving along at a slightly elevated angle as they pass over the solder wave. Dross is no problem in this process since it is not carried up on the solder wave. However, a great deal of dross forms on the solder in the pot. For this reason, some manufacturers add high-temperature oil coverings over the solder in the pot, which excludes the atmosphere and minimizes dross build-up. The oil, in addition to the dross, also does not carry up on the wave. Wave soldering is completely automated, so once the proper conveyor speeds, flux application, preheater controls, and solder temperature are set, the results obtained should remain consistent for similar pc board assemblies.

When the pc boards have been assembled and soldered, they may be tested for operational performance and modifications may be made if necessary.

PROBLEM SET

15.1 Lay out, shear, and bend an aluminum tray for holding a soldering iron tip cleaning sponge. The tray is to have inside dimensions of ¾ inch by 3½ inches by 3½ inches. Tabs are not required on the inside corners. These corners are sealed in Problem 18.3 to retain excess moisture. Break all sharp edges with the appropriate file.

15.2 Dip or hand solder the printed circuit device test jigs that were designed and fabricated in Problems 11.2, 13.3, and 14.2. The entire conductor pattern is to be tinned and all device pin locations numerically identified with dry transfers. A stand for mounting the test jigs is designed and constructed in Problem 18.4.

15.3 Solder the assembled pc board for the seven-fixed-interval UJT timer designed and fabricated in Problems 11.3, 13.5, and 14.3. Chassis mounting and interconnection wiring are done in Problems 18.5 and 19.5, respectively.

15.4 Solder the pc board for the series feedback regulator designed and constructed in Problems 11.4, 13.6, and 14.4.

15.5 Solder the pc board for the tachometer circuit designed and constructed in Problems 11.5, 13.7, and 14.5. Circuit board and meter installation are accomplished in Problem 18.6. Wiring is done in Problem 19.6.

16

Double-Sided Printed Circuit Artwork Design and Layout

16.0 INTRODUCTION

As electronic systems have become more complex, the requirements of higher-density packaging have partially been met with the widespread use of double-sided printed circuit boards.

Double-sided board fabrication requires the preparation of at least two pad masters, one for each side of the pc board. This, of course, adds to the time and costs of producing a double-sided board. Precise alignment of all terminal pads between both sides of the board is required for each lead access hole. This alignment problem is not a consideration in the preparation of single-sided boards since there is only one conductor pattern on one side of the board. In addition, double-sided boards require more steps in fabrication processing, which increases their cost as compared to single-sided boards.

Even though they are more costly and time-consuming to fabricate, double-sided pc boards allow for more components to be packaged in less space than single-sided boards and thus meet the demands for high-density electronic packaging.

16.1 BASIC DOUBLE-SIDED PRINTED CIRCUIT BOARDS

A double-sided pc board differs from a single-sided board in that both sides are processed with conductor patterns and terminal pad areas. The arrangement of conductor paths and pairs of terminal pads to form a double-sided pc board is illustrated in Fig. 16.1. As with single-sided boards, all components are mounted only on one side. This side is called the *component side* or *top*. Component leads pass through lead access holes through pairs of terminal pads on either side of the board. The component lead is *clinched* or bent over onto the terminal pad that is on the *conductor side* or *bottom* of the board. Note in Fig. 16.1 that the lead access hole drilled for the component lead also passes through a terminal pad on the component side of the board. Unlike single-sided layout, conductor paths may extend away from terminal pads on both sides of the board. Electrical connection between conductors on both sides of the board is made only when the component lead is soldered to both the conductor-side terminal pad and the component-side terminal pad. The component lead, then, is used as the jumper wire to make the electrical interconnection between both sides of the board. The conductor side of the board can be considered a conventional single-sided layout. The conductors and terminal pads on the component side serve to extend conductor path routing. With both sides of the board used to route electrical interconnections, less space is required for conductor paths on the bottom side, thus allowing for greater component density.

One of the considerations involved in the fabrication of a double-sided pc board is the component bodies that must rest on the conductor paths which are on the top of the board. Most component and device cases, such as resistors, capacitors, plastic-cased transistors, and integrated circuits, are electrically insulated from their leads. The bodies of these components may safely contact a conductor path without electrical interference. However, some metal-cased integrated circuits and transistors have one lead attached to the case. These packages must not be allowed to come into contact with any of the conductor paths on the top side of the pc board. Electrical contact is avoided by either mounting these components slightly above the board surface to create an insulating air gap or with the use of one of a variety of plastic insulating spacers, some of which are shown in Fig. 14.25.

16.2 DOUBLE-SIDED COMPONENT LAYOUTS

The generation of a component and conductor pattern layout for a double-sided pc board is initiated in a similar manner to that for a single-sided board, that is, with the selection of a suitable grid system and reduction scale. The selection criteria for a double-sided pc board is the same for a single-sided pc board, as detailed in Sec. 11.1. These selection criteria

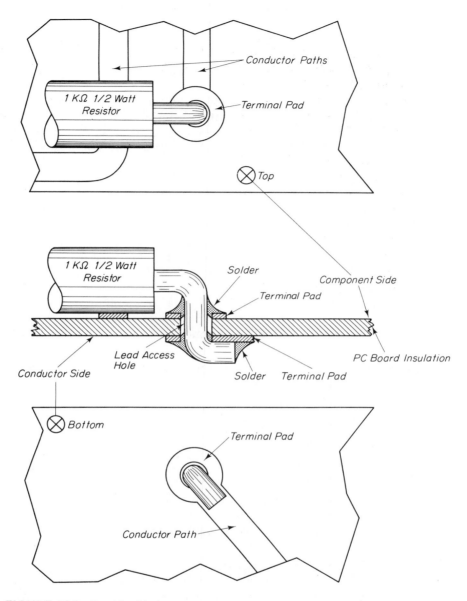

FIGURE 16.1. Double-sided pc board conductor path and terminal pad arrangement.

will not be repeated here. However, it should be emphasized again that a 4:1 scale layout on a 0.100-inch grid system should be used whenever possible.

For purposes of developing optimum component placement, it must be decided which of the three approaches — the schematic, central, or peripheral viewpoints — shall be employed. These component placement approaches are detailed in Sec. 11.2. One advantage of double-sided pc board layout is that it allows more flexibility in path routing because of the use of both sides of the board for this purpose. Thus, some of the major path-routing difficulties encountered in single-sided layout are largely overcome.

When drawing the composite view of component and conductor pattern layouts, lines representing conductor paths on the *opposite* side of the board may "cross" each other. Because the conductor routing paths on both sides of the board are drawn as a composite view, a system to avoid confusion and misinterpretation must be utilized. A color-coded system for this purpose has been adopted by much of the industry. When viewing the pc board from the top (component side), the components are first seen. Below the components, again on the *top* side of the board, are the conductor paths and terminal pads unique to that side. Still viewing the pc board from the top, the conductor pattern on the bottom of the pc board would be "seen" as though one were looking through the insulating material. While laying out a double-sided pc board, each side must be distinguishable on a single sheet of grid paper. Adopting the following color-coded system will, to a large extent, eliminate confusion:

1. The outlines of all component bodies are drawn in *black* lead pencil.

2. All items that are common to *both* sides of the board are also drawn in *black* lead pencil. These items are typically:
   a. Donuts (or teardrops) ⎫
   b. Special device outlines ⎬ requiring drill holes
   c. Mounting holes ⎭
   d. Corner brackets
   e. Reduction scale

3. All conductor paths for the top side should be drawn in *red* pencil. All literal and numerical information and component keying should be done in *red* pencil and *right-reading* if it is to appear on this side of the board.

4. All conductor paths for the bottom side should be drawn in *blue* lead pencil. All literal and numerical information and component keying should be done in *blue* pencil and *reverse-reading* if it is to appear on the bottom side of the board.

The use of a color system allows three levels of views to be drawn onto a single sheet of grid system without confusion. For example, a blue conductor trace crossing a red conductor trace does not produce an electrical connection, since each color represents conductor paths on opposite sides of the pc board which are insulated by the board material. The color-coding technique outlined is illustrated in Fig. 16.2. Because of the absence of colored illustrations in this book, blue traces are shown as

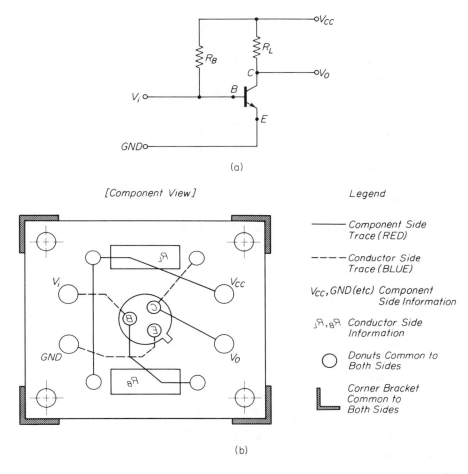

FIGURE 16.2. Double-sided path layout coding. (a) Basic transistor amplifier circuit. (b) Component layout showing composite of top and bottom views.

dashed lines and the red traces as solid lines to distinguish between paths appearing on opposite sides of the board. Figure 16.2 utilizes a basic transistor amplifier circuit to illustrate this double-sided layout technique.

16.3 DOUBLE-SIDED TAPED ARTWORK MASTER CONSTRUCTION TECHNIQUES

Before beginning the discussion of the preparation of the taped artwork master for double-sided pc boards, several important points must be emphasized. The objective of double-sided layout is to produce a taped artwork master for each side of the board with perfect registration of all donuts and drill locations on *both* sides. This is necessary to prevent drilling incorrectly through a pad on one side. Where there are few drill holes and they are small relative to the terminal pad diameter, this perfect registration problem is less severe when compared to the more complex designs required by high-density pc board packaging. High-density packaging demands that all the available space on a pc board be used as efficiently as possible. In addition, the IC patterns used extensively have very small terminal pad areas, owing to the fact that their on-center spacing is typically 0.1 inch. It can be seen that achieving perfect *top* to *bottom* registration requires more care as the packaging becomes more dense.

There are three popular methods used in the preparation of a double-sided taped artwork master. These methods are:

1. Single-acetate with color system (Fig. 16.3).
2. Two-acetate system (Fig. 16.4).
3. Three-acetate system or donut master technique (Fig. 16.5).

Each of these systems will be discussed together with their advantages and disadvantages.

The single-acetate color system, shown in Fig. 16.3, requires only one layer of acetate to which both sides of the double-sided pattern will be constructed. A single sheet of acetate is taped over the component and conductor pattern layout drawing onto the drafting board. All donuts and special layout patterns, such as those used for ICs and transistors, are laid out in black on one side of the acetate sheet. The conductor pattern on the component side is constructed with the use of red polyvinyl tape which is rigid and transparent. This tape will not bend or stretch and therefore changes in path direction cannot be made by contouring the tape. Changes in direction are made with the use of a variety of special precut angled corners. By splicing between the tape and angled corners, the desired changes in path direction may be achieved. An alternative method of changing path directions without the use of precut elbows is to cut and splice the tapes by overlapping to the desired angle. It can be seen that layout with the colored tapes is more time-consuming than artwork constructed of flexible black crepe tape which can easily be contoured. In addition, the cost of the colored tapes is higher than that for the black tape.

The conductor side is next constructed using blue tape. If this construction is to be made on the same side of the acetate as the component side, a problem will result where the blue tape must necessarily overlap the red tape. In the photographic reduction process, the areas where blue overlaps red results in a reduction of path width of the pattern in that area. This becomes more of a problem when extremely narrow tape is used. To overcome this problem, the blue taping should be done on the *opposite* side of the acetate sheet. However, this creates other concerns in construction. In order to place the blue tape on the opposite side, the acetate sheet must be removed from the component layout drawing and reversed. This creates a problem of alignment, since the component layout drawings were made from the component view. To overcome this alignment problem, a mirror image of the component layout drawing is required for correct orientation.

FIGURE 16.3. Single acetate red/blue color system. (a) Taped artwork for Figure 16.2, transistor amplifer. (b) Side view of acetate sheet with acceptable and unacceptable taping techniques.

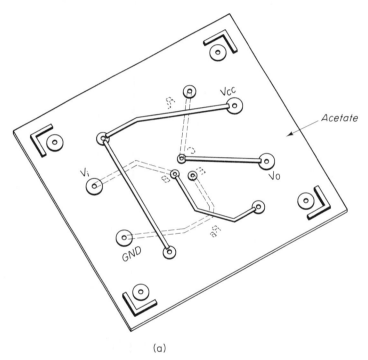

(a)

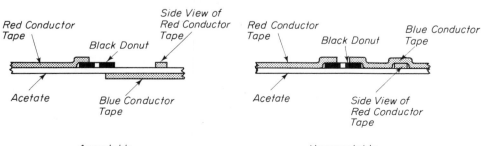

Acceptable

Unacceptable

(b)

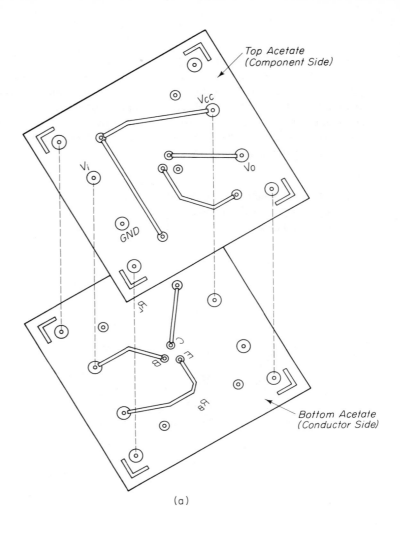

(a)

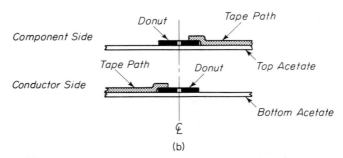

(b)

FIGURE 16.4. Two-acetate taping system for double-sided layout. (a) Separate top and bottom taped artworks for the transistor amplifier. (b) Side view of taped artworks showing correct donut alignment.

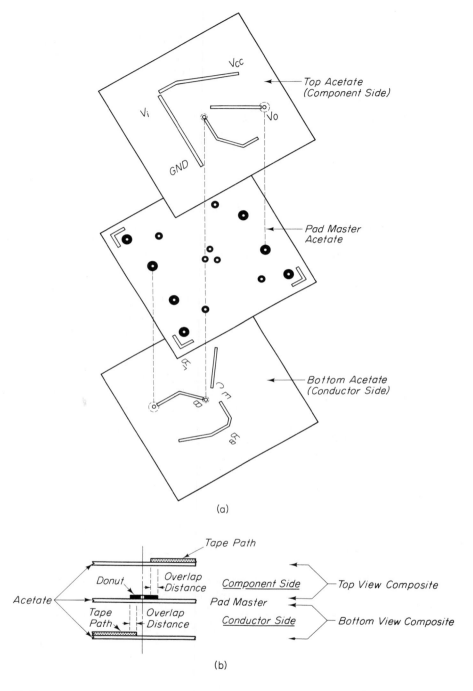

FIGURE 16.5. Three-acetate taping system for double-sided layout. (a) Separate top, pad master and bottom tapings for the transistor amplifier. (b) Side view showing correct tape overlapping with donut and path alignments.

Another problem that arises with the use of the single-acetate colored tape system is in the photographic process. A special blue filter must be attached to the camera, resulting in all blue lines being filtered out. Only red and black patterns will appear on the negative produced. Orthochromatic film, which can be processed in red safe light, is used for this purpose. To produce a negative showing the blue traces and black donuts, a red filter replaces the blue filter. With the use of panchromatic film, the red images are dropped and only the blue and black patterns appear on the negative. The use of panchromatic film greatly complicates the developing process, since this film must be processed in absolute darkness. In the absence of automated film processing equipment, developing time becomes a matter of many trials and errors. To overcome this, some graphic art tape manufacturers provide an amber/magenta color system to replace the red/blue system. This allows the total use of orthochromatic film in conjunction with a suitable filter system.

Through the use of the single-acetate color system technique, perfect alignment of artwork and resulting positive or negative photo tools can be realized since only one set of donuts is used for both sides of the board. However, the added costs of materials and the disadvantages discussed above render this technique the least cost-effective of the three listed.

The two-acetate system, shown in Fig. 16.4, makes use of black crepe tape and black donut and device patterns. As a result, no special lens filters or special film types are required for photographic processing. The fact that the two-acetate system is an extension of the single-acetate system will become apparent in the following discussion.

A single sheet of acetate is taped over the component and conductor pattern layout drawing onto the drafting board. All donuts and special terminal pad designs are first laid out. If this sheet is to be used to produce the artwork master for the conductor side of the pc board, all connections drawn in blue pencil would be routed using black tape with appropriate overlapping onto donuts and device pads. Finally, corner brackets, reduction scale, component keying, and *reverse-reading* literal and numerical information is added to complete the conductor side layout.

To produce the component-side artwork master, a second acetate sheet is taped over the completed conductor-side layout. The most critical phase of this process is the registration between the two artwork masters. All donuts and special device patterns must again be installed on this second acetate sheet with extreme care that they are aligned with those on the conductor-side layout. The placement of the donuts can be aligned easily with a reasonable degree of care. The alignment of the IC patterns, however, is much more difficult because there could be as many as 40 centers (40-pin DIP patterns) which must be simultaneously positioned exactly over the same style pattern on the conductor-side layout.

The more special IC patterns involved in the layout, the greater possibility of pattern misalignment.

The component artwork master is completed with the installation of the conductor-path taping for all the red paths appearing on the component layout drawing and with the addition of corner brackets, reduction scale, and numerical and literal information in a *right-reading* orientation.

This system is highly prone to misalignment, especially if extreme care is not exercised in firmly taping each configuration into its proper position. The two-acetate system is suitable for relatively low density packaging, in which few critical alignment situations are encountered.

A preferred method of preparing double-sided pc board artwork masters is the three-acetate or pad master system illustrated in Fig. 16.5. The pad master, which includes all donuts and special device patterns, corner brackets, mounting holes, and the reduction scale, is produced first. Since all the items that are common to both sides of the pc board are installed onto the pad master, they will not appear on the other acetate sheets, resulting in no expensive duplication of materials, less time required for layout, and no alignment problems.

On a second sheet of acetate, items unique to the conductor side are laid out. These include conductor paths drawn in blue pencil on the layout drawing and numerical information. The pad master aligned with the conductor-side taped master complete the composite artwork for the bottom side of the pc board.

On a third sheet of acetate, all items unique to the component side are laid out. Taping is provided for all red path traces on the component layout drawing in addition to the numerical and literal information. This layout, aligned with the pad master, completes the composite artwork for the top side of the pc board. Because the same pad master is used to produce *both* sides of the pc board, perfect alignment between both sides is easily achieved.

16.4 PAD MASTER FOR THE PREAMPLIFIER CIRCUIT

The pad master for the preamplifier circuit will be constructed using the three-acetate system. A sheet of acetate is positioned over the component layout drawing which has been taped to a drafting board. The acetate sheet should be 1 inch longer and wider than the component layout drawing. The acetate sheet is secured firmly in position by placing masking tape at each of the four corners. All device outlines, such as transistors and integrated circuit patterns, are first positioned in accordance with the underlying component layout drawing. The donuts are then positioned using the underlying drawing for correct placement. (Refer to Sec. 11.5 for techniques of device outline pattern and donut placement.) The pad master must include all device outlines and all donuts that are contained

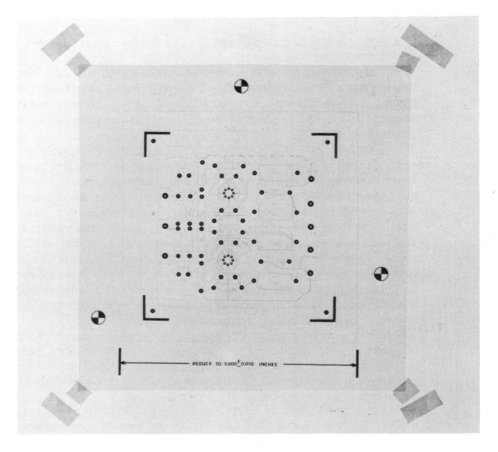

FIGURE 16.6. Completed pad master with the underlying component layout drawing for the preamplifier circuit.

on the component layout drawing in addition to four border delineation marks (corner brackets), three grid locators (targets) aligned with those on the drawing, and an accurate reduction scale. A completed pad master for the preamplifier circuit is shown in Fig. 16.6. Note that only items common to both sides appear on the pad master. No items unique to any one side (conductor path routing, component keying, literal and numerical information, etc.) are found on the pad master.

16.5 COMPONENT SIDE ARTWORK MASTER FOR THE PREAMPLIFIER CIRCUIT

To produce the artwork master for the component side of the preamplifier circuit, another sheet of acetate is placed directly over the pad master just completed. This sheet should be slightly larger than the pad master so that the masking tape used to secure it in place will not interfere with the tape holding the pad master. This will enable the component-side layout to be removed without upsetting the underlying pad master. On this component-side acetate sheet, all conductor path routing to complete this side is laid out. (Reference should be made to Sec. 11.5 for techniques for taping conductor paths.) Only those conductor-path traces drawn in red pencil on the component layout drawing will be taped. The tape should overlap the donuts and *not* butt them. No line drawn with blue pencil should be taped onto this artwork master. Upon completion, the taping of each path should be checked for accuracy in tape routing.

The component-side layout is a view of the pc board as it appears when viewed directly. Therefore, all literal and numerical information should appear *right-reading* on the acetate sheet. The finished layout of the component side of the amplifier pc board, together with the underlying pad master, is shown in Fig. 16.7. Component keying and labeling may be included on this side since their application to the acetate is simpler due to the fact that they appear *right-reading*. Before the component-side layout is removed from the pad master, three grid locators should be accurately positioned directly over those on the pad master. These will later be used for registration and alignment after the artworks have been removed from the drafting board and are to be photographed.

FIGURE 16.7. Completed component side taping with underlying pad master for the preamplifier circuit.

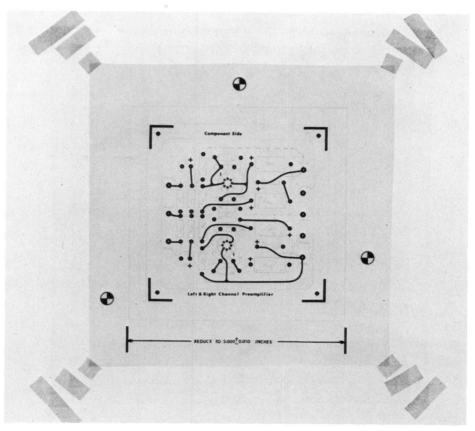

16.6 CONDUCTOR-SIDE ARTWORK MASTER
FOR THE PREAMPLIFIER CIRCUIT

After the component-side layout has been completed, it is removed and temporarily set aside being careful not to disturb the underlying pad master. Another sheet of acetate is positioned over the pad master to construct the conductor-side layout. This is done with the same procedure used for construction of the component-side layout except only those traces drawn in blue pencil on the component layout drawing are to be taped. When completed and checked, lettering is applied in a *reverse-reading* orientation, as is done for single-sided board layouts. Reverse-reading dry transfers are available *or* the acetate sheet may be removed from the pad master and right-reading lettering applied to the *back* of the sheet. The conductor-side layout for the preamplifier pc board, together with the underlying pad master, is shown in Fig. 16.8.

The pad master with the component-side layout and the conductor-side layout are shown in Fig. 16.9.

FIGURE 16.8. Completed conductor side taping with underlying pad master for the preamplifier circuit.

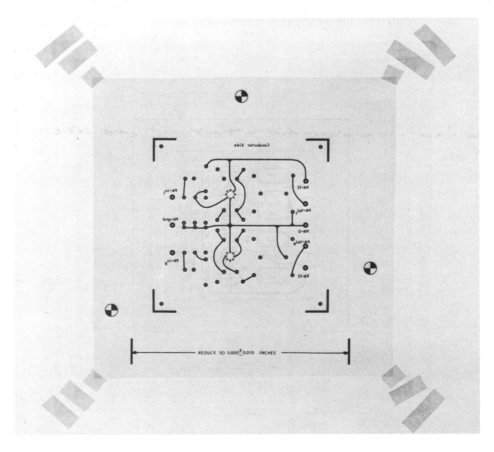

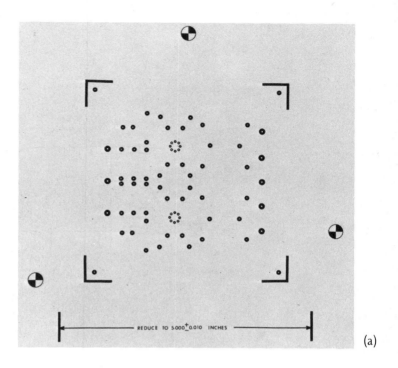

(a)

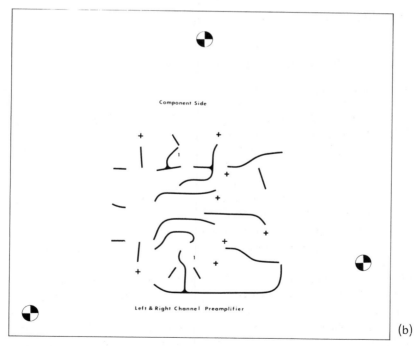

(b)

FIGURE 16.9. Individual completed layouts. (a) Pad master. (b) Component side. (c) Conductor side.

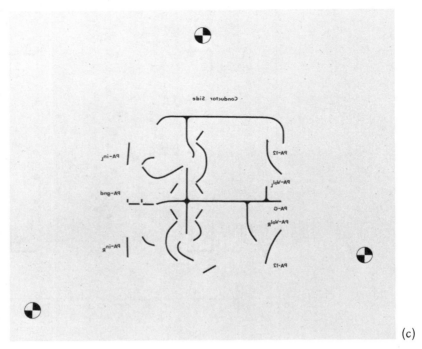

FIGURE 16.9. (cont.)

16.7 LAYOUT CONSIDERATIONS FOR PLATING

Section 17.12 describes the process of plating 60/40 tin–lead alloy onto the conductor path areas of both sides of the pc board. Many of the problems associated with an improper balance of plated metals can be attributed to the component layout drawing and the taped artwork masters. If a double-sided pc board is to be electroplated, it must meet the following specifications:

1. The total conductor pattern areas to be plated on each side of the board must be as close to equal as possible.
2. The conductor pattern on a given side of the board must be uniformly and evenly distributed throughout the board.

Strict adherence to the specifications above is essential if the correct 60/40 tin–lead alloy is to be deposited during the plating process. An imbalance of conductor pattern areas between the two sides results in the side with the less area to be lead-rich-plated and the side with more area to be tin-rich-plated. If the ratio of alloys plated on either side is significantly different from the optimum 60/40 ratio, proper soldering of the board will not be realized.

If the conductor pattern is not uniformly and evenly distributed across either side of the board, the area with the heavy concentration of copper conductors will be low in tin-content plating, while the area with the light concentration of copper conductors will be plated with an alloy having high tin content. This will again result in soldering difficulties.

One final provision that must be included in the conductor pattern layout phase for boards that will be electroplated is an area of solid copper just outside the corner brackets of the pc board. This area is called the *thief* or *robber* area and its purpose is to direct some of the high current density at the edge of the board toward itself during the plating process. This will result in a more consistent 60/40 alloy on both sides of the pc board. The thief area is added to the preamplifier pad master just outside the corner brackets with the use of appropriate-size tape. The processed negative should have a thief area between $\frac{1}{2}$ and 1 inch wide after reduction. The inclusion of this taped thief area on the pad master results in its appearance on *both* sides of the board.

The pad master for the preamplifier circuit with taped thief area is shown in Fig. 16.10.

FIGURE 16.10. The completed pad master with taped thief area.

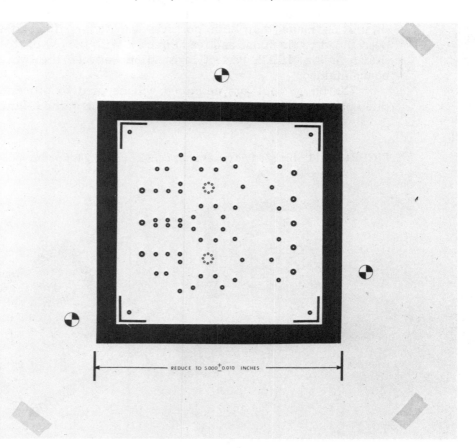

REDUCE TO 5.000$^{+}_{-}$0.010 INCHES

16.8 PHOTO TOOL FOR DOUBLE-SIDED PRINTED CIRCUIT BOARDS

For photographic reduction, the component-side layout is placed over the pad master on the copyboard and photographed. The component-side layout is removed and the conductor-side layout is placed over the pad master and another exposure is taken. Perfect registration is maintained with this system since the pad master is the same for both sides of the pc board.

The taped artwork masters are photographically reduced into two 1:1 scale negatives as described in Chapter 12. For processing a print–plate–etch pc board, two 1:1 scale *positives* are processed by contact printing, also discussed in Chapter 12. The finished positives after contact printing are shown in Fig. 16.11. To maintain front-to-back alignment during processing, a photo tool must be constructed with both positives. To do this, a scrap piece of pc board stock (the same thickness as that to be processed) is first cut 1 inch wide and slightly longer than the positives. One positive is first taped to one side of the scrap stock with masking tape to form a hinge allowing the corner brackets to extend below the hinge. This setup is then turned over onto a light table and the second positive is aligned with the first with the aid of the registration targets. Once aligned, the second positive is also taped to the hinge. A check should be made that both positives are *right-reading* after they have been taped to the hinge and read directly. With the use of this photo tool, shown in Fig. 16.12, perfect registration between the two positives can be maintained.

The photo tool just described will be used to process the double-sided pc board by the print–plate–etch technique in the following chapter.

FIGURE 16.11. Finished positives after contact printing for photo tool.

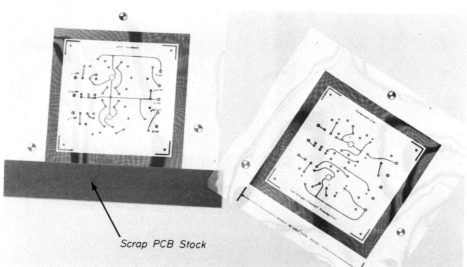

Scrap PCB Stock

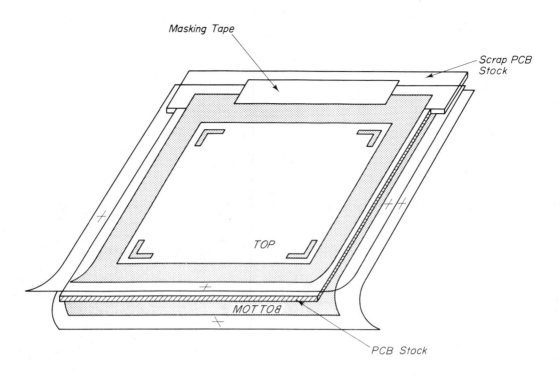

FIGURE 16.12. Photo tool set-up (conductor patterns not shown).

17

Double-Sided Printed Circuit Board Processing: Print-Plate-Etch Technique

17.0 INTRODUCTION

The fabrication techniques for single-sided printed circuit boards discussed in Chapter 13 employed a method called print-and-etch and is satisfactory for prototype work. This technique has been used for over 30 years and is the simplest of all methods employed in the manufacture of pc boards. To quickly review the print-and-etch method, you will recall that a light-sensitive liquid resist is applied to a clean, dry copper surface. This light-sensitive surface is exposed using a negative working tool of the desired conductor pattern. This image is then developed. The resist remaining on the pc board (the desired conductor pattern) prevented removal of the underlying copper during the etching process. After etching, the board is stripped of the resist, cleaned, and the exposed copper pattern protected by either an immersion tin or protective sealer.

Even though satisfactory boards are produced by the print-and-etch technique, there are some major drawbacks which must be considered. First, the uniform coating of liquid resist is extremely difficult to control resulting in exposure times becoming a matter of *art* rather than a *science*. A thin coating of resist, coupled with allowing impurities in the air to contact the resist during application, results in the necessity of excessive touch-up of pinholes. Second, the solvent-type developer required to process the liquid resist is becoming increasingly more difficult to properly dispose of because of environmental considerations. Finally, and perhaps the most serious drawback, the protective coatings (immersion tin or sealers) have a relatively short shelf life, resulting in the necessity to assemble and solder components as soon as possible after board fabrication.

Since the early 1970s, the printed circuit board industry has largely abandoned wet-film resist for the newer dry-film photo-resists. These dry-film resists overcome many of the above-mentioned problems encountered with the use of wet resists. Their thickness is more closely controlled. Uniform thickness results in more predictable exposure time and little, if any, pinhole touch-up is required. Dry-film resists are also available that can be developed and stripped in aqueous-based chemicals. These resists can be applied simultaneously to both sides of a double-sided pc board, lending themselves ideally to this application. Dry-film resists can more easily withstand almost any electroplating bath solution and, as a result, allow the application of the more durable electroplating of 60/40 tin–lead solder to improve the shelf life of the pc board.

The topics presented in this chapter are similar in sequence to those discussed in Chapter 13 but will be directed to the fabrication of double-sided pc boards employing the print–plate–etch technique of fabrication. The use of dry-film resists and solder plating techniques will be described in detail. It should be mentioned in the onset that although the sequential steps involved in this method are more intricate and the required equipment more expensive, the pc board fabricated is of the highest quality obtainable. Although this chapter discusses double-sided pc board fabrication exclusively, the processes described are easily adaptable to single-sided boards.

17.1 PROCESSES INVOLVED IN DOUBLE-SIDED FABRICATION

The basic steps to process a double-sided pc board by the print–plate–etch method are shown in Fig. 17.1. This flow chart shows the sequence of processes required. Subsequent sections in this chapter will expand on each step of each topic in this flow chart.

Comparison of the flow chart shown in Fig. 17.1 with the flow charts in Chapter 13 reveals many similarities. There are, however, many apparent differences between the two methods of pc board fabrication. The major differences are: (1) The bulk stock to be processed by the print–plate–etch method has copper on both sides, not only one side. Both sides must therefore be cleaned before dry-film resist is applied; (2) Because the pc board is double-sided, two positives are required. Positive photo tools at a 1:1 scale will be used. Developing a sensitized pc board after exposure through a positive photo tool will result in photo-resist being removed from all conductor paths and terminal pad areas. The result is that copper is exposed in all the pattern areas while the remaining resist covers all other areas; (3) The exposed conductor pattern

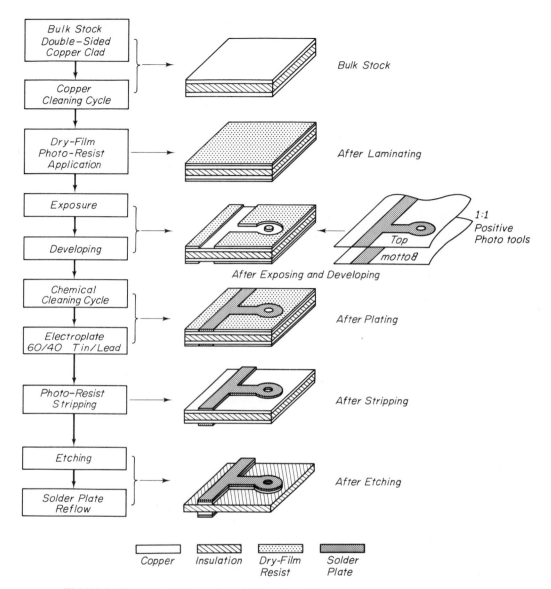

FIGURE 17.1. Basic flow chart with pictorial sequence of the print-plate-etch process for double-sided pc boards.

copper is then plated with a 60/40 tin-lead alloy. The resist is used as a plating "stop-off"; that is, the resist will prevent the plating from being deposited on all areas but the exposed conductor pattern. After the plating process, the resist is stripped from the board, exposing all of the copper areas to be removed during the etching process. The tin-lead plating over the conductor pattern will act as a stop-off to protect the underlying copper pattern from being etched. The etchant used will attack copper but not tin–lead alloy. Finally, the electroplated solder is "reflowed" or fused to the copper conductor pattern to complete the board fabrication process.

17.2 COPPER CLEANING FOR DRY-FILM APPLICATION

Optimum dry-film resist adhesion is accomplished only if the copper surface is properly cleaned. The cleaning cycle outlined in Fig. 17.2 will produce copper surfaces that are completely degreased, slightly acidic, and thoroughly dry. Careful attention to this cleaning cycle cannot be emphasized enough. Most problems that develop in subsequent processing steps can often be directly attributed to improper cleaning.

Cleaning is best accomplished in a sink, since cold tap water is required for rinsing. Care must be taken not to allow the pc board to be placed in the basin, however, because contaminants from the sink will redeposit themselves onto the copper surfaces. The stock should be either cupped in the hand or held by the edges. Fingers should not touch the copper surfaces since this will result in depositing body oils and greases onto the copper.

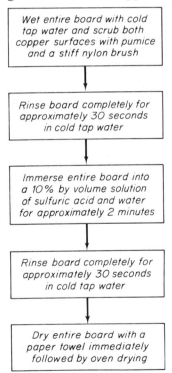

FIGURE 17.2. Mechanical and chemical cleaning procedure for bulk pcb stock prior to resist application.

Bulk stock sheared to size is first mechanically degreased with use of a nylon bristle brush and 3F (FFF) grade pumice. Pumice is a mild abrasive which will remove oxides, greases, oils, and particles from the copper surface. One side of the pc board is first flushed with cold tap water. The "beading" of the water that is observable on the copper surface is the result of surface contaminants that are present. A generous amount of pumice is sprinkled onto the copper surface and the entire surface is vigorously scrubbed with a wet nylon brush. Special attention should be paid to the edges of the pc board in this step of the cleaning process. The board and brush are rinsed in cold water and the scrubbing is repeated with more pumice until the copper surface is completely clean. This cleaning procedure is repeated on the other side of the pc board, being very careful to avoid touching the first side. After this mechanical degreasing process, the copper should be bright and shiny with no visible discoloration. The board is then rinsed thoroughly in cold tap water on both sides for at least 30 seconds to remove any pumice residue remaining on the copper surfaces. The copper is sufficiently degreased if, after rinsing, the remaining water sheets across the copper surfaces with no apparent dry areas (water breaks).

After rinsing, an acid dip is required to leave the copper surface in a slightly acid state to improve resist adhesion. The acid dip uses CP (chemically pure) or reagent-grade sulfuric acid. One part of acid is *slowly* added to 9 parts of cold tap water. This mixture will yield a 10% by volume acid dip solution. *Caution: Never add water to acid.* This causes violent eruptions to occur. The container used for the dip tank can be of Pyrex or an acid-resistant plastic such as polyethylene or polypropylene. Household plastic gloves should be worn whenever handling acid solutions. The pc board is dipped into the acid solution for approximately 2 minutes while gently rocking the container (Fig. 17.3). The pc board is then removed from the acid dip and completely rinsed on both sides for at least 30 seconds and then immediately dried with clean paper towels. The board is then placed vertically into a forced-air oven for approximately 15 minutes at a temperature of 150° F (66° C). This prebake cycle just prior to resist application accomplishes two functions: (1) the elevated temperature will remove any moisture from the board, which would adversely affect resist adhesion, and (2) by raising the temperature of the copper surfaces to 150° F, improved resist adhesion results. Timing and temperature in this heating phase is somewhat critical. Prolonged exposure to heat will result in surface oxides to be redeposited onto the copper surfaces. Although dry-film resist can be applied successfully over a *light* oxide film, this should be avoided. In addition, increasing the oven temperature to shorten the prebake cycle should also be avoided, since higher temperatures do not improve resist adhesion and can degrade the copper bond.

After this prebake cycle, the board is ready for dry-film resist application.

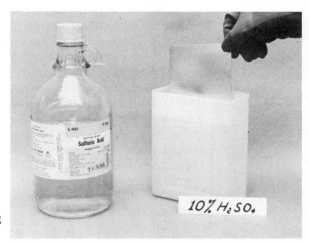

FIGURE 17.3. Copper surface being treated in acid bath.

17.3 DRY-FILM PHOTO-RESIST

The use of dry-film photo-resists in the manufacture of printed circuit boards is a relatively new process used extensively since the early 1970s. A typical composition of the dry-film resist is shown in Fig. 17.4. The resist is a light-sensitve photopolymer, dry in nature and available in thicknesses of from 0.5 to 2.5 mils. Available widths range from 3 to 25 inches and on continuous rolls of 100, 250, and 400 feet. The resist is sandwiched between two layers of plastic. One side is protected with a polyester cover sheet and the other side is covered with a polyethylene separator sheet. These plastic sheets are intended to protect the dry film from handling damage and also to provide a gastight seal against the oxygen in the air, which would quickly degrade the film.

As with any other light-sensitive resist, correct handling is crucial. Because these dry films are sensitive to light in the ultraviolet range, they should be stored and used only in an area illuminated with gold fluorescent tubes or lighting that does not emit ultraviolet light.

FIGURE 17.4. Composition of dry film resist.

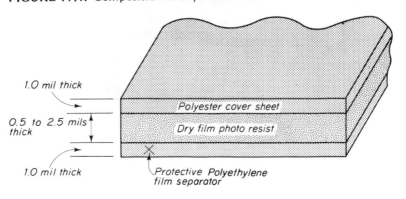

1.0 mil thick

0.5 to 2.5 mils thick

1.0 mil thick

Polyester cover sheet

Dry film photo resist

Protective Polyethylene film separator

Dry films are also sensitive to extreme heat and should therefore be stored at temperatures below 70°F (21°C).

Dry-film resists are categorized as either solvent- or aqueous-developing types. With the demands being placed on the industry for more and better environmental controls, the aqueous types are being used more extensively.

Although dry-film resists are available from several manufacturers, all are similar in makeup and also in the fabrication processes required. For our purposes, we have elected to use Laminar A, which is an aqueous-developing dry film, green in color, manufactured by the Dynachem Corporation of California. This resist will withstand the 60/40 tin–lead plating solution which will be used in the plating phase of the pc board described later in this chapter.

The optimum size of resist selected for our purposes is a thickness of 1 mil and a width of 8 inches, which will cover the bulk stock described in this chapter.

17.4 DRY-FILM LAMINATOR

Dry-film photo-resists are applied to the clean copper surfaces with a *laminator*. A typical laminator capable of processing boards up to 12 inches wide is shown in Fig. 17.5. This laminator applies the dry film simultaneously to both sides of the pc board. Dry-film resists are applied under pressure at elevated temperatures with a specific feed-through rate.

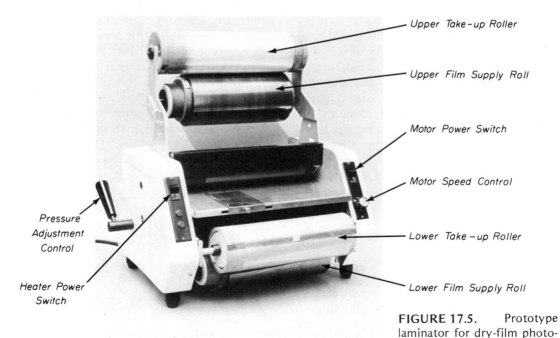

FIGURE 17.5. Prototype laminator for dry-film photo-resist application.

A typical laminator consists of (1) a pair of hard-rubber pressure rollers, (2) a pair of take-up rollers, (3) a pair of heat shoes, (4) a pair of temperature sensors for each shoe, (5) a variable-pressure adjustment control, and (6) a variable-motor-speed control to adjust the feed rate of the dry film.

Dry-film threading of the laminator for double-sided pc boards is shown in Fig. 17.6. Two rolls of dry film, one for each side of the pc board, are used. The protective polyethylene separator sheets are removed from both rolls of dry film and fed into both the upper and lower take-up rollers. The upper and lower rolls of dry film are fed between the hard-rubber rollers with exposed dry film in contact with each other. The opposite sides of both rolls of film, together with the polyester cover sheets, are in contact with both heat shoes. As the pc boards are fed between the rolls of dry film, the take-up rollers remove, on a continuous basis, the polyethylene separator sheets just prior to the film contacting the pc board. The film is heated as it passes over the heat shoes and is laminated onto the copper surfaces by the pressure of the hard-rubber rollers as the pc board moves through the laminator.

FIGURE 17.6. Typical threading diagram for laminator.

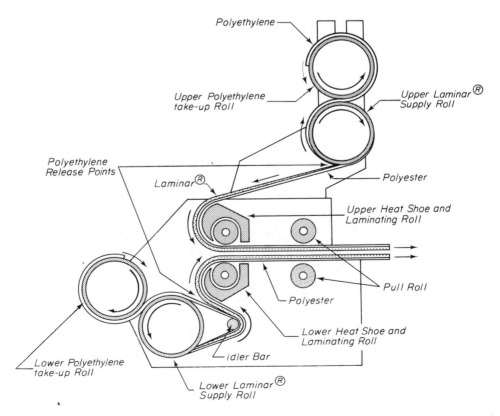

Successful application of Dynachem Laminar A to prepared copper surfaces requires the following laminator specifications: (1) heat shoe temperature set to 235 ±10° F, (2) laminating feed rate of approximately 5 linear feet per minute, and (3) moderate pressure setting of the hard-rubber rollers.

With the laminator's parameters correctly set, the application of the dry-film photo-resist onto the pc board can now be performed.

17.5 DRY-FILM PHOTO-RESIST APPLICATION

In Sec. 17.2, the pc board was cleaned, placed into an acid dip, towel-dried, and prebaked in an oven. These heated boards should be laminated immediately upon removal from the oven. Cold boards would act as heat sinks and remove heat from the dry film, resulting in poor resist adhesion. For this reason, the oven should be placed close to the laminator so that no delay will result between removing the board from the oven and placing it into the laminator.

With the laminator controls set to the settings specified in Sec. 17.4 and allowing the laminator time to reach the preset temperatures, approximately 1 foot of dry film is run through the laminator. This is a precautionary measure that should be done each time the laminator has remained idle for any length of time. The reason for doing this is that the first foot of film on the rolls has been in close contact with the heat generated by the heat shoes of the laminator and, as a result, has been degraded by the long exposure to elevated temperatures. Film damaged from excessive heat results in *heat fog*. A pc board processed with heat-fogged film will not process successfully in the subsequent steps.

After the foot of film has passed through the laminator and with the film feed still on, the board is removed from the oven (held by the edges only) and fed immediately into the laminator by pushing it gently up to the moving rollers. The board contacting the rollers will be pulled into the laminator, where heated dry-film resist will be applied simultaneously to both sides of the copper surface. This process is shown in Fig. 17.7a.

As the board with the applied resist leaves the laminator, it is removed by cutting off the film either with the aid of the slide cutter mounted at the rear of the laminator or with a single-edge razor blade (Fig. 17.7b). Efficient laminating can be realized if two people operate the unit: one person feeding the boards into the laminator while the other cuts and removes the processed boards.

A laminated double-sided pc board is shown in Fig. 17.8. Notice that the board is sandwiched between two layers of dry film, each protected by a polyester cover sheet. After the board has been laminated, it is extremely important not to disturb or remove the polyester cover sheets (even at the corners). These sheets serve as "gas seals," preventing

(a)

(b)

FIGURE 17.7. Technique for pc board laminating. (a) PC board fed into laminator. (b) Board removal using rear film cutter.

FIGURE 17.8. PC board laminated with dry-film photo-resist.

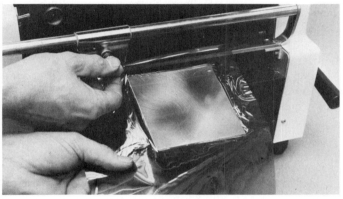

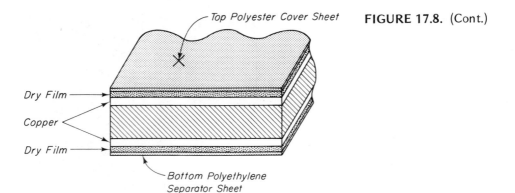

Top Polyester Cover Sheet

FIGURE 17.8. (Cont.)

Dry Film

Copper

Dry Film

Bottom Polyethylene
Separator Sheet

oxygen in the atmosphere from contacting the underlying resist film and causing it to polymerize (i.e., change it chemically into heavier molecules). The excess film is carefully trimmed as close as possible to the board edges, using a sharp razor blade. This technique is shown in Fig. 17.9.

The temperature of the laminator must be carefully maintained at $235 \pm 10°F$. At elevated temperatures, resist blistering occurs, and at temperatures lower than the specified range, a breakdown in resist adhesion to the copper surface can result. If the laminator is in continuous use, careful monitoring of the two thermometers is essential to minimize problems.

After laminating, a normalizing period of 10 minutes before exposure is recommended by the manufacturer. This 10-minute period allows the board and dry film to return to room temperature, resulting in an improvement of the adhesion properties of the film. Because the film is extremely sensitive to ultraviolet light, the normalizing interval must be in a gold-light area.

FIGURE 17.9. Trimming of excess dry film from edges of laminated pc board.

17.6 EXPOSURE OF DRY-FILM PHOTO-RESIST

The dry-film resist is now ready to be exposed to an ultraviolet light source. The amount and quality of ultraviolet light contacting the resist is critical to obtain results that are consistently reliable. Too short an exposure time will result in an image that is not completely developed. Too long an exposure time will result in resist brittleness and improper resist adhesion to the copper surface. Resist adhesion problems usually become apparent in the cleaning cycle prior to plating and during the plating operation itself.

Proper exposure time for a particular resist and exposure unit must be determined empirically. The procedure for determining proper exposure time for any exposure unit is detailed in Appendix XIV.

Any point-light source rich in ultraviolet light is suitable for exposure of Laminar A resist. Excellent results are obtainable with an exposure unit equipped with high-pressure mercury vapor lamps. The exposure unit shown in Fig. 17.10 is a small prototype unit suitable for double-sided exposures. This unit consists of two 400-watt high-pressure mercury vapor lamps, one above and one below the vacuum frame. Also included in this unit are the controls to activate the vacuum pump, a vacuum gauge, and a resettable timer to control the exposure time. The technique of exposure is somewhat a function of the equipment available. The following explanation relates directly to the exposure unit shown in Fig. 17.10 but is similar to most fixed-light-source units. The main power is

FIGURE 17.10. Prototype exposure unit for double-sided pc boards.

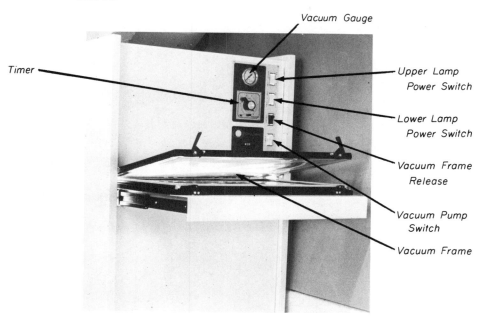

397

turned on and both upper and lower lamps are energized. The unit is then allowed to warm for at least 30 minutes before exposing any boards. A warm-up time is required of all exposure units equipped with high-pressure mercury vapor lamps to allow the lamps to operate at full light intensity.

After the laminated pc board has normalized to room temperature, it is placed between the top and bottom positives of the photo tool, which was discussed in Chapter 16. The setup for exposure is shown in Fig. 17.11a. The photo tool and pc board are placed onto the glass portion of the vacuum frame (Fig. 17.11b). The plastic cover is then lowered and locked into place over the setup. Engaging the vacuum pump will create a vacuum between the glass and the plastic sections of the frame forcing both the top and bottom positives into immediate contact with the polyester cover sheets of the dry film. A good seal is indicated by a reading of 25 psi for the unit illustrated in Fig. 17.10. A good seal having been achieved, the frame is pushed into the exposure unit between the upper and lower lamps and exposed for the preset interval on the timer control. For the exposure unit shown, an exposure time of 1 minute and 30 seconds, as determined with the step table, results in correct exposure for Laminar A photo-resist.

FIGURE 17.11. Preparation for exposure. (a) Placing pc board into photo tool. (b) Photo tool placed into vacuum frame of exposure unit.

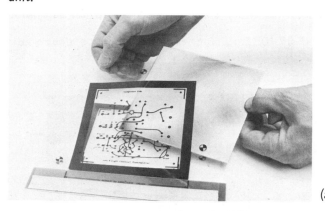

(a)

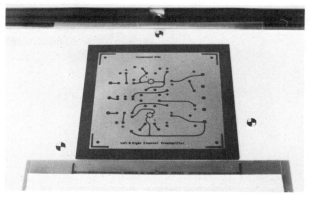

(b)

As with any negative–acting resist, exposure of the film to ultraviolet light causes the following to occur. All those areas of the resist protected from the light (those areas appearing as black conductor patterns on the positive) will remain unchanged chemically and will be completely removed in the developing phase. Those areas that are exposed to the light (all transparent areas) are polymerized and will not be removed in the developing phase.

After the film is properly exposed, the vacuum frame will be automatically ejected from the unit. The vacuum pump is then turned off and both the exposed film and the photo tool are removed from the unit.

At this point, a faint image of the conductor pattern is discernible through the transparent gas seal on the resist. This is the result of a visible indicator dye that was added to the film by the manufacturer. Again, the exposed pc board needs to be allowed to normalize to room temperature for approximately 10 minutes. The protective gas seal should remain in place and undisturbed during the normalizing period.

One side of a correctly exposed double-sided pc board with its protective polyester cover sheet partially removed (after the 10-minute normalizing period) is shown in Fig. 17.12. This board is now ready for developing.

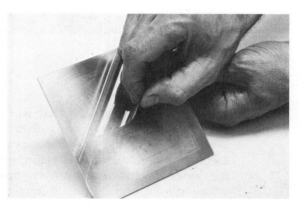

FIGURE 17.12. Removal of protective polyester cover sheet after exposure.

17.7 SPRAY DEVELOPING

Spray developing is accomplished by spraying, under high pressure, a mildly alkaline developer solution onto the exposed pc board for a specific length of time and at a specific temperature. This temperature should be within the range 80 to 90°F (27 to 32°C). Higher developing temperatures will decrease developing time, but it may degrade the quality of the resist pattern remaining on the pc board. Too low a temperature will increase the required developing time. The prototype spray developer shown in Fig. 17.13 will completely develop a pc board in approximately 45 seconds at a temperature of 85°F (29°C) with new developer solution. Development time is determined by the "freshness" of the developing solution and its temperature. Optimum developing can be determined

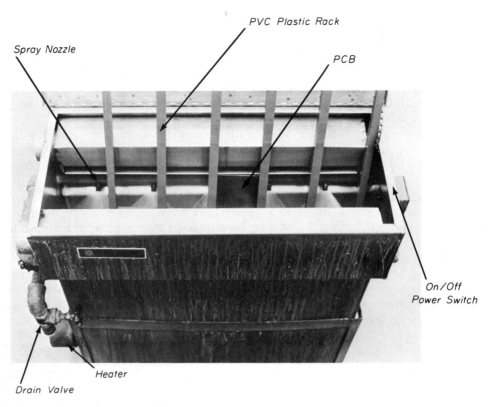

Spray Nozzle

PVC Plastic Rack

PCB

On/Off
Power Switch

Heater

Drain Valve

FIGURE 17.13. Spray developing a pc board.

visually — all of the resist that was not exposed to the ultraviolet light is completely removed, leaving clear copper. A properly developed pc board is shown in Fig. 17.14. In mass-production conditions, an additional 15 seconds is added to the developing time to compensate for the fact that the more boards that are developed, the more resist goes into the developing solution resulting in longer time.

A developer that can be used for Laminar A resist is a 1.5% by weight solution of trisodium phosphate (TSP) and tap water. For this solution, 0.125 pound (57 grams) of TSP per gallon of water is required. This solution is cost-effective but results in a relatively slow developer. Improved developing efficiency can be achieved with Dynachem's concentrated developer solutions for Laminar A resist. These developers are under proprietary trade labels of Concentrate Developer KB-1A and KB-1B. An example of preparing developing solution will make use of the spray developer shown in Fig. 17.13. This developer has a sump capacity of 10 gallons minimum and 15 gallons maximum.

To prepare the developing solution, 12 gallons of cold tap water are placed into the developer sump. One-half gallon ($\approx$ 1890 milliliters) of Concentrate Developer KB-1A and $\frac{1}{6}$ gallon ($\approx$ 630 milliliters) of

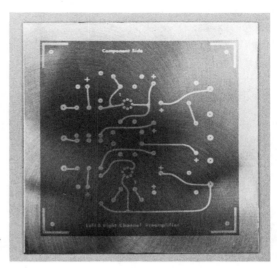

FIGURE 17.14. PC board after developing.

Concentrate Developer KB-1B are added to the water. To suppress the formation of excessive foam during the developing operation, the addition of Dynachem Antifoam No. 50 at the rate of $\frac{1}{2}$ milliliter per gallon of developer will complete the solution. To the approximately 13 gallons (exactly $12\frac{2}{3}$ gallons) of solution in the sump, $6\frac{1}{2}$ milliliters of Antifoam No. 50 is added. This solution will result in rapid and effective pc board developing. It should be mentioned at this point that with this developing solution, if developing time exceeds $1\frac{1}{2}$ minutes at a spray temperature of 85° F, this is an indication that the solution has been saturated with developed Laminar A. It should be discarded and a fresh batch of developer prepared.

In preparation for developing, the immersion heater temperature control is adjusted to 85° F and at least 1 hour should be allowed to pass for the solution to reach this operating temperature. After the solution has reached its operating temperature, the board to be developed, with both protective polyester cover sheets removed, is placed in a suitable rack and suspended into the spray. A slow up-and-down motion improves the uniformity of developing, as does rotating the entire rack 180° in the spray at least once for each batch of boards to be developed.

After developing, the board is immediately removed from the rack and given a rinse under a cold-water tap for at least 30 seconds on each side. Any developer solution remaining on the pc board after rinsing is neutralized with a 30-second dip in a diluted sulfuric acid solution. A 5% by volume solution is recommended. This solution is prepared by slowly adding 5 parts of CP sulfuric acid to 95 parts of tap water. After this neutralizing dip, the board is rinsed for 30 seconds and then allowed to air-dry.

The next phase will be to chemically clean the developed conductor pattern and thief area and to plate these areas with 60/40 tin–lead.

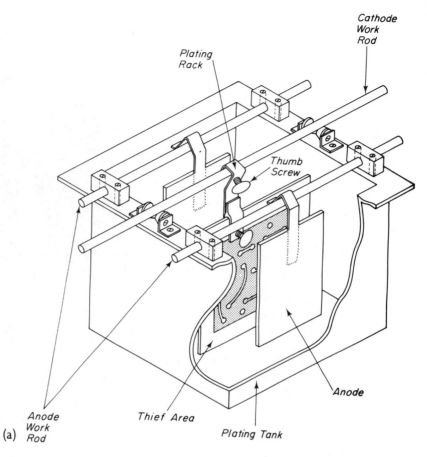

FIGURE 17.15. Plating rack and accessories. (a) Plating and rack layout (b) Commercial plating rack. (c) Simple laboratory style plating rack.

17.8 PLATING RACK DESIGN

A rack for plating printed circuits holds either one or more boards in a fixed position in the plating solution. It must be capable of handling the large currents required in the plating process and not be easily affected by the plating solution. A basic plating and rack layout is shown in Fig. 17.15a. The rack contacts the cathode work rod of the plating tank and is held securely with a wing nut or thumb screw. Tight contact between the thumb screw and cathode reduces resistance and holds the rack rigidly to the rod. The lower section of the rack connects to the thief area of the pc board and is held firmly by another thumb screw or set screw.

Plating racks are often fabricated from square copper rod stock, because this material is capable of handling 1000 amperes per square inch of cross-sectional area. A commercially fabricated plating rack capable of handling at least four boards at one time is shown in Fig. 17.15b. Because most of the rack assembly will be in contact with the plating solution, it is protected with a pestasol coating to prevent plating buildup. The contact points for the boards are fabricated of 300 series stainless steel. A smaller, single board rack is shown in Fig. 17.15c. It is constructed of copper stock and stainless steel screws. Subsequent discussions on plating will make use of this rack.

17.9 CLEANING CYCLE FOR 60/40 TIN-LEAD ELECTROLYTIC PLATING

Quality plating of 60/40 tin–lead alloy onto the exposed circuit pattern can only be achieved if standard plating practices and effective cleaning cyles are employed. The cleaning cycle must properly prepare the copper surface to accept the plating while at the same time not allow the resist (stop-off) remaining on the board to be damaged. The cleaning cycle shown in Table 17.1 is recommended for 60/40 tin–lead plating where Dynachem Laminar A dry-film resist is used.

The cleaning phase should be set up in an area with a sink available having hot and cold running water. The chemicals and solutions required for board cleaning can be prepared in clean laboratory beakers or small plastic tanks (polyethylene or polypropylene). A cleaning line is shown in Fig. 17.16.

The board to be cleaned is first racked, being careful that a good electrical contact is made between the rack and the thief area of the board. The board is then processed through the following baths in the order presented:

Step 1. A proprietary solution of Dynachem LAC-41 will be used to remove oxides and greases from the exposed copper area of the conductor pattern and the thief area. This solution is selected because it will not adversely affect the resist or its bond to the copper surface. To

TABLE 17.1

Recommended Cleaning Cycle for Dynachem
Laminar A Photo-resist

1. Dip board into 20% by volume solution of Dynachem LAC-41 and water for 5 minutes at $160 \pm 10^\circ F$ ($71^\circ C$). (Slowly agitate.)
2. Rinse board in warm tap water for 1 minute.
3. Rinse board in cold tap water for 1 minute.
4. Place board into ammonium persulfate, $(NH_4)_2 S_2 O_8$, and etch-dip for 2 minutes at $70^\circ F$ ($21^\circ C$). (Slowly agitate.)
   Etch-dip solution – $1\frac{1}{2}$ pounds (680 grams) ammonium persulfate per gallon of water.
5. Rinse board in cold tap water for 1 minute.
6. Dip board into 10% by volume solution of sulfuric acid and water for 2 minutes. (Slowly agitate.)
7. Rinse board in cold tap water for 1 minute.
8. Dip board into 10% by volume solution fluoboric acid and water for 2 minutes. (Slowly agitate.)
9. Electroplate.

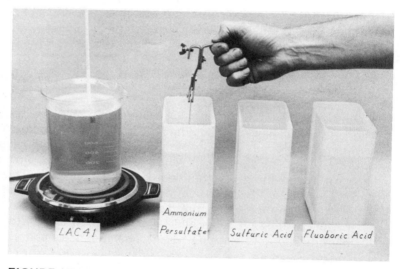

FIGURE 17.16. Cleaning line (water rinses not shown).

prepare this cleaner, 1600 milliliters of water is placed into a 2000-milliliter beaker. To this beaker, 400 milliliters of concentrated LAC-41 solution is added and stirred. This yields a 20% by volume working bath of cleaning solution. The beaker is then placed onto a hot plate and the solution brought up to and stabilized at a temperature of $160^\circ F$ ($71^\circ C$). A laboratory thermometer is used to monitor this temperature. The racked board is then placed into the heated solution for 5 minutes, during which time the board is slowly rocked back and forth to allow fresh solution to pass across the exposed copper areas.

Step 2. The board is removed from the LAC-41 solution and rinsed completely under a warm-water tap for at least 1 minute.

Step 3. While still at the sink, the cold-water tap is turned on and the board is rinsed for an additional minute. This assures that all the cleaning solution is removed from the copper surfaces.

Step 4. An ammonium persulfate (AP) etching bath is the next phase of the cleaning cycle. This bath is prepared by adding 12 ounces of AP crystals to 2000 milliliters of warm water. To achieve uniform etching, it is absolutely essential that the board be slowly rocked back and forth continuously for approximately 2 minutes, or until a visual inspection of the conductor paths shows a smooth mat finish that is salmon in color.

Step 5. The board is removed from the etching bath and rinsed under a cold-water tap for at least 1 minute to remove the majority of the ammonium persulfate residue from the board surface.

Step 6. To remove all traces of ammonium persulfate residue and to neutralize the copper surfaces, the board is dipped into a diluted solution of sulfuric acid and rocked slowly for approximately 2 minutes. This solution is prepared by adding 200 milliliters of CP sulfuric acid to 1800 milliliters of water. This mixture will yield a 10% by volume solution of sulfuric acid.

Step 7. The board is removed from the sulfuric acid bath and rinsed under cold water for 1 minute.

Step 8. A final acid dip in fluoboric acid is the last step in the cleaning process. Because the tin-lead plating bath is extremely susceptible to contaminants, especially sulfates from the previous acid dip, this fluoboric acid dip is extremely important. This 10% by volume fluoboric acid bath is prepared by adding 200 milliliters of concentrated fluoboric acid to 1800 milliliters of water. The board is placed into this bath and slowly rocked back and forth for approximately 2 minutes.

Step 9. After the final acid dip, the board is transferred directly into the plating tank for the tin-lead plating process.

It is recommended that all cleaning solutions used be prepared with either CP (chemically pure) or reagent-grade acids. In small quantities, all solutions used in the cleaning line should be discarded daily to minimize the possibility of cleaning problems. The use of large quantities makes daily discarding and preparation time-consuming and economically unfeasible. To determine the effectiveness of the cleaning solutions and procedures a modified water-break test should be used periodically. A test pc board (no pattern necessary) is passed sequentially through all the cleaning steps as outlined above. After the board is removed from the final fluoboric acid bath, the acid should "sheet" across the entire copper surface for a minimum of 30 seconds with no detectable "water breaks," which is an indication of the presence of contaminants not successfully removed. A cleaning cycle that passes this test is effective in preparing the copper surfaces for quality plating.

17.10 TIN–LEAD PLATING TANK AND ACCESSORIES

In the previous section, the copper pattern to be plated with 60/40 tin–lead alloy was properly cleaned in preparation for electroplating. Prior to the discussion of the plating technique, however, the construction of a prototype plating tank will first be presented.

Because the tin–lead alloy will be electroplated from a fluoborate plating solution, the recommended material for the tank is stress-relieved polypropylene or polyethylene plastic. Glass-silicated materials or titanium are not suitable for fluoborate solutions and their use should be avoided.

A tank size of 12 inches long by 12 inches wide by 12 inches deep will be adequate for most prototype work. This tank will have a solution capacity of approximately 6 gallons when the plating solution is 9.75 inches from the bottom. The exact volume can be calculated by dividing the number of cubic inches of solution by 231 cubic inches per gallon. The following example will demonstrate this calculation.

Example: Determine the exact volume of plating solution for a 12 inches by 12 inches by 12 inches tank.

Volume of solution (cubic inches) = length $\times$ width $\times$ depth of solution

$$= 12 \text{ inches} \times 12 \text{ inches} \times 9.75 \text{ inches}$$
$$= 1404 \text{ cubic inches}$$

$$\text{Solution volume (gallons)} = \frac{\text{Volume of solution (cubic inches)}}{231 \text{ cubic inches/gallon}}$$

$$= \frac{1404 \text{ cubic inches}}{231 \text{ cubic inches/gal}} = 6.08 \text{ gallons}$$

A 12-inch-deep tank with approximately 10 inches of plating solution (2 inches of clearance to the rim of the tank) will allow the insertion of the anodes and the racked pc board to be plated without causing the solution to spill over the top of the tank.

The tank is fitted with two anode work rods and one cathode work rod. The anode work rods are made of $\frac{3}{8}$-inch-diameter solid copper rod stock. This diameter will easily support the weight of the tin–lead anodes as well as handle the currents used. These anode work rods are held securely into position with pairs of Lucite or plastic blocks fastened onto opposite ends of the tank rim. The rods are positioned over the tank parallel to one another and 1 inch from the sides of the tank. The cathode work rod made from the same $\frac{3}{8}$-inch-diameter copper stock is positioned parallel to the anode work rods and centered between them. The spacing between either anode work rod and the cathode work rod is 5 inches. The cathode work rod rests on a pair of plastic rollers attached to opposite ends of the tank rim and centered between the Lucite anode rod support blocks. This allows movement of the cathode rod parallel to the stationary anode rods. The cathode work rod must be capable of back-and-forth travel to provide agitation of the pc board in the plating solution. This type of rod motion will produce "knife agitation" of the racked pc board through the solution and is recommended for tin–lead alloy plating provided that the rate of motion is not violent. For tin–lead alloy plating, the recommended distance of travel of the pc board in the solution is from 3 to 6 feet in 1 minute. To realize this optimum range of travel distance, a high-torque motor operating at 12 rpm is mounted to the tank rim. A drive wheel of approximately 2 to 3 inches in diameter is assembled to the motor's shaft and the cathode work rod is linked to the drive wheel at a distance of 1 inch from the center of the motor's shaft. The linkage of the cathode work rod to the motor *must* be made with a nonconductive material, preferably plastic. As the motor rotates the drive wheel through one revolution, the cathode work rod will be pushed forward, pulled back, and then pushed forward again for a total distance of 4 inches. With a 12-rpm motor, the total distance traveled will be 48 inches or 4 feet in 1 minute. This distance is within the recommended range for "knife agitation" used for tin–lead alloy plating. A typical plating tank fabricated as described in this section is shown in Fig. 17.17.

The negative terminal of a plating power supply is electrically connected to the cathode work rod. Both anode work rods are electrically connected together (anode strap) and then connected to the positive terminal of the power supply. All electrical connections should be made with AWG No. 12 stranded wire protected with a chemical-resistant plastic insulation.

The plating power supply should have the following characteristics:

1. Continuous adjustment of output voltage from 0 to 6 volts dc.
2. Current capacity in excess of plating current requirements (a maximum of 10 amperes is acceptable for the plating tank illustrated).
3. Percent ripple at full load current should be equal to or less than 5%.
4. A meter to monitor the dc output voltage at the + and – terminals.
5. A meter to monitor the load current.

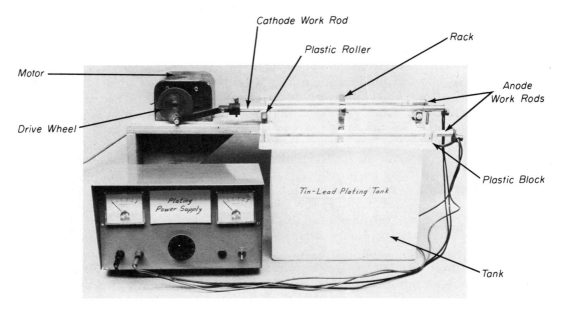

FIGURE 17.17. Typical plating tank and accessories.

FIGURE 17.18. Schematic diagram of plating power supply.

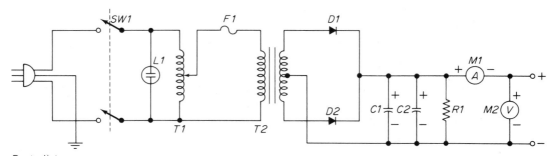

Parts list

SW1 — DPST switch
L1 — Neon lamp
T1 — 2 amp, O to 120 Vac, 0.25 kVA autotransformer
F1 — 1 amp, 250 V, Slo–blo fuse
T2 — Stancor P–6464 6.3 VCT at 10 amp transformer
D1, D2 — IN3890 12A at 100 PIV
C1, C2 — 21,000 μF at 40 WVdc
R1 — 1 kΩ, $\frac{1}{2}$ W, 10%
M1 — O to 10 A meter
M2 — O to 10 V meter

The circuit schematic diagram for a typical plating power supply is shown in Fig. 17.18.

The type of material used for the anode, together with its shape and surface area, will affect the quality of the alloy that is plated onto the pc board. Because the tin-lead alloy to be plated requires a metal ratio of 60% tin and 40% lead, anodes with this same ratio should be utilized. The anodes must also be free of oxides and have low metal impurities. Anode material that meets all these specifications is high-purity Vaculoy. Anodes made with this material will corrode evenly in the plating solution, produce less sludge, and improve bath stability.

Anodes are available in various shapes and in a wide range of sizes (Fig. 17.19). The standard shapes are *flat*, *oval*, *round*, *sawtooth*, and *gear-tooth*. A flat anode is preferable in a small plating tank since it results in more uniform current density. The oval, sawtooth, and other special shapes are designed to increase the surface area of the anode. The criteria for selecting an anode is its surface area relative to the area of the pc board to be plated. For the plating of tin-lead alloy, an anode-to-cathode surface area ratio of approximately 2:1 is recommended. This means that the surface area of the anode facing the pc board in the plating tank should have a surface area approximately twice as large as the one side of the pc board that it is facing. You will recall from Chapter 16 that uniform plating on double-sided pc boards will result if the overall exposed conductor path area is equal on both sides of the pc board. Double-sided plating is accomplished with the use of two anodes, one positioned on each side of the pc board, both having identical surface areas. This will simplify and optimize the plating process and a uniform coating of 60/40 tin–lead alloy will be plated on both sides of the pc board.

The anodes must be provided with a hook whose length will allow the anode to extend from the anode work rod down into the plating solution yet allow the top of the anode to protrude slightly above ($\frac{1}{4}$ inch) the surface of the solution. The length of the anode should be such

FIGURE 17.19. Standard anode shapes.

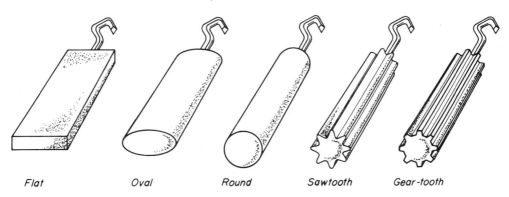

| Flat | Oval | Round | Sawtooth | Gear-tooth |

that it extends from 1 to 2 inches below the bottom edge of the pc board, which is suspended from the cathode work rod. This will result in a higher current density on the thief area of the pc board and tends to improve the uniformity of alloy composition in the conductor-path area.

To illustrate the calculation of the correct size of an anode for a specific application, the preamplifier layout shown in Fig. 16.11 will be used. The area of exposed copper must first be accurately determined for each side of the pc board to be plated. The thief-area calculation presents no special problem, since it is normally a square or rectangular border outside the conductor pattern area. The thief area is calculated as follows (Fig. 17.20). The area of the bulk stock is first determined in square inches. The thief area is then obtained by subtracting the available conductor pattern area from the bulk stock area. In equation form, with reference to Fig. 17.20:

Thief area (square inches per side) = $(A \times B) - (C \times D)$

(where dimensions A, B, C, and D are in inches). For the preamplifier circuit, the dimensions are:

$$A = 5.0 \text{ inches}$$
$$B = 5.0 \text{ inches}$$
$$C = 4.0 \text{ inches}$$
$$D = 4.0 \text{ inches}$$

Therefore,

Thief area (square inches per side) = (5 by 5 inches) – (4 by 4 inches)

= (25 square inches)
– (16 square inches)

= 9 square inches

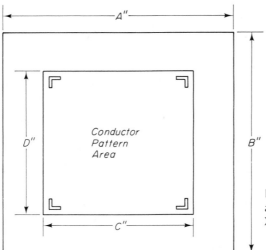

FIGURE 17.20. Calculation of thief area. Thief area = $(A \times B) - (C \times D)$.

The determination of the area of the conductor pattern is not quite as simple. To calculate the copper surface area of the conductor pattern that will face each anode, the area of each of the shaped configurations (i.e., donuts, IC patterns, tape, etc.) used in the layout, reduced to a 1:1 scale, must first be determined. Table 17.2 is provided to aid in this calculation. Many standard patterns used in pc layouts are listed together with their areas in square inches. In addition, several common conductor path widths are given, together with their square inch area per linear inch. The information given in Table 17.2 is for a 1:1 scale, (i.e., the same size as appears on the pc board). If, for example, a $\frac{1}{4}$-inch-diameter donut was used on a 4:1 layout, the actual 1:1 scale diameter on the finished pc board would be $\frac{1}{16}$ (0.0625) inch.

TABLE 17.2

Areas of Some Standard PC Patterns

Patterns		Approximate Copper Area (Square Inches)
Terminal	0.150	0.018
Pad	0.125	0.012
Diameters	0.100	0.008
(Inches)	0.09375	0.007
8-pin mini-DIP IC		0.035
14-pin DIP IC		0.06
16-pin DIP IC		0.07
8-pin round IC		0.038
10-pin round IC		0.039

Conductor Width (Inches)	Area (Square Inches) per Linear Inch	Area (Square Inches) per 10 Linear Inches
0.020	0.020	0.2
0.025	0.025	0.25
0.030	0.030	0.3
0.040	0.040	0.4
0.050	0.050	0.5
0.060	0.060	0.6

The following procedure is used with Table 17.2 to determine the square-inch area of a conductor pattern for *one* side of a double-sided layout: (1) count the number of donuts of each size used in the layout and multiply that number by the appropriate square-inch area listed for the terminal pad areas—this will determine the total area of just the donuts; (2) count and add the total area of each special layout configuration, such as IC patterns; and (3) determine the total length in inches of conductor tape and multiply this length by its area per linear inch. The addition of items 1, 2, and 3 will produce the total conductor area for one side of the pc board.

An example of the foregoing procedure for the *preamplifier circuit* will illustrate its use:

Pattern	Number		Area of Each		Total Area (Square Inches)
$\frac{1}{8}$-inch donuts	8	×	0.012	=	0.096
$\frac{3}{32}$-inch donuts	52	×	0.007	=	0.364
8-pin ICs	2	×	0.038	=	0.076
$\frac{1}{32}$-inch tape ($\approx$0.030)	16.8 inches in length	×	$\dfrac{0.030 \text{ sq in.}}{\text{linear inch}}$	=	0.504
				Total	1.04 square inches

To calculate the total area to be plated, simply add the thief-area calculation to the copper surface area of the conductor pattern. For the preamplifier layout:

Total area to be plated = thief area + conductor pattern area

$$= (9 \text{ square inches}) + (\approx 1.00 \text{ square inch})$$
$$= 10 \text{ square inches}$$

This total area to be plated can now be used to determine the anode area required for plating. This calculation will also be used in Sec. 17.12 to determine the plating power supply current density for optimum plating.

The two restrictions which are placed on the selection of the anode size is that (1) it has a surface area from one to two times larger than that of the pattern to be plated and (2) it is from 1 to 2 inches longer than the pc board bulk stock. Selecting an anode whose length is 6 inches (1 inch longer than the 5-inch bulk stock) will meet this criteria. If the anode width is chosen to be 3 inches, the anode surface area will be 6 inches × 3 inches = 18 square inches, which is within the range of from one to two times larger in area than the area of the pattern to be plated. Since each side of the preamplifier layout has approximately equal areas, two 3- by 6-inch anodes will meet all of the requirements for plating the preamplifier pc board.

Prior to the placement of the anodes into the plating tank, proper procedures must be followed to avoid contaminating the plating solution. Tin–lead fluoborate plating solutions are extremely susceptible to organic contaminations. For this reason, the anodes must be properly cleaned and correctly *bagged*. Anode degreasing is accomplished by scrubbing the surface with a brass bristle brush and cold tap water. The anode is then rinsed thoroughly and dipped into a 15% by volume solution of fluoboric acid for approximately 10 minutes. After the acid dip, the anode is ready to be bagged.

The anode bags used should be made of either polypropylene or dynel (*but never cotton*) and should be 2 inches wider and 2 inches longer than the anode. The bag must be *leached* (purified) of all organic contaminations that it was exposed to in its manufacturing process. Leaching is accomplished by boiling the bag in water for at least 1 hour at 150°F (66°C). The bag is then removed, rinsed in cold water, and tied around the cleaned anode. The anodes are now ready to be installed into the plating tank (Fig. 17.21).

(a)

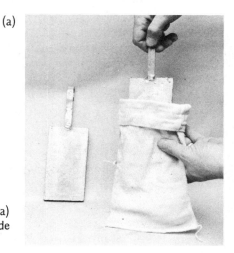

FIGURE 17.21. Anode bag preparation. (a) Anode and leached anode bag. (b) Anode installation into plating tank.

(b)

In addition to the anode preparation, the plating tank itself must be thoroughly leached of all contaminations prior to the addition of the plating solution. Tank leaching should be done with all the hardware and electrodes removed from the tank. Leaching of the tank need be done only before the plating solution is added to the tank *for the first time*. Detailed information for leaching the tank is given in Appendix XV. The student should refer to this appendix before proceeding to the next section.

17.11 PLATING SOLUTION

After the plating tank has been properly leached, the preparation of the plating solution may begin. For our purposes, a high-throwing power formulation will be used to electrodeposit 60/40 tin–lead alloy onto the copper conductor pattern surface. The solution described is a high-acid, low-metal (tin and lead) mixture formulated under the trade name HI-THRO, manufactured by the Allied Chemical Corporation in Morristown, N.J., under U.S. Patent 3,554,878. The plating bath is made up of the following chemicals and compounds:

1. Stannous fluoborate concentrate, 51.0% by weight
2. Lead fluoborate concentrate, 51.0% by weight
3. Fluoboric acid, 49% by weight
4. Boric acid
5. Stabilized liquid peptone solution (available from Allied Chemical Corporation)
6. Water

Stannous and lead fluoborate are used to provide the source of metal in the plating bath. Because variations in the metal content of the plating bath will adversely affect the content of the alloy deposited, both the stannous and lead fluoborate additives must be carefully controlled when preparing the bath so that it will contain 60% by weight of tin as a metal and 40% by weight of lead as a metal.

The fluoboric acid aids in increasing the conductivity of the bath and improves the grain quality of the deposit by making it finer and smoother. Peptone also improves the fine-grained deposit and inhibits the formation of "trees," which are small growths of deposit extending away from the conductor pattern. Boric acid aids in maintaining stability by minimizing the decomposition of the fluoborates in solution.

The chemicals and compounds necessary to prepare a 10-gallon bath of HI-THRO tin–lead fluoborate plating solution are listed in Table 17.3. The table includes the required amounts and the order of mixing the ingredients.

TABLE 17.3

Formulation of 10 Gallons of Tin–Lead Plating Solution

Order of Mixing Ingredients	Compound	Amount	
5	Stannous fluoborate concentrate, 51%	0.45 gal	1703 mL
3	Lead fluoborate concentrate, 51%	0.21 gal	795 mL
4	Fluoboric acid, 49%	6 gal	22.7 L
2	Boric acid	1.85 gal	839 g
6	Stabilized peptone solution	0.17 gal	643 mL
1, 2, 7	Water (deionized)	3.2 gal	12.1 L
		Total 10.03 gal	3.8 L

For purposes of preparing a plating bath for the 6-gallon tank previously discussed, Table 17.4 lists the amount of each ingredient and the order of mixture. The required amount of each of the ingredients was derived by taking $^6/_{10}$ (0.6) of the amounts listed in Table 17.3.

The 6-gallon plating solution is prepared by first placing one-half of the required amount of water (0.96 gallon or 3633 milliliters) into the plating tank. To 1500 milliliters of hot water, 1.11 pounds (503.5 grams) of boric acid is added and stirred until it is completely dissolved. This solution is then added to the plating tank and stirred. The following are then stirred into the plating tank in the order given: 0.126 gallon (477 milliliters) of lead fluoborate solution; 3.6 gallons (13.63 liters) of fluoboric acid; 0.27 gallon (1022 milliliters) of stannous fluoborate solution; and 0.102 gallon (386 milliliters) of stabilized peptone. The remaining water (2134 milliliters) is added to the tank to complete the bath solution. The solution is thoroughly stirred and allowed to settle overnight before use.

TABLE 17.4

Formulation of 6 Gallons of Tin–Lead Plating Solution

Order of Mixing Ingredients	Compound	Amount	
5	Stannous fluoborate concentrate, 51%	0.27 gal	1022 mL
3	Lead fluoborate concentrate, 51%	0.126 gal	477 mL
4	Fluoboric acid, 49%	3.6 gal	13.63 L
2	Boric acid	1.11 lb	503.5 g
6	Stabilized peptone solution	0.102 gal	386 mL
1, 2, 7	Water (deionized)	1.92 gal	7.27 L
		Total 6.018 gal	22.78 L

The height of the solution in the tank should be marked on the *outside* to note any decrease due to evaporation. Any significant decrease of solution should be replaced so as to return it to the original level in the tank. The bath should be maintained at room temperature and loosely covered with sheet plastic when not in use to protect the solution from dust and other foreign matter.

All plating solutions require periodic chemical analysis. In alloy plating, the ratio of metals deposited must be maintained within narrow limits. The techniques of chemical analysis for the type of bath just described are beyond the scope of this book. However, since the amount of solution described in Table 17.4 is relatively small and not significantly expensive, the entire bath should be discarded if plating problems are encountered. A fresh batch of solution can be used successfully for a reasonable length of time if extreme care is exercised to minimize the introduction of organic contaminations into the solution. Detailed chemical analysis techniques may be obtained from Allied Chemical Corporation for HI-THRO plating solution.

17.12 60/40 TIN–LEAD PLATING

After the pc board has been cleaned and prepared chemically (as described in Sec. 17.9), it is removed from the final fluoboric acid rinse (also described in that section) and placed into the plating tank. The pc board is then lowered into the solution and the rack is secured onto the cathode work rod. The motor is turned on to produce the knife agitation and finally the power supply is adjusted for the correct current density. This current setting determines the proper alloy plating and needs to be properly set. The manufacturer recommends that the optimum current density is 20 amperes per square foot (≈ 140 milliamperes per square inch). Simply stated, the power supply should deliver a current of 20 amperes for every square foot of copper that is exposed to the plating solution. In Sec. 17.10, it was calculated that the thief area and the conductor pattern on each side of the pc board had a total surface area of 10 square inches. Therefore, the total surface area to be plated is:

$$\text{Total surface area (square inches)} = \text{surface area of one side in square inches} \times 2 \text{ sides}$$

$$= 10 \times 2 = 20 \text{ square inches}$$

The plating power supply therefore must be set at:

$$\text{Current} = \text{current density} \times \text{total surface area}$$

$$= \frac{140 \text{ milliamperes}}{\text{square inch}} \times 20 \text{ square inches}$$

$$= 2800 \text{ milliamperes} = 2.8 \text{ amperes}$$

It will be noted that when the power supply is set at 2.8 amperes, the output voltage is low (in the order of 1 volt or less) for this plating solution.

The thickness of the plating onto the pc board is a function of the current density and the plating time. For the HI-THRO alloy bath, 60/40 tin–lead is plated at a rate of 0.0001 inch (0.1 mil) every 2.3 minutes. To achieve a plating thickness of 0.3 mil requires 6.9 minutes, and 0.5 mil requires 11.5 minutes of plating. A minimum of 0.3 mil of plating thickness is required for the tin–lead alloy to provide an effective etching resist. A plating time of 7 minutes has been established as optimum to provide the pc board with the minimum required plating thickness.

After the pc board has been plated, the rack is removed from the cathode work rod and the power supply is turned off. The pc board is then thoroughly rinsed in cold tap water for at least 1 minute and then dried. Inspection of the plating should reveal a smooth, fine-grained deposit with a light dull-gray surface having no apparent dark streaks. A properly plated pc board is shown in Fig. 17.22.

FIGURE 17.22. Correctly plated pc board.

17.13 RESIST STRIPPING

The Laminar A resist previously applied to the pc board may now be removed by chemical stripping since it has served its purpose. The Dynachem Corporation provides an alkaline solution that will remove the resist by simple immersion. For Laminar A resist, 400K stripper is available as a concentrate. The stripping solution is prepared by mixing 1 part of 400K stripper to 9 parts of water in a Pyrex dish or beaker. The stripping solution is then heated on a hot plate to 130°F (54°C). The temperature should be checked with a laboratory thermometer. Immersion of the pc board into the stripper for approximately 2 minutes will completely remove the resist. The board is then removed from the stripper and rinsed first in hot tap water and then in cold tap water for at least 1 minute. It is recommended that the board surface be gently rubbed over with a nylon bristle brush while under the hot-water rinse to ensure complete removal of stubborn resist residue.

17.14 ETCHING

The processed pc board is now ready to be etched. The copper conductor pattern is protected with a 0.3-mil-thick layer of 60/40 tin–lead alloy while the remainder of the copper surfaces is unprotected and removed by etching. Etching is accomplished in a similar manner, as described in Chapter 13. However, ferric chloride, which was used in Chapter 13 as the etching solution, cannot be used, as it will attack the tin–lead alloy as well as the exposed copper. The tin–lead alloy plating acts as the etch "resist" and an etchant solution must be used that will be effective in attacking the exposed copper without adversely affecting the plating. Three common etching solutions can be used for this purpose: alkaline ammonia, chromic–sulfuric acid, or modified ammonium persulfate. Both the alkaline ammonia and the chromic–sulfuric acid solutions produce noxious and poisonous fumes, especially when heated. When using these solutions as etchants, extreme caution and good ventilation are required. For these reasons, alkaline ammonia and chromic–sulfuric acid are limited to industrial applications. The modified ammonium persulfate solution is much safer to use and is the best choice for prototype applications. This etchant solution is made by dissolving 2 pounds (0.91 kilogram) of ammonium persulfate crystals per gallon (3.785 liters) of deionized water at a temperature of 120°F (49°C). A mercuric chloride catalyst solution is then added in the amount of 1 milliliter per gallon. The mercuric chloride catalyst solution is made by dissolving 0.27 gram of the compound to 10 milliliters of deionized water. This amount of catalyst is sufficient for 10 gallons of etching solution. Finally, phosphoric acid is added in the amount of approximately 50 milliliters per gallon of etching solution. The formulation of 10 gallons of modified ammonium persulfate etchant is shown in Table 17.5.

TABLE 17.5

Formulation of 10 Gallons of Modified Ammonium
Persulfate Etchant

Compound	Amount for 10 Gallons of Etchant
Deionized water	10 gal
Ammonium persulfate crystals	20 lb (18.2 kg)
Mercuric chloride catalyst solution	10 mL
Phosphoric acid	500 mL

Etching double-sided pc boards is best accomplished with the use of a spray etcher, such as that shown in Fig. 17.23. This etcher allows for both sides of the pc board to be etched at the same time. The spray etcher is operated at a temperature of from 110 to 120°F (43 to 49°C). The time required to accomplish the etching is a function of the temperature and the amount of copper in the solution. A 1-ounce copper board should satisfactorily be etched in 3 to 5 minutes. As more boards are etched, additional copper goes into the solution and a point will be reached where etching time will exceed 6 minutes. At this point, the etching solution should be discarded and a fresh batch prepared. Where small volumes are involved, it is advisable to prepare a fresh batch daily.

After the board has been etched, it is rinsed in cold tap water and thoroughly dried with paper towels. The etched pc board is shown in Fig. 17.24.

FIGURE 17.23. Spray etcher for both single and double sided pc boards.

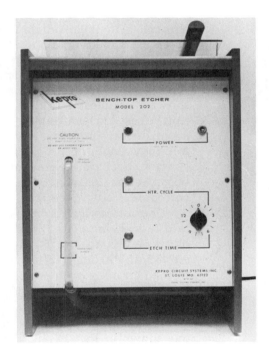

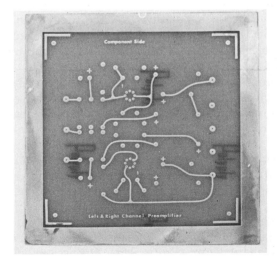

FIGURE 17.24. Properly etched pc board.

17.15 SOLDER FUSION

After the pc board has been etched, it is now ready for final processing: drilling, shearing, component assembly, and component soldering. However, an interim step is recommended at this stage of processing. This process is called *fusion*, sometimes referred to as reflow, and involves heating the 60/40 tin–lead alloy electroplated onto the copper surface to a temperature sufficient to melt the plating and fuse it to the underlying copper. The following advantages are realized through this process: (1) improved solderability, (2) extended shelf life, (3) elimination of solder overhang which may have resulted after the etching process, (4) a check to observe if de-wetting of the copper surface is a problem, and (5) a rough check on the metal ratio of plating. Items 1, 2, and 3 are self-explanatory. Items 4 and 5 will be discussed in detail. During the plating process, the 60/40 tin–lead metals have been plated onto the copper surface. The plating has not been alloyed with the copper but is adhered through a metallic bond. For alloying to occur, the solder must be elevated to its melting temperature.

If the solder has been plated to a clean copper surface, the fused solder will appear as bright and shiny over the entire conductor path areas. After fusion, an indication of an improper bond is a nonuniform surface showing nodules of solder. This de-wetting condition is a probable indication of an unclean copper surface and the board should be rejected to avoid further problems during the component soldering process, whether it be by hand or dip methods. The fusion process, then, is a quality-control check on the cleaning process (detailed in Sec. 17.9). If de-wetting occurs after fusion, the cleaning process needs to be improved and/or corrected.

Fusion can also be used as a means of roughly checking the tin–lead ratio being plated. Reference to Fig. 15.1 shows that at temperatures of above 361 to 400°F, the range of tin–lead solder alloys that will melt are 55/45 through 60/40 to 80/20. If the plating has not melted and formed an alloy with copper after the fusion process, this is an indication that the metal content of the plating bath is out of balance and should be replaced.

Fusion is accomplished in an infrared drawer-type oven, similar to the one shown in Fig. 17.25. This style of oven consists of a wire belt drawer with infrared heating elements positioned above and below, a preheat timer, a high-intensity timer, a cycle button, and an exhaust blower system.

The recommended time settings for $1/16$-inch pc board stock are 12.5 seconds for the preheat timer and 6.5 seconds for the high-intensity timer.

Before the pc board is placed into the oven drawer, it must be fluxed with a solder-fusing liquid. This liquid is applied to all areas of the pc board by hand brushing, spraying, or roller coating. The fusing liquid should be a mildly acidic flux that is completely water soluble. One such flux is Solder Flow No. 2111A available from Glo-Quartz® Ovens, Inc.

FIGURE 17.25. Bench type solder fusion oven. Courtesy of Glo-Quartz® Ovens, Inc.

The purpose of the flux is to prevent air from contacting the molten solder, to minimize the formation of oxides, to hold any impurities in suspension so that they will not interfere with the alloying process, and to reduce the possibility of de-wetting.

After the pc board has been properly fluxed, it is inserted into the drawer of the oven. The oven door is closed and with the two timers set, the cycle button is activated. During the preheat cycle (≈ 12.5 seconds) the board is elevated at a temperature of approximately 200°F (93°C) to reduce the possibility of thermal shock. At the end of the preheat cycle, the high-intensity cycle (≈ 6.5 seconds) automatically increases the board's temperature to approximately 400°F (204°C), melting the plated alloy and fusing it to the copper surface. The board is allowed to cool for a few minutes before removing it from the oven.

Because the solder plating on the surface of the copper is very thin, there is a good possibility that the solder will not completely flow over the edges of the copper surfaces during the reflow process. Etching causes these edges to become visible because only the top surfaces of the conductor pattern have been plated. Although those unplated copper edges will not affect the performance of the finished pc board, they can be plated to improve the overall appearance of the board. This is done by dipping the board into an immersion tinning bath between the etching and reflow processes. The tinning will ensure complete wetting of all edges. A typical immersion tinning bath is described in Sec. 13.5. After the board has been etched and thoroughly rinsed, it is dipped into the immersion tinning bath for approximately 10 minutes. A layer of tin, approximately 40 millionths of an inch thick, will be deposited onto all surfaces of the conductor pattern. The board is removed from the tinning bath and thoroughly rinsed in cold tap water, dried, fluxed, and then reflowed. After the tinning process, the entire conductor pattern will have a dull-gray appearance. This will change to a bright, shiny finish after the reflow process.

Because Solder Flow No. 2111A is a mildly acidic solution, the residue remaining on the copper surfaces must be completely removed. This is best accomplished by briskly brushing both sides of the pc board in warm water using a nylon bristle brush. A final rinse under a cold water tap for 30 seconds on each side followed by paper-towel drying completes this phase of the pc board processing.

17.16 FINAL PROCESSING SEQUENCES

The final sequence of operations to complete the pc board is cutting to size, drilling, component and hardware assembly, and soldering. Because the operations are discussed in detail in previous sections as they relate to single-sided pc boards, they will be briefly mentioned here as they pertain to double-sided boards.

For prototype work, shearing the pc board to the *inside* edges of the corner brackets is usually satisfactory. Shears such as those shown in Chapter 4 are suitable for this purpose. Shearing leaves the edges of the board rather rough. These rough edges are easily smoothed by placing a fine file onto a workbench and pushing the board edges across the file. About six passes across the file for each edge are sufficient. Once all edges have been filed, they are polished by firmly rubbing them with No. 00 steel wool. When the edge color matches that of the board surface, optimum smoothness is achieved. All steel wool and dust particles may be removed with a soft dry cloth.

With the board cut to finished size, the drilling operations are then performed. A high-speed drill press and carbide drill bits are used. (The proper use of these tools is described in Chapter 13.) Recall that special precautions were taken in the graphic layout process (Sec. 16.3) to ensure that the pads on both sides of the board are exactly registered so that no misalignment of drill holes will occur. Extra care in aligning the drill bit for double-sided boards should be taken, since any hole drilled on one side will have the same alignment with the underlying pad on the opposite side. The criteria for determining properly drilled holes are shown in Fig. 17.26.

After the board has been drilled, hardware requiring staking or swaging should be installed before any components are assembled onto the board. (Staking and swaging are discussed in Chapter 14.) Similar to single-sided boards, components that are not sealed or have materials that can be adversely affected by flux-cleaning solvents should not be initially installed. (Refer to Chapter 14 for a detailed discussion on component assembly.) After all noncritical components have been assembled on one side, the opposite side is coated with flux and dip-soldered. The board is cleaned of all flux residue and the final component assembly may be completed. (Refer to Chapter 15 for detailed discussions on proper soldering procedures.)

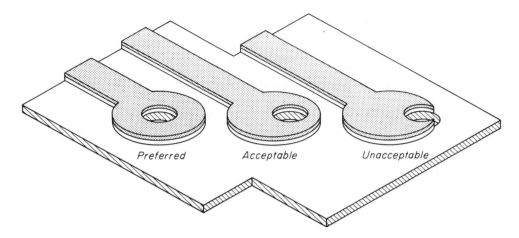

FIGURE 17.26. Drill hole criteria for acceptability.

Electrical connections between both sides of the board just assembled and dip soldered must be made. One method of forming these connections is to hand-solder all component leads having conductor paths leaving their terminal pads. (It is unnecessary to solder pads on the component side of the board that have no conductor paths leading to or from them.)

Hand-soldering the component leads on the component side of the pc board to form the required electrical connection can be avoided with the use of special interconnecting techniques for double-sided boards. These can be classified as either *electroplated* or *mechanical* interconnections. An electroplated interconnection between board sides is called a *plated through-hole*. A very thin layer of copper is first electrolessly plated onto the wall surface of all drilled holes. A greater thickness of copper is required in the hole than is possible to achieve through electroless plating. Therefore, a thicker electroplated copper surface is provided over the thin layer of electroless copper to achieve reliable connections. The results of such plating are shown in Fig. 17.27. To produce these plated through-holes requires a multiprocess involving cleaning, electroless copper plating, and then copper electroplating procedures that are not practical for prototype work.

An alternative and more economically feasible method for providing electrical connections between both surfaces of a pc board is by mechanical means. These may be in the form of a *clinched jumper*, or *roll* or *flare flange eyelet*. These types of interconnections are shown in Fig. 17.28. Clinched jumper wires require soldering to both sides of the board to achieve sound mechanical and electrical connections. The roll or flare eyelets do not necessarily require soldering but should be soldered to ensure positive and reliable electrical connections. Solder rings may be used

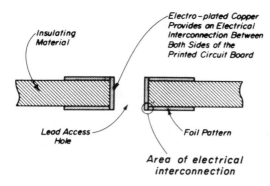

FIGURE 17.27. Copper-plated through-hole-type interfacial connection.

for this purpose. The proper size of either eyelet type is selected with the aid of Fig. 17.28 and manufacturer's information such as that provided in Table 17.6. Dimension B in Fig. 17.28 is determined by the board thickness, whereas dimensions A and C depend on the diameters of the terminal pad and hole into which the eyelet will be inserted. The technician must anticipate the use of eyelets in the pc board design and provide a terminal pad having a diameter on the processed board approximately 0.020 inch larger than dimension A. In addition, the drilled hole in the board should be 0.005 to 0.010 inch larger than dimension C for proper fit. When eyelets are not to be soldered, the roll flange type is generally chosen to ensure a sound electrical connection.

When the appropriate eyelet has been selected, it is installed by the swaging methods previously discussed. An arbor press with the appropriate anvil and setting tool are used to mechanically roll or flare the edge of the eyelet against the copper foil. Soldering both sides of the eyelet to the foil establishes highly reliable electrical connections between surfaces of the board. The cross-sectional views of eyelets properly swaged or soldered to the pc board are shown in Fig. 17.28 together with a properly formed and soldered jumper.

FIGURE 17.28. Methods of interfacial connections for double-sided printed circuit boards. (a) "S"-type clinched jumper. (b) Roll flange eyelet. (c) Flare flange eyelet.

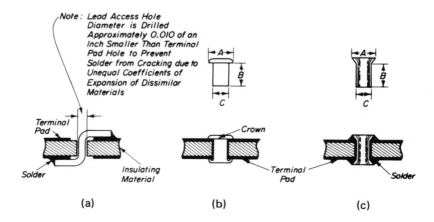

TABLE 17.6

Dimensional Information for Roll and Flange Eyelets

A	B	C	Drill Size	Board Thickness
Roll Eyelets				
0.105	0.093	0.059	Use	1/16
0.105	0.125	0.059	#51	3/32
0.105	0.156	0.059	Drill	1/8
0.150	0.093	0.089	Use	1/16
0.150	0.125	0.089	#41	3/32
0.150	0.156	0.089	Drill	1/8
0.200	0.093	0.121	Use	1/16
0.200	0.125	0.121	#30	3/32
0.200	0.156	0.121	Drill	1/8
Flange Eyelets				
0.095	0.093	0.059	Use	1/16
0.095	0.125	0.059	#51	3/32
0.095	0.156	0.059	Drill	1/8

All dimensions given in inches.

With the eyelets properly mounted, a component lead passing through the eyelet need only be soldered to the conductor side of the board (by hand or dip methods) to form an electrical connection to both sides.

A completely assembled and properly soldered pc board is shown in Fig. 17.29.

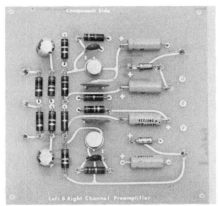

FIGURE 17.29. Completely processed and assembled preamplifier pc board.

18

Chassis Hardware and Assembly

18.0 INTRODUCTION

With the completion of all subassemblies (the pc boards in the amplifier project), the assembly of hardware and components onto the main chassis element may begin. In addition to the boards, this involves the mechanical securing of jacks, connectors, fuse holders, terminal strips, switches, potentiometers, lamps, meters, power cords, and transformers.

Before beginning to assemble parts, the overall task should be examined. If there are few components and parts to be assembled and if they are small, random mounting will pose no particular problem. If, however, the assembly involves many component parts with a wide range of sizes and weights, some order will be necessary to minimize needless effort and reduce the possibility of damaging delicate components.

Small, sturdy components such as fuse holders, terminal strips, lugs, grommets, and connectors should be mounted first. Transformers, filter chokes, and other large or bulky components should be among the last to be mounted, to avoid having to work with a heavy, bulky chassis during the initial assembly phase. Fragile components such as lamps, meters, sensitive relays, and pc boards should be the last to be mounted. As working space diminishes with the addition of components, extreme care must be exercised to avoid damaging parts with assembly tools.

Mounting electronic components and hardware to a metal chassis may appear to be a simple task. However, efficient and quality workmanship will result only if the four major factors of packaging are considered. These are: (1) mechanical security, (2) appearance, (3) accessibility, and (4) cost. Each factor is discussed in this chapter in conjunction with specific information on common and special fastening devices. Also considered are correct tool selection and assembly techniques for securing components and hardware.

18.1 MACHINE SCREWS

Machine screws are among the most common types of fasteners used in electronic packaging since they provide sound mechanical security and are easy to assemble and remove. Machine screws are designated by (1) *head style*, (2) *drive configuration*, (3) *diameter*, (4) *threads*, (5) *length*, (6) *material*, (7) *finish*, and (8) *fit*.

Some of the common head configurations are the *fillister*, *binder*, *pan*, *round*, *flat*, *oval*, and *truss* styles. These are shown in Fig. 18.1a. Although somewhat of a personal preference, the head style selected is generally determined by appearance, application, and mechanical security.

FIGURE 18.1. Machine screw styles and drives. (a) Machine screw-head styles. (b) Head-style drive configurations. (c) Drive configurations requiring special driving tools. (d) Machine screw-thread measurements.

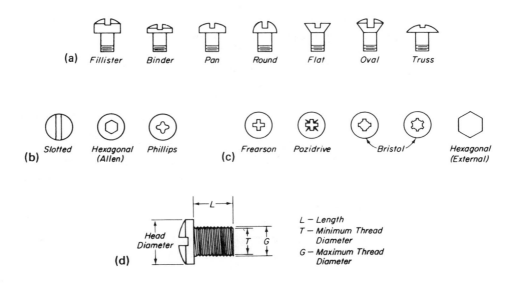

Fillister, binder, and pan heads are usually preferred over round heads for appearance sake. These head styles are mechanically superior to the round head because there is more material at the driver blade pressure points. These head styles have approximately the same-size sealing plane (surface contact area) for a particular screw size, although the fillister style has the smallest head diameter. Both the flat and oval heads require that the work surfaces be countersunk prior to assembly. The flat head is used for flush mounting and has sealing plane angles of 82 or 100 degrees. Neither the flat nor the oval heads are recommended for use with thin sheet metal, wherein countersinking might weaken the assembly. The oval head is often used with a cup-style washer for securing control panels to racks. Countersinking in this case is unnecessary since the shape of the washer provides the required seating. This arrangement tends to provide both a decorative appearance as well as excellent mechanical security. The truss head style, common to sheet-metal screws, has a larger sealing plane that provides greater surface contact area with the work.

The drive configuration refers to the style of head for tool access. Those most common to electronic assemblies are the (1) *slotted*, (2) *Phillips*, and (3) *hexagonal socket* or *Allen* design. These are shown in Fig. 18.1b. The slotted style is the most widely used. The Phillips head has the advantage over the slotted type in that it provides four positive-drive pressure points completely within the perimeter of the head. It is often used when the screw is visually obstructed during assembly, since it allows the screwdriver to be self-aligned. In addition, the recessed style of the Phillips head greatly reduces the possibility of the driver's slipping from the screw head, which could damage the work surface. The Allen configuration allows for even more positive tool positioning and gripping and is commonly used on set screws for securing knobs to control shafts. Other less common drive configurations that may be encountered are shown in Fig. 18.1c. Each of these requires special driving tools. The various driving tools common to electronic assembly will be discussed in Sec. 18.4.

The maximum thread diameter (dimension G in Fig. 18.1d) varies from 0.086 to 0.216 inch for general electronic assembly. The minimum diameter, shown as dimension T in Fig. 18.1d, will vary from 0.0628 to 0.1619 inch. Small sizes of machine screws are generally designated by *gauge number*. For the thread diameters referred to, the gauge numbers vary from Number *2*, having a diameter of 0.086 inch, to Number *12*, having a diameter of 0.216 inch. Larger thread diameters are given in fractions of an inch. Smaller diameter screws, with gauge numbers of *1* or *0*, are not common in electronic assembly. A listing of some common thread gauge numbers together with their decimal equivalents is shown in Table 18.1. A more complete listing is provided in Appendix VII. One way of determining the gauge number for an unknown screw size is simply to measure the thread diameter with a micrometer and compare the reading thus obtained with the corresponding gauge number in Table 18.1.

TABLE 18.1

Machine Screw Sizes Most Frequently Used in Electronics

Gauge Number	Decimal Equivalent in Inches	Threads Per Inch	
		UNC	UNF
2	0.0860	56	64
4	0.1120	40	48
6	0.1380	32	40
8	0.1640	32	36
10	0.1900	24	32
12	0.2160	24	28

Another method is to use a *screw gauge*, shown in Fig. 18.2. The threads of the unknown screw are placed into the tapered V-shaped slot and moved downward until the threads contact both edges of the slot. The correct gauge number is the one that is aligned most closely with the center of the screw.

Machine screws are manufactured with *right*- and *left*-hand threads. Right-hand threads are secured by clockwise (cw) rotation and extracted by counterclockwise (ccw) rotation. Left-hand threads are rotated in the opposite directions and are used only in special cases wherein vibration or rotation of the system would tend to loosen the conventional right-hand thread.

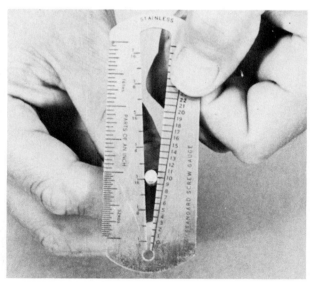

FIGURE 18.2. Screw diameter gauge.

Machine screws are also designated by the number of threads per inch. The simplest method of determining this is by placing a scale against the threads and counting the number of threads in 1 inch. For large screws with few threads per inch, this method is suitable. However, the smaller-diameter screws having many threads per inch require a different technique. The *thread gauge*, shown in Fig. 18.3, is employed for this purpose. This gauge contains as many as 30 blades, each stamped with a number designating the number of threads per inch. The gauge is positioned with a blade in contact with the threads. The threads per inch are determined when one blade exactly meshes with the threads to be measured.

For a particular screw gauge or diameter, the number of threads per inch will depend on the screw thread system used. Although there are other systems of screw designations, machine screws having either *Unified Coarse* (UNC) or *Unified Fine* (UNF) designations are predominantly used for electronic assembly. Referring to Table 18.1, a UNC screw having a gauge No. 4 diameter has 40 threads per inch, whereas a UNF screw with the same gauge number has 48 threads per inch. It is apparent that the UNC system has fewer threads per inch than the UNF system. The coarseness of the screw selected will depend on its application. Generally, UNC threads are preferable for rapid assembly since thread meshing is a less exacting process. This coarser screw is also used when threading into soft material. Since a machine screw of a given size with a coarse thread has a smaller minimum diameter than the fine thread, it will mesh deeper into the material surrounding the tapped hole. This advantage greatly reduces the possibility of thread shearing when an axial load is placed on the screw. If vibrations are a consideration, fine threads are preferable.

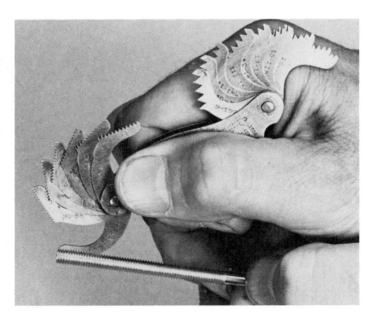

FIGURE 18.3. Thread gauge.

Because of their greater minimum diameter, they can be torqued more than comparably sized coarse thread screws and will not shake loose as readily.

For all head styles except the flat and oval designs, the thread length is measured from the sealing plane to the end of the screw (Fig. 18.1d). Flat-style and oval-style screw lengths are measured from the top of the sealing plane to the end. Screw lengths between $\frac{1}{8}$ and $\frac{5}{8}$ inch are available in increments of $\frac{1}{16}$ inch. Lengths from $\frac{5}{8}$ to $1\frac{1}{4}$ inches are graduated in $\frac{1}{8}$-inch increments, whereas those from $1\frac{1}{4}$ to 3 inches are graduated in $\frac{1}{4}$-inch increments.

Screws for electronic assembly are made from *steel, brass, aluminum,* or insulating materials such as *nylon.* The finishes available include *black anodizing, cadmium, nickel, zinc,* or *chrome* platings. The finish selected will depend on the appearance and the necessary corrosion protection.

Screws are also designated with a specific relationship (fit) between mating threads of nuts or tapped holes and machine screws. There are four classes of fit between threaded parts: *Class 1* is an extremely loose fit with considerable "play" between the threaded parts; *Class 2,* sometimes designated as a "free fit," can be readily assembled with finger pressure; *Class 3* has more of a snug fit than Class 2 but requires only finger pressure for initial assembly; *Class 4* is characterized by the tightest available fit, requiring screwdrivers and wrenches for assembly of the threaded parts. Selecting the most appropriate class of fit will depend on such factors as (1) speed of assembly, (2) type and style of washer employed, and (3) the amount of looseness between threaded parts that can be tolerated. For most electronic applications, the Class 2 fit is standard.

Machine screws may be completely identified by the following abbreviated description:

$$\#6\text{-}40 \times 3/8 \text{ UNF} - 2A \text{ LH Pan Hd. Steel C.P.}$$

An explanation of this example follows:

#6	—Major thread diameter (equivalent to 0.138 inch, refer to Table 18.1).
40	—Number of threads per inch.
3/8	—Length of thread in inches measured from the sealing plane to the screw end.
UNF	—Type of thread (Unified Fine). If the type was UNC, this designation would normally be omitted and assumed to be understood.
2A	—The 2 represents the class of fit between threaded parts and the A designates that the thread is external. A B suffix designates an internal-type thread for machine nuts.

LH — Left-hand thread. If a right-hand was specified, no designation is given since it is assumed to be understood.

Pan Hd. — Head style.

Steel — Material.

C.P. — Plating (cadmium-plated).

Clearance hole sizes for several machine screws are given in Appendix VIII. The clearance hole size is determined by the maximum diameter of the screw thread. This type of hole is drilled when components and hardware are to be assembled with screws and nuts. The diameter of a clearance hole must be sufficiently larger than the thread diameter to allow complete unobstructed passage of the screw through the hole. The amount of clearance will depend on the materials to be fastened. However, the values given in Appendix VIII are average sizes, which are satisfactory for most applications. A clearance hole together with its tolerance is shown in Fig. 18.4.

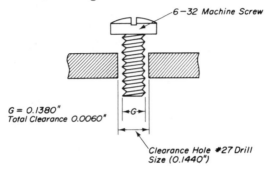

6–32 Machine Screw

G = 0.1380"
Total Clearance 0.0060"

←G→

Clearance Hole #27 Drill
Size (0.1440")

FIGURE 18.4. Clearance hole for a 6–32 machine screw.

18.2 NUTS AND WASHERS

Nuts used with machine screws are designated by (1) *style*, (2) *chamfer*, (3) *hole diameter and threads*, (4) *fit*, (5) *material*, (6) *finish*, (7) *width*, and (8) *thickness*. Nuts common to electronic assembly are shown in Fig. 18.5a. The hexagonal style is by far the most widely used and is generally preferable to the square nut for reasons of appearance and ease of assembly. This nut may have *single chamfer*, *double chamfer*, or *single chamfer and washer face* (Fig. 18.5b). When a single chamfer nut is used, the flat side should be against the work to provide the best appearance with no exposed sharp edges. If the work surface is finished, double chamfer or washer face nuts should be used. When tightening the nut against the work surface, the chamfer or washer face will prevent any visible surface marring, such as the circular abrasions formed by the points of a single chamfer square or hexagonal nut.

The hole diameter and thread type are, naturally, consistent with the mating screw. This is also true for class of fit, material, and finish. For

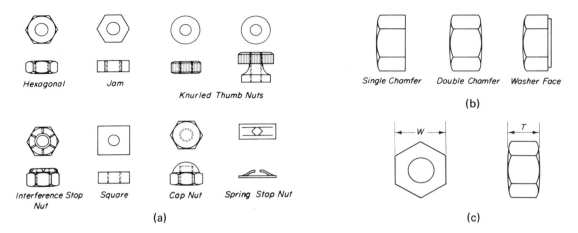

FIGURE 18.5. Machine screw nuts and fasteners. (a) Nuts common to electronic assembly. (b) Hexagonal nut face styles. (c) Width and thickness measurements for hexagonal nuts.

each size and style of nut, different widths (measured across pairs of opposite flats) and thicknesses are available. These external dimensions are shown in Fig. 18.5c. Tabulated sizes for common hexagonal nuts are given in Appendix IX. Nuts with larger widths are normally used for ease of handling and rapid assembly whereas those with smaller widths are selected if there is a possibility of interference with other hardware or if assembly clearance is limited.

The special-purpose nuts, also shown in Fig. 18.5a, have styles that dictate their application. A description and application of each follows:

Jam nut — basically the same as a hexagonal nut except that it is from $\frac{2}{32}$ to $\frac{3}{8}$ inch thinner, depending on the hole diameter. This nut may be used alone but it is more often employed under a regular hexagonal nut for a more secure assembly. When this nut is tightened, its flats align with those of the hexagonal nut, providing the desired locking action.

Interference stop nut — used to prevent loosening due to vibration. It has an internal unthreaded plastic or fiber collar. When the nut is tightened, the screw threads are forced into the inside surface of this tapered collar. Being elastically deformed, the collar squeezes against the screw threads, providing a secure locking action and eliminating the need for a lock washer.

Square nut — often used in securing channel stock in steel cabinets or racks. The flat of the nut often fits snugly or jams against the inside member when being tightened, thus requiring merely a screwdriver for assembly.

Knurled thumb nut — allows for quick finger tightening. It is commonly used on terminal posts such as those on some primary cells and telephone terminal blocks.

Cap nut—used primarily for purposes of appearance. It completely covers the end of the screw and its shape eliminates sharp corners or edges. Chrome-plated cap nuts add a decorative appearance to a cabinet.

Spring stop nut (speed nut)—pressed onto the screw with only minimal final tightening required. It is self-retaining as well as self-locking. Because it greatly reduces assembly line operations, the speed nut is often used in radio and television receivers. It may also be used to secure smooth studs or tubing.

Washers are classified as either *flat* or *lock type* (Fig. 18.6). Flat washers are used in contact with a screw head or the chamfer of a nut to provide more surface area to distribute the forces imposed on the screw and nut. Lock washers are used to secure square or hexagonal nuts from loosening due to vibrations or handling. For electronic assembly, the *tooth-type* lockwasher is used extensively since it provides maximum gripping action. Normally, the use of one lock washer is sufficient. However, under severe vibrations lock washers should be placed under both the nut and the screw head. The tooth-type lock washer should not be reused when removed. The gripping edges of the teeth are easily dulled and the spring action becomes fatigued. These washers have from 10 to 24 combined points of contact, depending on the number of teeth. *Internal* tooth washers are preferred over the *external* type from the standpoint of appearance since they cause no visible surface marring when assembling components such as the potentiometer shown in Fig. 18.7a. External tooth washers are generally used with flat washers for mounting components having oversized clearance holes. This arrangement is shown in Fig. 18.7b.

The *spring-type* or *split* lock washer, also shown in Fig. 18.6, uses the temper of the steel to maintain locking action. As the washer is compressed, the spring tension exerts a force that prevents counterrotation.

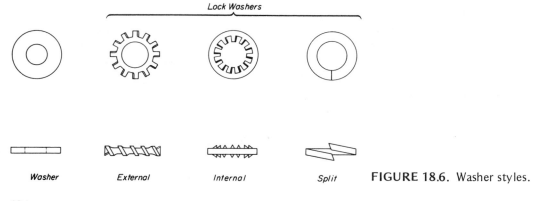

FIGURE 18.6. Washer styles.

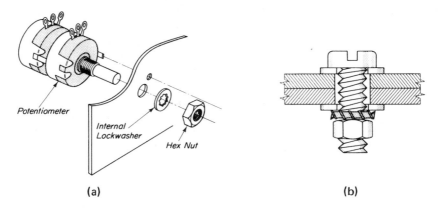

FIGURE 18.7. Washer assembly techniques. (a) Washer assembly for potentiometer mounting. (b) Washer assembly for oversized clearance holes.

This type of washer is less effective than the tooth type because the locking action results only from spring tension.

Flat steel washers are used in conjunction with tooth-type lock washers when securing soft materials, whereas resilient washers, such as lead or fiber, are used to provide a cushion effect when assembling brittle material.

The inside diameter of a washer must coincide with the clearance hole diameter of the screw with which it is associated. The outside diameter will depend on the application. Recommended washer sizes are given in Appendix X.

18.3 THREAD-FORMING AND THREAD-CUTTING SCREWS

Thread-forming and thread-cutting screws are widely employed when the use of nuts or the tapping of machine threads is not practical. These screws do not require a prethreaded hole for fastening. Manufactured from hardened steel, they form or cut their own threads as they are turned and driven into a hole. They cannot be driven into a material harder than their own. These screws are available in slotted, hexagonal, and Phillips drives with head styles common to machine screws, in addition to the *washer* head style shown in Fig. 18.9. Three of the popular thread designs are shown in Fig. 18.8. These are the *sheet metal* (thread forming), *machine screw thread forming*, and *self-tapping* (thread cutting) types. The sheet metal screw, shown in Fig. 18.8a, when forced into the proper-size predrilled hole, will form the metal around the hole into a threaded shape matching its own thread. This screw has a widely spaced thread extending to a point. The machine screw thread-forming type, shown in Fig. 18.8b, has a slightly tapered end but does not extend to a point. It has standard machine screw threads, which require greater force

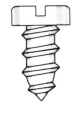 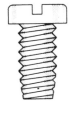 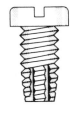

FIGURE 18.8. Thread-forming and thread-cutting sheet metal screw styles. (a) Sheet metal. (b) Machine. (c) Self-tapping.

(a) *Sheet Metal.*　　　**(b)** *Machine.*　　　**(c)** *Self-tapping.*

to drive than a sheet metal screw. This screw is used in place of the sheet metal screw when the possibility of loosening from vibrations exists. The self-tapping screw, shown in Fig. 18.8c, has one or more slots in the threaded end for chip relief. As this screw is driven into a hole, it cuts threads in basically the same manner as a tap. (Taps are discussed in Sec. 18.8.) The self-tapping screw is not used on sheet metal but is intended for fastenings to castings and soft materials such as Bakelite, copper, aluminum, and plexiglass.

Sizes of thread-forming and thread-cutting screws are similar to those for machine screws. To result in optimum mechanical security, the size of the predrilled hole is important. The size of this hole will depend on the *screw size, gauge of the metal,* and *type of metal.* Generally, for the same size screw, a larger hole must be provided in harder materials. For example, if a No. 6 self-tapping screw is to be used to secure sheet metal whose thickness is 0.060 inch, a No. 32 drill bit is recommended for steel, whereas a No. 36 would be necessary for aluminum. Information on hole sizes for thread-forming and thread-cutting screws is provided in Table 18.2.

Thread-forming and thread-cutting screws do not provide the mechanical security obtainable with machine screws with lock washers and nuts. However, when vibration is not a factor, they overcome accessibility problems in areas where nuts cannot be installed. Access or cover plates are an example of this application.

The proper securing of two pieces of sheet metal with a sheet metal screw is shown in Fig. 18.9. The hole drilled in the metal closest to the screw head must be a clearance hole to ensure positive metal-to-metal

Serration to Ensure Positive Gripping

Washer Head Style Sheet Metal Screw

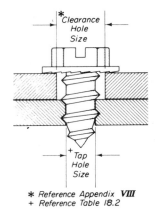

* *Clearance Hole Size*

+ *Tap Hole Size*

* *Reference Appendix* **VIII**
+ *Reference Table 18.2*

FIGURE 18.9. Recommended method for sheet metal assembly.

TABLE 18.2

Recommended Drill Sizes for Sheet Metal Screws

Thread Forming Machine and Sheet Metal Screws							*Self-Tapping (Thread Cutting) Screws*							
	Metal Gauge Numbers							*Metal Gauge Numbers*						
Screw Size	*Aluminum*						*Screw Size*	*Aluminum*						
	28	26	24	22	20	18		24	22	20	18	16	14	12
4	45	44	42	42	42	40	4	45	44	44	44	44	43	43
6	40	39	39	39	38	36	6	38	38	37	37	37	36	35
8	34	34	33	32	32	31	8	33	32	32	31	30	29	28
10	32	31	30	30	30	29	10	28	28	27	27	27	27	26
12	29	28	27	26	25	24	12	21	21	20	20	20	19	17
	Steel							*Steel*						
4	44	44	42	42	40	38	4	43	42	42	41	39	38	37
6	39	39	39	38	36	35	6	36	36	35	34	32	31	30
8	33	33	33	32	31	30	8	32	31	31	30	29	28	25
10	31	30	30	30	29	25	10	27	27	26	24	24	22	20
12	28	26	26	25	24	22	12	19	19	19	18	16	14	13

These drill sizes are intended only as a guide and may vary with application.

contact when the threading is completed and the sealing plane of the screw head is flush with the metal. For sheet metal applications, screws with larger sealing planes, such as the washer-head type shown in Fig. 18.9, should be used.

18.4 SCREWDRIVERS AND NUTDRIVERS

Screwdrivers are the most common tools used for securing the screws discussed in Sec. 18.3. They are available in a variety of *handles, shank lengths* and *shapes*, and *driver styles*. A popular-style screwdriver is shown in Fig. 18.10a together with the nomenclature of its parts. The handle is usually made of hardwood or plastic. In addition, a rubber grip may be provided for additional electrical protection, positive gripping, and comfort. For those screwdrivers without a rubber grip, large flutes are formed in the handle for improved gripping. The shank is manufactured from a tempered steel alloy specifically designed to withstand distortion caused by the excessive torque to which these tools are subjected. The handle

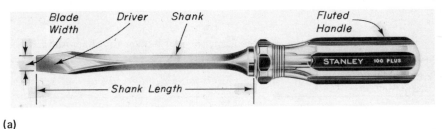

(a)

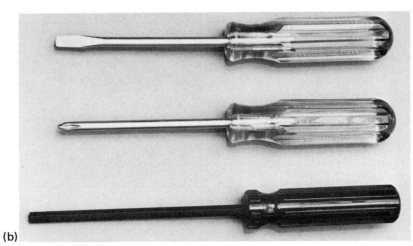

(b)

FIGURE 18.10. Screwdriver styles. (a) Screwdriver nomenclature. Courtesy of Stanley Tools. (b) Standard, Phillips, and hexagonal drivers.

and shank, however, will not withstand severe forces when the screwdriver is misused, such as for prying or bending. The shank may have either a round or square cross section. The square design is usually found on larger-size screwdrivers, where severe torque is expected. Shank lengths normally vary from 3 to 16 inches. For general electronic assembly, shank lengths up to 10 inches find wide application.

Driver styles are available for all screw-head drive configurations shown in Fig. 18.1b and c. The three most common types are the *standard*, *Phillips*, and *hexagonal* drivers. These are shown in Fig. 18.10b.

The standard driver has a tapered tip design that fits into the drive slot of a screw head. Common blade widths are $3/32$, $1/8$, $5/32$, $3/16$, $1/4$, and $3/8$ inch. The proper blade width should exactly equal the length of the drive slot to take advantage of the maximum surface contact area for optimum drive efficiency and security. The ideal blade-width criteria for slotted screw heads are shown in Fig. 18.11a and b. It can be seen in these figures that the blade width should exactly equal the head diameter. When selecting the most suitable screwdriver to avoid damage to the screw or the work surface, the width of the blade should closely match the slot length but should not extend beyond the perimeter of the screw head.

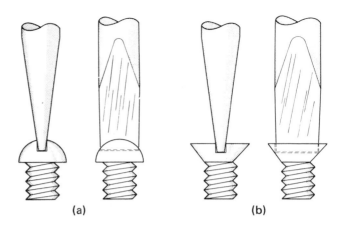

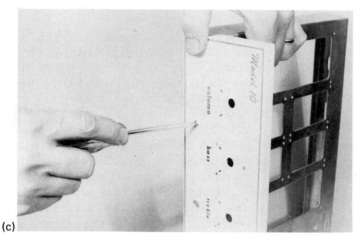

(a)

(b)

(c)

FIGURE 18.11. Ideal blade-width criteria for slotted-head screws and proper driver positioning. (a) Proper blade fit for round-head screws. (b) Proper fit for flat-head screws. (c) Perpendicular shank positioning.

Phillips screwdrivers are designed to fit the tapered cross-slots on screw heads made for this type of drive (Fig. 18.1b). Since more contact points are available with this design than with the slotted type, more driving force can be applied with less possibility of slippage. This style is commonly available in five number sizes; 0, 1, 2, 3, and 4. These will fit a large variety of Phillips screw-head sizes. Table 18.3 shows the screwdriver number and the screw size to which each will accommodate. Those most frequently used in electronic assembly are numbers 1, 2, and 3. This screwdriver must be held in exact alignment with the drive configuration of the screw head to be effective. The Phillips screwdriver properly positioned for use is shown in Fig. 18.11c.

TABLE 18.3

Phillips Screwdriver Selection

Number Tip Size	Machine Screws		Sheet Metal Screws
	Flat or Oval	*Round*	*Flat, Round, Oval, or Binder*
0	0,1	0,1	------
1	2-4	2-4	2-4
2	5-9	5-10	5-10
3	10-16	12-16	12-14
4	above 16	above 16	------

The Frearson or Pozidrive (a trade name of Phillips International) styles (see Fig. 18.1c) are similar to the Phillips drive except that they allow for a much more precise fit between the driver and the screw head. This fit enables the screw to be supported unaided on the end of the driver, which can be a time-saving advantage in assembly work in which accessibility is limited.

The *Allen* drive, as well as the special *Bristol* type (whose screw heads are shown in Fig. 18.1), also have a distinct advantage over the slotted style in that they minimize the possibility of the driver's slipping out of the screw head and damaging the work surface. The Allen drive is encountered more often in electronics assembly than the Bristol style. A selection of Allen drives from $3/64$ to $3/8$ inch is suitable for most electronic applications. These sizes vary by $1/64$ inch for sizes from $3/64$ to $3/32$ inch (measured across opposite flats as shown in Fig. 18.12), by $1/32$ inch for $3/32$ inch to $1/4$ inch and by $1/16$ inch for $1/4$ to $3/8$ inch sizes.

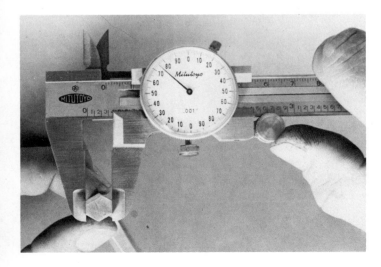

FIGURE 18.12. Method of determining size of Allen drive.

For extremely small set screws, the Bristol drive is often preferable to the Allen style since its multiple spline design prevents rounding of the driving surfaces of either the driver or the screw head through continued use. Bristol drives are available with four or six flutes.

For small and delicate work, jewelers' screwdriver sets, shown in Fig. 18.13a, may be used. The barrel of the handle is knurled with a swivel top finger rest. This configuration allows the driver to be turned using the thumb and forefinger while applying pressure with the index finger (Fig. 18.13b).

Screwdrivers for special applications are shown in Fig. 18.14 and are available in all the standard drive configurations. The *"stubby"* design, shown in Fig. 18.14a, is used if there is limited access to the screw head. These drivers are available in lengths from 1 to 2 inches.

Offset drivers, such as those shown in Fig. 18.14b, are also used for work in restricted areas where even the stubby style would not fit. When the driver is of the blade type, alternate ends of the offset screwdriver must be used successively in order for the screw to be completely

FIGURE 18.13. Jeweler's screwdrivers for delicate work. (a) Jeweler's screwdriver set. No. AA 0.025″, No. A 0.040″, No. B 0.055″, No. C 0.070″, No. D 0.080″, No. # 0.100″. Courtesy of L.S. Starret Company, Athol, Massachusetts. (b) Properly held jeweler's screwdriver.

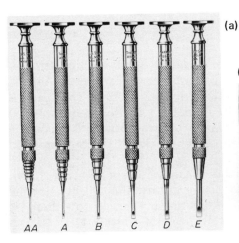

(a)

(b)

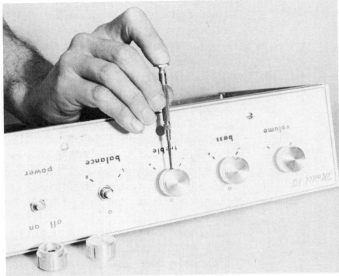

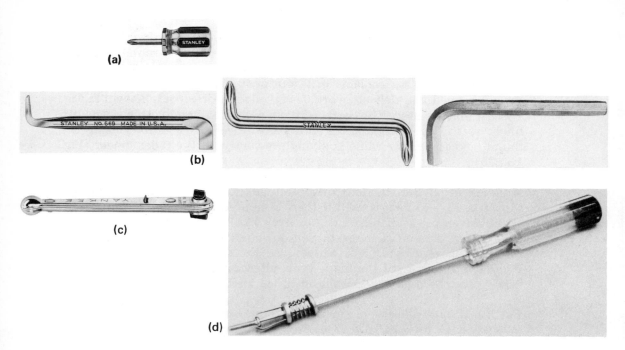

FIGURE 18.14. Screwdrivers for special applications. (a) "Stubby."
(b) Standard, Phillips, and Allen offset drivers. (c) Offset ratchet. (d)
Driver with screw-holding mechanism. (a), (b), and (c) Courtesy of
Stanley Tools.

tightened by alternate partial turns. When using the offset Phillips or Allen
drivers, it is unnecessary to use alternate ends. An improvement on the
offset driver is the *ratchet* design shown in Fig. 18.14c. This tool is
similar to a ratchet-type socket wrench. A screw is tightened or loosened
by a back-and-forward motion of the handle. A two-position lever on the
driver end of the handle is set to obtain either clockwise or counterclockwise ratchet driving action.

Screwdrivers having special screw-holding devices, such as that shown
in Fig. 18.14d, retain the screw head firmly against the driver, allowing
the full length of the shank to be used in gaining access to otherwise
impossible assembly positions. The screw head is grasped by a spring-
loaded dual leaf device. When this assembly is forced against the driver tip,
the spring compresses and the leaves spread outward. The screw head is
then inserted against the driver and the leaves cup the head under its
sealing plane as the spring assembly is released and forced toward the
handle.

The proper operating position of any screwdriver is always in alignment with the axis of the screw. This position minimizes the possibility
of slippage damage and provides maximum surface contact area between
the driver and the screw head.

Nutdrivers are used in a similar manner to screwdrivers but are designed to accommodate hexagonal-style machine nuts or screw heads. A set of nutdrivers is shown in Fig. 18.15a. They are available in sizes 6, 7, 8, 9, 10, 11, 12, 14, 16, and 18. The number represents the size in 32nds of an inch. For example, a number 6 will fit a $^{6}/_{32}$ ($^{3}/_{16}$) hexagonal nut or screw head as measured across opposite flats. Sizes 6, 8, 10, 11, and 12 are most frequently used in electronics assembly. Sizes above $^{1}/_{2}$ inch are useful for the larger nuts associated with potentiometers and switches.

FIGURE 18.15. Hexagonal nutdrivers. (a) Nutdrivers are available with individual handles or interchangeable handles. (b) Hollow-shank nutdriver. Courtesy of Xcelite, Inc. (c) Combined use of nutdriver and screwdriver.

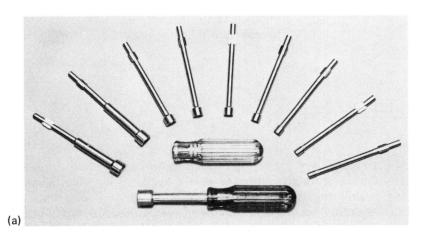

(a)

(b)

(c)

The sizes of nutdrivers are often stamped into the shank or colored handles are provided for quick identification. A shank length of 4 inches is suitable for most applications. Nutdrivers are available with hollow shanks for threading nuts onto long machine screws that extend more than $\frac{1}{4}$ to $\frac{3}{8}$ inch into the driver (Fig. 18.15b).

A screwdriver in conjunction with a nutdriver should be used to assemble and disassemble hardware whose fasteners require considerable force. This technique is shown in Fig. 18.15c.

The hexagonal *nut starter*, such as the type shown in Fig. 18.16, is an extremely useful aid where accessibility is difficult. It is worn over the index finger to hold the nut in position until the screw can be started with a screwdriver. These tools are available to accommodate $\frac{5}{32}$ - to $\frac{3}{8}$ -inch nuts.

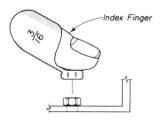

FIGURE 18.16. Nut starter for assembly in limited-access areas.

18.5 WRENCHES

Wrenches are necessary for larger nuts found on some potentiometers and switches when a suitable nutdriver is not available or when accessibility does not permit its use. The most common types of wrenches used in electronic assembly are the *box*, *open-end*, and *adjustable* wrenches shown in Fig. 18.17.

The box wrench, similar to a nutdriver, requires complete access over the nut since it encircles and grips the nut at its points. This wrench is a 12-point drive as opposed to the 6 points of a nutdriver. If used to secure a hexagonal nut, all 6 of the points will be contacted, virtually eliminating the possibility of shearing or rounding the points. The box wrench is preferable over other types of wrenches because it is much less likely to gouge the work surface. The box wrench should fit snugly onto the nut. All sizes listed for the nutdriver are also available for the box wrench, in addition to larger sizes that are not normally required in electronic assembly. Lengths of 4 to 6 inches are suitable for most assembly applications. The proper use of a box wrench used to secure a potentiometer to the front panel of the amplifier is shown in Fig. 18.18a. Masking tape may be applied to the panel where the edge of the wrench would contact the work surface around the nut to eliminate the possibility of damaging the finish.

Open-end wrenches are primarily intended for use with square nuts. The proper-size wrench can be determined when the entire length of the inside faces of the jaws fit snugly against any two opposite flats of the

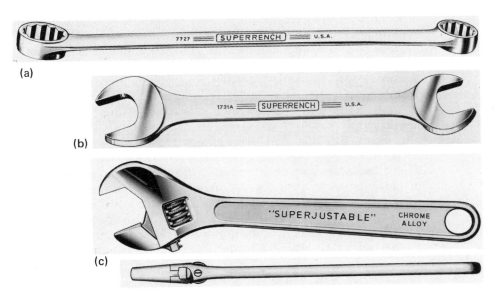

FIGURE 18.17. Wrenches. (a) Box. (b) Open-end. (c) Adjustable.
Courtesy of J. H. Williams & Co.

FIGURE 18.18. Wrenches in assembly. (a) Box wrench. (b) Correct
position of jaws when tightening right-handed threads.

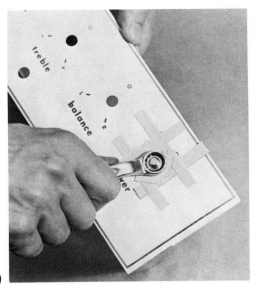

(a)

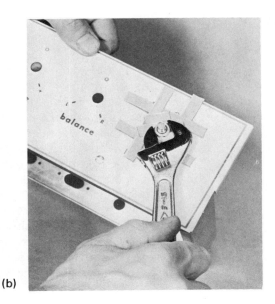

(b)

nut. This ensures maximum surface contact and minimizes slippage and rounding of the nut. The jaws of this wrench are generally angled at 15 degrees to the handle center line. This design makes it possible to work in confined areas where the nut can only be turned partially. Open-end wrenches for electronic assembly are available in a complete range of sizes from $\frac{3}{16}$ to 1 inch and between $3\frac{1}{2}$ to 10 inches long. Although available in sizes much larger than 1 inch, these are seldom required in electronic applications.

Both the box and open-end wrenches require a complete set to accommodate all possible nut sizes that may be encountered in assembly. The adjustable wrench eliminates the need for complete sets since the distance between jaws can be changed to any size within its capacity. It does not, however, provide as secure a grip as either the box or open-end wrench. The adjustable wrench is similar to the open-end wrench with the 15-degree angle except that only one jaw is stationary. The second jaw can be moved by means of a thumb-operated *adjusting screw* located within the head of the wrench. To adjust the wrench for proper fit, the jaw is first opened sufficiently to allow both jaws to easily slip over opposite flats on the nut. The movable jaw is then closed until both jaws are gripping firmly against the flats. It is important that when positioned, the *pulling* force should be applied to the stationary jaw. If excessive force is applied to the movable jaw, the adjusting screw may be damaged or the wrench could slip, damaging the work surface or surrounding parts. The correct position and adjusted fit for the adjustable wrench used to tighten a right-hand threaded nut is shown in Fig. 18.18b.

18.6 POP RIVETS

An alternative method of securing sheet metal or hardware by means other than machine screws and nuts or sheet metal screws is with the use of *rivets*. The standard rivet is a metal stud with a flared head. The rivet shaft is placed through the clearance holes of the metal parts to be fastened. The head is then placed on an anvil and the shaft end is uniformly peaned over with a rivet-setting tool and hammer. This procedure is shown in Fig. 18.19. Since this process is time-consuming and requires skill and free access to both work surfaces, it is not a popular means of fastening in prototype work. *Pop riveting*, on the other hand, is rapid, requires little skill, and free access to only one side of the work is necessary. The only tool used for this process is a hand-type riveting gun. These advantages make this method of fastening convenient in prototype work, even though it does not provide the rigid mechanical security and overall uniformity of appearance obtainable with a standard rivet.

A typical pop rivet, shown in Fig. 18.20a, consists of a *truss style head*, *sealing plane*, *hollow shaft*, *stem*, and *button head*. The stem is tapered near the button head. When a tensile force is applied to the stem, the button head is drawn up into the hollow shaft. This action flares the

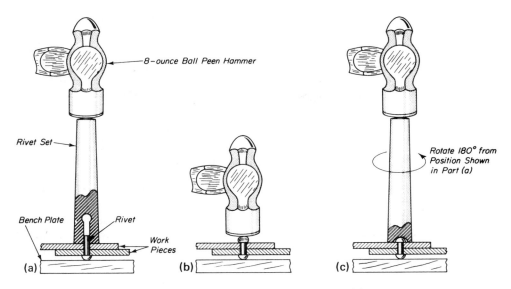

FIGURE 18.19. Sequence for standard rivet installation. (a) Draw (setting workpieces tightly together against rivet head). (b) Upset (striking rivet to slightly flatten or mushroom the shank end). (c) Head (using shallow hole in rivet set to form second head).

FIGURE 18.20. Pop rivet and its assembly. (a) Pop rivet assembly and nomenclature. (b) Pop rivet gun positioned against rivet head. (c) Installed pop rivet.

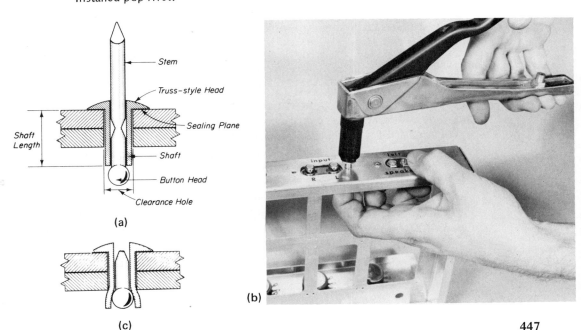

end of the rivet against the metal, thus providing a positive fastening of the metal parts. When a pulling force of from 500 to 700 pounds is reached, the stem will break at the tapered section. The sheet metal or hardware to be fastened must be provided with clearance holes to accept the shaft of the rivet. A clearance of 3 to 5 thousandths of an inch is sufficient. The shaft is then pressed through the holes with the sealing plane flush against the metal surface. The pulling force is applied to the stem through the locking action of the pop rivet gun *nose piece* (stem collet). The nose piece is placed over the stem and pressed flush against the head of the rivet (Fig. 18.20b). As the operating lever of the gun is squeezed, the stem of the rivet is gripped, drawing the button head upward into the hollow shaft. Continued squeezing of the operating lever will create the desired flaring action. The stem is broken when sufficient pulling force is reached. The operating lever may have to be actuated more than once to complete this operation. When the lever is again released, the broken stem drops out of the nose piece. An installed pop rivet is shown in Fig. 18.20c.

Rivets are a permanent type of fastener intended for use if components, sheet metal, or hardware are not to be removed once fastened. This fact must be considered when using rivets since they are difficult and time consuming to remove. If it is necessary to remove a rivet, drilling through the rivet center with a drill bit size which is at least equal to the clearance hole is required.

Pop rivets of the style described are available in steel or aluminum with shaft diameters of $\frac{1}{8}$, $\frac{5}{32}$, and $\frac{3}{16}$ inch and shaft lengths of $\frac{1}{8}$ to $\frac{5}{8}$ inch. The clearance hole drill numbers for these three diameters are Nos. 30, 20, and 11, respectively. The stems of the three available shaft diameters require their own nose piece on the pop rivet gun. These nose pieces thread into the gun assembly and are easily removed with an open-end wrench.

18.7 PLIERS FOR HARDWARE ASSEMBLY

The two types of pliers most commonly used in electronic assembly are the *long nose* and the *gas* pliers. These are shown in Fig. 18.21. Both styles consist of a *pivot, jaws,* and *handles.*

The long-nose pliers consist of a pair of long, narrow tapered jaws. These jaws are smooth or may have serrations near the tips for improved gripping. These pliers are designed to aid in assembly for holding or positioning small parts in tight spaces where finger accessibility is limited. They are not intended to withstand severe gripping or twisting forces and should not be used for tightening machine nuts. Misusing the tool in this way could cause the jaws to spring out of alignment, rendering the pliers useless. Long-nose pliers holding a machine nut for assembly are shown in Fig. 18.21a.

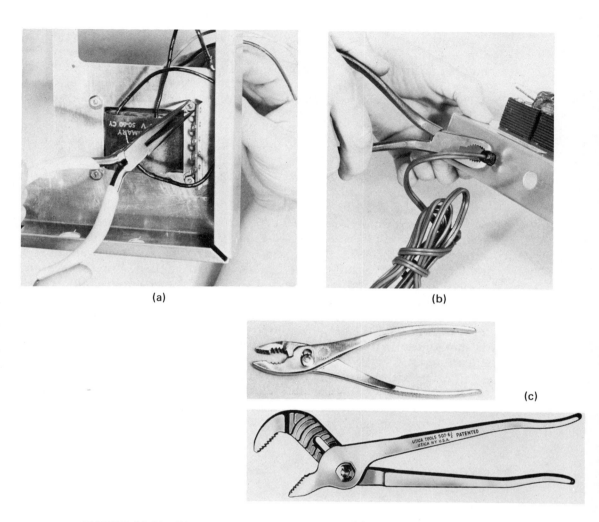

FIGURE 18.21. Pliers for hardware assembly. (a) Long-nose pliers aid in assembling small parts. (b) Gas pliers used to install power cord strain relief. (c) Slip and rib-joint pliers. Courtesy of Utica Tool Company, Inc.

Gas pliers have a much more rugged design than long-nose pliers and have coarse serrations over the complete inside surface of each jaw. These pliers are used primarily for holding round stock, such as inserting a power cord strain relief bushing, as shown in Fig. 18.21b. Excess gripping force on soft material should be avoided because the serrated jaws will mar the surface. For this reason, gas pliers should be used with extreme care. A minimum of two layers of masking tape wrapped around the stock will help prevent the jaws from marring.

Although many other styles of pliers are available, such as the *slip-joint* and *rib-joint* types, shown in Fig. 18.21c, they are designed for holding heavy stock in which marring of the surface is not a factor. They should not be used in place of any of the drivers, wrenches, or pliers discussed in this chapter. Special pliers for cutting wire will be considered in Chapter 19.

18.8 TAPS AND DIES

Taps and *dies* are *thread-cutting* tools constructed of hardened steel. The tap is used to cut internal threads into predrilled holes for accepting screw threads. The die will cut external threads on rod stock for assembling nuts. Normally, the thickness of the metal used to form chassis is too thin to allow an adequate number of threads to be tapped. For screws to be mechanically secure, a metal thickness that will allow a minimum of three full threads to be cut is required. For this reason, components and hardware mounted to chassis use either machine screws and nuts, sheet metal screws, or rivets. In applications in which one of these three methods is not desirable, special fastening techniques, discussed in Sec. 18.9, can be employed. However, the use of taps and dies is sometimes the only practical alternative to make an effective and neat fastening.

A tap with its nomenclature is shown in Fig. 18.22a. Taps are available in common machine screw sizes and are designated identically. For example, a hole formed for a 4-40 machine screw is cut with a 4-40 tap. Three basic hand taps are shown in Fig. 18.22b, the (1) *taper* tap,

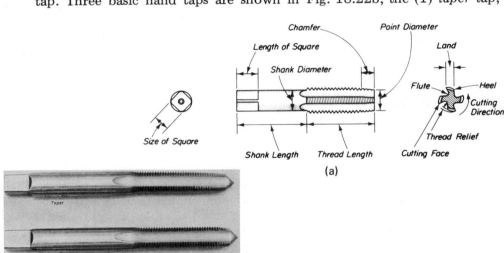

FIGURE 18.22. Tap nomenclature and styles. (a) Machine screw tap nomenclature. (b) Tap styles.

(2) *plug* tap, and (3) *bottoming* tap. The distinction in each style is in the amount of chamfered taper at the front end of the threaded section. As can be seen in Fig. 18.22b, taper taps have the most prominent taper and are preferable for starting threads. For this reason, they are sometimes referred to as *starter* taps. The taper tap is used for threading *through* holes (passing completely through the work). For *blind* holes (terminating at some depth within the work) tapping involves a sequence of operations. The thread is first started with a taper tap. Owing to its chamfered taper, the last several threads toward the bottom of the hole cannot be cut. For this reason, the plug tap is used next to cut all but the last few threads. To thread a blind hole completely to the bottom, a bottoming tap is finally used to form the remaining threads.

The design of a tap is similar to that of a drill bit. It is fluted to allow entry of lubricants as well as to provide space for chips to accumulate while thread cutting. The number of flutes will vary depending on the size of the tap, with the smaller taps having 3 flutes and the larger ones having 4.

Hand tapping can only be accomplished with a tap secured in a *straight adjustable* tap wrench or a *chuck handle "T"* tap wrench. These are shown in Fig. 18.23.

The straight adjustable tap wrench, shown in Fig. 18.23a, has two jaws that grip the square drive end of the tap. By rotating the handle attached to the adjustable jaw, the tap is firmly secured for use.

The chuck handle "T" tap wrench, shown in Fig. 18.23b, grips the tap in the jaws of its chuck, which is rotated to secure the tap in place.

The tapping operation begins by drilling a hole whose diameter is equal to the minimum diameter of the required screw (dimension T in Fig. 18.1d). The correct tap drill sizes for most common screws are given in Appendix XI for various metals.

When the correct size tap hole has been formed and the taper tap is secured into a tap wrench, a lubricant is applied directly to the tap. The tapered end of the tap is positioned into the hole perpendicular to the work surface. When a great deal of tapping is to be done, small hardened

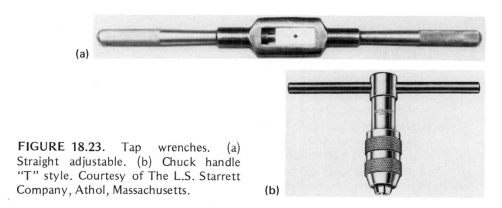

(a)

(b)

FIGURE 18.23. Tap wrenches. (a) Straight adjustable. (b) Chuck handle "T" style. Courtesy of The L.S. Starrett Company, Athol, Massachusetts.

steel blocks (*tapping blocks*) with predrilled clearance holes for various tap sizes are extremely useful. With the tap positioned in the tapping block, proper tool alignment will be achieved when the block is in contact with the work surface and their holes are aligned. With the tap positioned, it is gently rotated clockwise (for right-handed threads) with a uniform downward pressure. After the first half-turn, the tap is rotated in the reverse direction for at least one-quarter turn. The purpose of this technique is to break the chips generated and to clear them into the flutes, where they fall toward the bottom of the hole. This procedure is continued throughout the tapping process. After the first two or three full turns, downward pressure on the tap is no longer required since it reaches a point where it threads its own way into the work. When working with hardened materials, considerable resistance to the tap may be encountered. If an unusually high spring effect is felt at the handle of the tap wrench, this is an indication that the tap is jammed in the work. The tap is extremely brittle and is easily broken. When this excessive resistance to thread cutting is encountered, the cutting rotation must immediately be reversed to break the bound chips and additional lubricant added to the cutting area by way of the flutes. For through holes, a fine wire may be passed through each flute to help dislodge the broken chips. This is not recommended for blind holes since the chips may be compressed, causing the tap to jam in either direction. When working on blind holes, the tap must be removed periodically so that the chips may be shaken or blown out. It may be that forward rotation may be limited to something less than a one-half turn with a necessary reverse turn of one-half to three-quarters. Each technician must develop his or her own procedures based on personal experience. It must be emphasized that a broken tap is virtually impossible to remove without damaging the work surface. When working with a specific metal for the first time, it is advisable to experiment on a piece of similar scrap stock to gain experience in the necessary sequence of forward and reverse rotation.

The purpose of the lubricants introduced through the flutes is to reduce friction between the tap and the work and to prolong tool life. For this reason, lubricants should be used freely. A table of cutting lubricants for several common materials is given in Table 6.1.

Although dies are available in several styles, the *split die* is the most commonly used and will be discussed exclusively. This die is shown in Fig. 18.24a. The center of the die has cutting threads that are interrupted by holes along the perimeter of the threads, giving the appearance of a clover. The purpose of these holes is for chip relief. On one side of the die, the threads are chamfered outward and away from its center line to facilitate alignment with the work. An adjustment screw is located in the perimeter of the die to obtain the desired fit. This is a trial-and-error process. If too much stock is cut where a tight fit is desired, the threaded portion is cut off and the adjusting screw tightened. On the other hand,

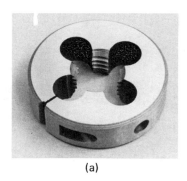

(a)

(b)

FIGURE 18.24. Die tools for cutting external threads. (a) Split die.
(b) Die stock with die inserted.

if not enough stock is cut to provide a loose fit, the adjusting screw is loosened and the die is run over the first set of threads to cut them deeper.

The die is used in conjunction with a handle termed a *die stock*, shown in Fig. 18.24b. The die is positioned with its unchamfered thread closest to the inside shoulder of the die stock and held firmly in place with a set screw that aligns with a recess provided along the outer edge of the die. With the die secured in the die stock, the work is held in a vise or other suitable clamp. The chamfered threads of the die are properly aligned when the handles of the die stock are perpendicular to the work. Similar to the tap, the die is rotated first with a forward one-half turn followed by a reverse one-quarter turn to clear chips. A properly installed die is shown in Fig. 18.24b. Lubricants recommended for tapping are also used for dies.

Dies, as well as taps, are designated in sizes identical to those of machine screws. An 8-32 die will cut 32 threads per inch on stock whose diameter before threading is 0.164 inch (refer to Appendix VII).

Forming threads by hand is one of the most difficult machining operations because of chip removal and tool alignment. The important factors to consider for obtaining the desired threads are (1) proper size of predrilled hole or diameter of stock, (2) correct size and type of tap or die, (3) recommended lubricant, and (4) proper alignment between the tool and the work.

18.9 SPECIAL FASTENERS AND HARDWARE

There arise many instances in which the conventional fastening methods do not completely satisfy the assembly requirements because of many reasons, such as accessibility, orientation, frequent removal, and location. Many special fasteners and hardware are commercially available that find wide application in electronic assembly. Some of the more common of these will be discussed here.

An alternative method of fastening components to sheet metal or securing chassis elements is with the *self-clinching nut*, shown in Fig. 18.25a. This type of nut provides a means of securing to sheet metal with a machine screw thread and is capable of being used with metal stock as thin as 0.032 inch. Self-clinching nuts are available in *steel*, *stainless steel*, and an alloy of *aluminum* and *Monel*. The knurled clinching ring grips the sheet metal, causing it to be pressed around the tapered base forming a permanent bond as the nut is forced into the metal. These nuts are available in thread sizes from 2-56 to ½ -20.

To attach the self-clinching nuts into the sheet metal, a hole from 0.166 inch for the 2-56 nut to 0.656 inch for the ½ -20 size is required. The tolerance for this hole must be within +0.003 inch to –0.000 inch for maximum gripping of the clinching ring and tapered base into the sheet metal. The base of the nut is inserted into the hole with the clinching ring against the work surface. The nut is then pressed into the metal with a hydraulic press. The installed self-clinching nut, shown in Fig. 18.25b, is a raised design. Flush-mounted, threaded stud, and solder terminal designs are also available.

If a package is designed for continual access to a portion of the circuitry within an enclosure for test or service, *panel locks* are often specified. One such arrangement is shown in Fig. 18.26. This arrangement provides excellent security and rapid access.

The layout design may specify a single wire or cable to pass through a hole formed in the metal chassis element. For this requirement, a *grommet* should be used to prevent damage to the wire installation. A

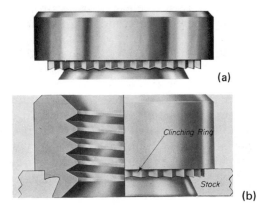

(a)

Clinching Ring

Stock

(b)

FIGURE 18.25. Self-clinching nut. (a) Raised self-clinching nut style. (b) Installed self-clinching nut. Courtesy of Penn Engineering & Manufacturing Corp.

FIGURE 18.26. Quick couple slide lock.

grommet is a rubber insulator that "cups" around the hole in the chassis and prevents wires from rubbing against the metal edges. A grommet being inserted into the amplifier's main chassis element is shown in Fig. 18.27a. An access hole must be provided in the chassis that is equal to dimension A in Fig. 18.27b. Some common size soft-rubber grommets together with the necessary size of access hole for each are listed in Table 18.4. Although dimension A is the grommet size designation, the selection of the proper size grommet depends on the minimum required size of dimension B.

If the packaging design requires that a machine screw or other metal stud be insulated from the chassis, *fibre shoulder washers*, such as those shown in Fig. 18.28a may be used. Two washers are necessary, as shown in Fig. 18.28b, to ensure complete electrical insulation of the stud from the surrounding metal edges of the access hole. The major diameter of the

FIGURE 18.27. Grommets are inserted into holes to cover sharp edges. (a) Grommet installation technique. (b) Important dimensions for grommet installation.

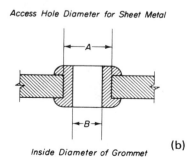

Access Hole Diameter for Sheet Metal

Inside Diameter of Grommet   **(b)**

TABLE 18.4
Rubber Grommet Dimensions

Grommet Size Designation	A in inches	B in inches
1/4	1/4	1/8
5/16	5/16	3/16
3/8	3/8	1/4
1/2	1/2	3/8
5/16	5/16	3/16
3/4	3/4	9/16
9/16	9/16	7/16
5/8	5/8	1/2
7/8	7/8	3/8
7/8	7/8	5/8
1	1	11/16

FIGURE 18.28. Fiber shoulder washers. (a) Common shoulder washers. (b) Recommended method of assembly. (c) Important shoulder washer dimensions for installation.

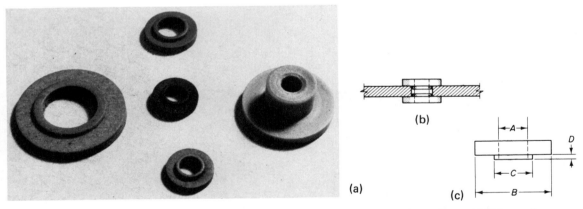

stud used determines the size of dimension A in Fig. 18.28c, whereas dimension B is the outside diameter of the washer. Dimension C is the outside diameter of the shoulder for which the access hole in the metal must be formed, and dimension D corresponds to the shoulder height. If the height of the shoulder is more than one-half the thickness of the metal, a flat fibre washer may be necessary in place of a second shoulder washer to ensure flush mounting and total insulation. The appropriate size shoulder may be determined from Table 18.5.

TABLE 18.5

Fiber Shoulder Washer Dimensions

Screw Size	I.D. A	O.D. B	Shoulder	
			Diameter C	Height D
3	0.110	0.250	0.187	0.031
4	0.136	0.250	0.187	0.031
5	0.136	0.312	0.187	0.031
6	0.140	0.375	0.237	0.031
8	0.172	0.375	0.246	0.031
10	0.196	0.375	0.308	0.031
12	0.250	0.500	0.312	0.028
1/4	0.265	0.500	0.365	0.031
5/16	0.375	0.750	0.500	0.031
3/8	0.385	0.625	0.500	0.031

Dimensions A, B, C, D are in inches.

The use of machine screws and nuts, as used to mount the seven pc boards in the amplifier shown in Fig. 18.31, is one acceptable mounting arrangement. However, large areas of metal must be removed from the chassis. The sheet metal work could be reduced by the use of *standoffs*, *spacers*, and *circuit board supports*, which often find application in prototype as well as production assemblies. This hardware is shown in Fig. 18.29. Straight standoffs and spacers are made of plastic for applications requiring electrical insulation and are available in lengths from ⅛ to several inches. Standoffs, shown in Fig. 18.29a, are threaded at both ends for mounting and will accept a screw size range of from No. 4 to No. 10. *Hinged* standoffs are available in cadmium-plated brass. Both ends of this style standoff are threaded for mounting to accept screw sizes from 4-40 to 8-32. Spacers, shown in Fig. 18.29b, are unthreaded, with clearance holes for screw sizes from No. 4 to No. 10.

By drilling holes into the four corners of the pc board, straight standoffs or spacers may be used between the board and chassis to support the board away from the chassis surface. Generally, if the screw length required for this assembly exceeds ¾ to 1 inch, the threaded standoff and two screws are used. The length of standoff or spacer is determined by the tallest projection or component extending from the side of the pc board that faces the chassis. A minimum of ⅛ inch must be added to this dimension to ensure adequate clearance.

The hinged standoff is generally selected to make circuit testing or servicing easier. Two corners of the pc board are fastened to hinged

(a)

(b)

(c)

FIGURE 18.29. Hardware for mounting printed circuit boards. (a)
Threaded standoffs. (b) Spacers. (c) Nylon printed circuit board
support.

standoffs and the two opposite corners are secured to standard standoffs.
The height of the hardware must be selected to allow the board to rest
parallel with the chassis surface with sufficient clearance. For servicing,
the two screws at the corners of the board secured to the standard standoffs are unfastened. The board can then be lifted upward to an angle
of 90 degrees by the pivoting action of the hinged standoffs, thereby
exposing wiring and components for testing and repair. This method of
mounting not only provides easy access for repair, but also can reduce the
overall package size.

Circuit board supports, shown in Fig. 18.29c, are available in nylon
and other plastic materials. They consist of two snap-lock arrangements,
one at each end of the support. For mounting, one end of the support
snaps securely into a 0.187-inch-diameter hole drilled into the chassis.

These supports provide another convenient method of mounting pc boards to chassis. The board is secured to the upper end of the support by a spring-type locking action. Support heights range from $\frac{3}{16}$ to $\frac{7}{8}$ inch. Circuit board supports can be mounted on all four corners of small boards. More may be used on larger boards for greater support. They have the advantage of allowing boards to be removed and quickly replaced and eliminate the need for removing large areas of metal from the chassis. These circuit board supports can be used in conjunction with connectors mounted for electrical interconnections between the board and chassis, which adds to the versatility of this method of mounting.

Printed circuit board guides, shown in Fig. 18.30, are grooved runners made of plastic and designed for use with edge connectors. These guides ensure alignment between the edge of a pc board and a connector and also provide additional support for the board. Generally fastened to the chassis with machine screws and nuts, pc board guides are also available with an adhesive backing for rapid attachment. The grooves in these guides are designed to accept all standard board thicknesses.

The final method of fastening to be considered is the use of adhesives, particularly *epoxy resins*. High-strength adhesives find wide application in electronic packaging since they provide a number of advantages over other methods of bonding. Some of these advantages are elimination of machine operations, provision of electrical insulation and low thermal resistance between parts joined, and uniform stress distribution over the entire bonded area. Although the use of adhesives requires thorough preliminary cleaning of parts and often heat and pressure during the curing period, the advantages of this method of bonding far outweigh the time necessary for its proper use.

FIGURE 18.30. Printed circuit grooved guides.

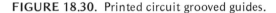

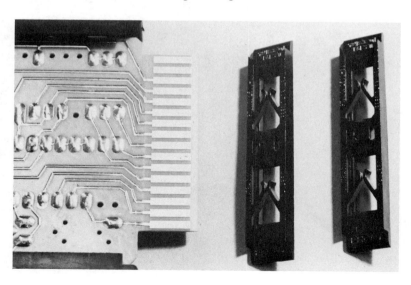

As mentioned in Sec. 14.7, devices may be mounted on heat sinks with epoxy. For this application, the thermal resistance and coefficient of thermal expansion of the adhesive should approximate that of aluminum and copper for satisfactory results.

Epoxy resins must generally be mixed with a *catalyst* (hardener) prior to use. The parts are measured by volume or weight. Manufacturers' mixing instructions must be consulted for the correct amounts for a specific epoxy since there is a wide variation among the many different adhesive types. The *working time* (time between mixing and initial hardening) must also be predetermined for the specific epoxy to avoid mixing more than can be used during the prescribed time.

For proper bonding, the surfaces must be thoroughly cleaned. Fine emery cloth to remove oxides may be used followed by degreasing with a solvent such as toluene or trichlorethylene. After the epoxy is properly mixed, a thin coating is applied to each of the mating surfaces and pressed together to squeeze out air bubbles and excess adhesive. Improved bonding to aluminum results if the surface is anodized prior to bonding. Depending on the epoxy used, curing time will vary from 5 minutes to 24 hours at room temperature. These periods can be reduced if the bonded surfaces are baked in an oven but care must be exercised not to exceed the temperature range of the epoxy. In addition, baking is not recommended when bonding devices to heat sinks because of the possibility of damaging components.

FIGURE 18.31. Completed mechanical assembly of all amplifier parts to chassis and front-panel elements.

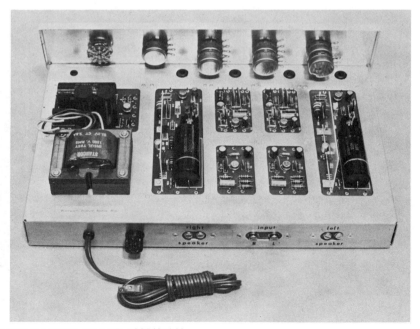

Epoxy is used for bonding when the stresses encountered will be only tensile or shear. Since the shelf life of many adhesives is often less than 1 year, the expiration date must not be exceeded for sound bonds. In addition, adhesives should be stored in a cool area to ensure longest possible shelf life.

The hardware and pc boards of the amplifier have been assembled using materials and techniques discussed in this section. The amplifier completely assembled is shown in Fig. 18.31.

Now that assembly is completed, wiring of the boards and chassis mounted hardware are necessary to complete the amplifier. The next chapter will consider wire and wiring assembly techniques.

PROBLEM SET

18.1 Design and fabricate a steel tapping block having approximate dimensions of ½ inch by 2 inches by ¾ inch. Referring to Appendix VIII, determine the clearance hole sizes for gauge numbers 4, 6, 8, and 10 machine screws. Drill four evenly spaced holes to the sizes found in the table. Number stamp the holes for the corresponding gauge. Break all sharp edges with the appropriate file and give all surfaces a light coat of oil to prevent rusting.

18.2 Using an appropriate epoxy, bond the two halves of an ink eraser to an aluminum handle such as that shown in Fig. 18.32. The handle is fabricated from 16-gauge aluminum having blank dimensions of ¾ inch by 10 inches and bent around a ⅝-inch-diameter dowel.

FIGURE 18.32.

18.3 Select an appropriate epoxy and seal the four corners of the soldering iron tip cleaning sponge tray constructed in Problem 15.1.

18.4 Design and construct a test jig stand for mounting printed circuit device test jigs designed and completed in Problems 11.2, 13.3, 14.2, and 15.2. Break all sharp edges with the appropriate files. Identify each test jig position on the stand with appropriate dry transfer letters and numbers.

18.5 Mount the lamp (L1), on-off switch (S8), and battery clip for the seven-fixed-interval UJT timer printed circuit designed and fabricated in Problems 11.3, 13.5, 14.3, and 15.3 onto the chassis fabricated in Problem 8.5. The pc board may be mounted onto the chassis with standoffs or by flush mounting using the appropriate size mounting hole. Consideration must be given to providing access to the pc board-mounted time interval switch. Wiring among chassis-mounted components and hardware and the pc board is done in Problem 19.5.

18.6 The meter for the tachometer printed circuit designed and fabricated in Problems 11.5, 13.4, 14.5, and 15.5 may be mounted onto a panel of the same design as that shown in Fig. 5.22. The pc board is to be mounted with proper length spacers and mounting screws to the rear of the meter, which in turn is secured to the panel. Wiring for the installation of the tachometer is done in Problem 19.6.

19

Hook-up Wire and Interconnection Techniques

19.0 INTRODUCTION

One of the least emphasized and perhaps the most important aspect of packaging is the selection of the most suitable hook-up wire for a specific assembly. From the standpoint of reliability, this selection must be based on such factors as *conductor material and plating, conductor size, type of insulation, flexibility, environmental factors*, and *current and voltage limits*, as well as *cost*. This chapter provides guidelines to aid in evaluating and selecting wire. Coupled with the problem of selecting are the methods for *interfacing interconnections*. Since there are a variety of standard interconnections (in addition to many special types), this chapter also includes pertinent information for evaluating the suitability of various connection methods.

Specific information also is provided here on common conductor materials and platings and American Wire Gauge (AWG) classifications of size, along with a comparison of solid wire with various types of stranded wire. Typical types of insulation are examined with respect to their application, including color-coding methods. Additional information is provided on current, voltage and mechanical and environmental criteria to allow for a total understanding of wire selection from the many types commercially available. Special wire types also are discussed.

Methods of interconnections are introduced, and the tools used for wire preparation and mechanical assembly are described, such as *strippers*, *diagonal cutters*, and others. *Wrap-around* and *feed-through* connections as they apply to swaged solder-type terminals, eyelet-type terminals, and solder lugs are discussed and expanded to include special hardware connections, such as solder pots, plugs, and jacks. Finally, solderless wiring techniques, such as solderless or crimp-type connectors and wire-wrap connections are set forth.

19.1 CONDUCTOR MATERIALS AND WIRE CONFIGURATIONS

Most of the wire used for electronic interconnections is fabricated from *annealed* (softened) copper. Copper used for this purpose is termed *Electrolytic Tough Pitch* (E.T.P.). Although other materials, such as aluminum, steel, and silver, are employed in wire construction, they are not common to electronics fabrication. Alloys of copper have been developed, however, which display superior characteristics when compared to pure copper. Although pure copper possesses excellent electrical conductivity characteristics as well as being malleable and ductile, it has the pronounced disadvantage of being highly susceptible to fractures under vibration and flexing conditions. Modern high-temperature annealing techniques, however, have made it possible to develop high-strength alloys with improved fatigue life. Some of these common high-strength alloys are *cadmium–copper*, *chromium–copper*, and *cadmium–chromium–copper*. One disadvantage in using these alloyed wires is that they are available only with coatings of silver or nickel, which are both expensive and difficult to solder. Tin-coated high-strength alloys are, unfortunately, not readily available.

Copper wire is usually coated, because pure copper oxidizes quickly when exposed to the atmosphere. *Tin* is the most common coating, although *silver* and *nickel* coatings are used in special applications. Tin-coated wire, recommended for applications wherein environmental temperatures will not exceed 150°C (302°F), is the least expensive of the three coatings mentioned. In addition, tin coating improves solderability. Silver coating, recommended for use in temperatures ranging from approximately 150 to 200°C (302 to 395°F) greatly improves solderability but tends to corrode at elevated temperatures. Silver is the most expensive coating of those mentioned. Nickel coating is capable of oxidation protection to temperatures of up to 300°C (575°F) but requires activated fluxes and high-temperature soldering techniques.

Tin-coated wire is by far the most commonly used in the electronics industry. Tin is applied to the bare wire by either *dipping* or *electroplating*. Dipping is the least expensive method and is used if uniform plating thickness is not critical. Electroplated tin should be specified if close tolerance automatic stripping is to be employed, since uniform plating is achievable by this method.

Wire used in the electronics industry can be divided into two groups: *solid* and *stranded*. Generally, solid wire that is comparable to a specific size of stranded wire is less expensive to manufacture and subsequently less expensive to purchase. Solid wire, however, does not possess the flexibility characteristics or the fatigue life of stranded wire. Whereas solid wire tends to fracture even under mild flexing, stranded wire remains highly flexible. Because much of the electronic equipment produced today is exposed to some form of vibration under normal use or has wiring flexibility requirements during assembly, stranded wire is the type most often specified.

Solid wire finds extensive application in the fabrication of leads for small components, such as resistors, capacitors, and solid-state devices. Hook-up applications of solid wire are generally limited to conditions wherein flexibility is not a criterion or lead runs are to be short, rigid terminations such as the direct connection of two closely spaced swaged turret-type terminals on a pc board. Solid, uninsulated tinned copper wire used as a short jumper or as a common circuit point is termed *bus* wire.

When longer lengths of bus wire are needed, it is usually recommended that the wire be straightened prior to its use. The best way to straighten wire is to secure one end of it in a vise and the other end in the chuck of a hand drill (egg beater). All that is usually required to straighten the wire is 8 to 10 complete turns of the drill in one direction followed by an equal number of turns in the opposite direction while applying continuous tension to the wire.

Stranded wire is classified as either *bunch* or *concentric* (Fig. 19.1). Bunch-type stranded wire is constructed of several small-diameter solid wires "bunched" together without regard to symmetry. Although the cost of bunch wire is low and it is extremely flexible, it does not have a consis-

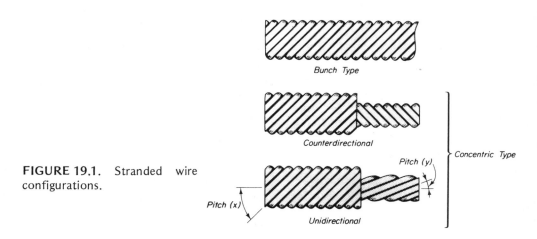

FIGURE 19.1. Stranded wire configurations.

tent diameter. Because of this disadvantage, it is not recommended, because modern insulation stripping tools may nick or break individual strands, which will weaken the wire as well as reduce current capacity.

Concentric-type stranded wire overcomes the disadvantages of bunch wire in that its consistent circular cross section makes it more suitable to manual as well as automatic stripping techniques. *Counterdirectional* concentric stranded wire, shown in Fig.19.1, is a multilayer wire with alternate layers of strands rotating or twisting in an opposite direction to the next succeeding lower layer. *Unidirectional* concentric stranded wire is also a multilayer wire, but each layer of strands twists in the same direction but with a different degree of *pitch* (angle at which strands cross the axis of the overall wire) as they are twisted. Unidirectional stranded wire is more flexible than the counterdirectional type, but its diameter is not so precisely uniform.

Solid and stranded wire each have their specific size designation systems. Solid wire is designated in terms of its cross-sectional area and diameter by *gauge number*. The most commonly used designation is the *American Wire Gauge* (AWG) standard system. Appendix XII lists the AWG numbers of solid bare copper wire with its equivalent cross-sectional area and diameter for gauge Nos. 00 to 50.

To better understand the information provided in wire tables, certain relationships need to be examined. In the electrical field, characteristics of different wire gauges and lengths are, for convenience, compared to a standard so that different wire sizes may be readily compared. The accepted standard is a piece of solid annealed copper wire with a diameter of 0.001 inch (1 mil) and a length of 1 foot whose resistance is 10.37 ohms at 20°C. (This standard is often referred to as the *mil-foot*.) Since most electrical conductors have round cross sections, a system of circular measure has been established to reduce the tedious task of comparing conductor characteristics. The parameters involved in such a comparison are resistance, length, and cross-sectional area. These relationships are expressed mathematically as:

$$\frac{\text{resistance wire I}}{\text{resistance wire II}} = \frac{\text{length wire I}}{\text{length wire II}} \times \frac{\text{area wire II}}{\text{area wire I}} \tag{19.1}$$

The expression for determining the area of a circle is $A = \pi/4 \, (\text{diameter})^2$. When the area of two conductors is compared, as above, $\pi/4$ is a constant for both wire I and wire II. Therefore, Equation (19.1) may be simplified as follows:

$$\frac{\text{resistance wire I}}{\text{resistance wire II}} = \frac{\text{length wire I}}{\text{length wire II}} \times \frac{(\text{diameter})^2 \text{ wire II}}{(\text{diameter})^2 \text{ wire I}} \tag{19.2}$$

The term $(\text{diameter})^2$ represents the cross-sectional area of a round wire. This term is designated as the *area* in *circular mils* (CM), where the area now is equal to the square of the diameter ($A = d^2$). All wire tables refer to diameters in mils and to cross-sectional areas in circular mils.

To determine the resistance of copper wire, the following relationship is used:

$$\text{resistance} = \frac{10.37 \times \text{length (feet)}}{\text{diameter}^2 \text{ (CM)}} \qquad (19.3)$$

If, for example, it is necessary to find the resistance of 50 feet of AWG No. 18 wire, Appendix XII shows that the area of this wire is 1624 CM. Substituting these known values into Equation 19.3 yields

$$R = \frac{(10.37)(50)}{1624} = 0.32 \text{ ohm} \qquad (19.4)$$

The wire diameter in mils can be converted to inches merely by moving the decimal point three places to the left (multiply by 0.001). For example, AWG No. 18 wire has a diameter of 40.30 mils, which is equal to 0.0403 inch.

Because one of the most important criteria for selecting wire is its current-carrying capacity, wire tables provide the cross-sectional areas of each wire gauge expressed in CM. The fourth column in Appendix XII lists the maximum recommended current capacity in amperes for common wires used in electronics. These values were obtained by the "rule of thumb" relationship that each 500-CM cross section will safely accommodate 1 ampere of current. For example, AWG No. 10 wire, having a cross section of approximately 10,000 CM, can handle 20 amperes. AWG No. 13 wire, which is 3 gauge numbers higher, reduces the cross-sectional area by one-half to approximately 5000 CM and therefore is capable of handling one-half the current, or 10 amperes. Again, AWG No. 23 (10 gauge numbers higher than No. 13), with its cross section of approximately 500 CM, is capable of handling 1 ampere. From these relationships, the following guidelines can be used: For each *decrease* or *increase* of three gauge numbers, the CM area doubles or halves, respectively, as does its current-carrying capacity. Correspondingly, a gauge number decrease or increase by a factor of 10 will increase or decrease the cross section by 10 or $\frac{1}{10}$, respectively, as will its current capacity. (Refer to Appendix XII.)

Common gauges of hook-up wire used in the electronics industry range from AWG No. 10 to No. 26, and their current-carrying capacities, from approximately 20 amperes to $\frac{1}{2}$ ampere.

The AWG standard numbering system was developed principally for bare solid-copper wire. This system is also used to designate stranded wire together with an additional number to specify the number of strands composing the wire. For example, in the designation *7/34*, the first number specifies that there are a total of 7 strands of wire and the second

7 Strand

12 Strand

FIGURE 19.2. Cross-sectional views of stranded wire.

19 Strand

37 Strand

number indicates that each strand is AWG No. 34. The cross-sectional views of various stranded wire configurations are shown in Fig. 19.2. To determine the equivalent AWG solid-wire size to a specific stranded wire, the cross-sectional area of each strand is multiplied by the number of strands in the overall wire. For example, to determine the equivalent solid-wire size of No. 7/34 stranded wire, the cross-sectional area of No. 34 is found to be 40 CM (from Appendix XII). The 7 strands constitute a total cross section of 7 × 40, or 280 CM. Appendix XII reveals that its closest equivalent is AWG No. 26 solid wire. Therefore, both No. 7/34 and AWG No. 26 have a maximum recommended current-carrying capacity of approximately 0.5 ampere. Table 19.1 lists some of the common stranded-wire types with their equivalent AWG numbers.

TABLE 19.1

Stranded-Wire Configurations
to AWG Numbers

Equivalent AWG Number	Stranding
10	37/26
12	19/25
14	19/27
14	37/29
16	19/29
16	27/30
18	7/26
18	19/30
18	27/32
20	7/28
20	19/32
22	7/30
22	19/34
24	7/32
24	19/36
26	7/34
26	19/38

To determine the AWG wire size of either solid or stranded wire, a *wire gauge*, such as that shown in Fig. 19.3a, may be used. Each slot in the gauge is accompanied by the corresponding AWG number size and a relief hole having a diameter greater than the slot width. The purpose of the relief hole is to allow for proper measurement "feel" and also for damage-free removal of the wire from the gauge after measurement. The proper method of positioning the gauge and wire to be measured is shown in Fig. 19.3b. Notice from the figure that when measuring stranded wire, the insulation is not completely removed as it is for solid wire. Rather, a section of insulation is separated to expose the conductor. This section should not initially be cut closer than 1 inch from the end of the wire to prevent the individual strands from separating or flattening as the wire is placed in the gauge. When removing the section of insulation, the underlying wires must not be twisted, for the diameter of the conductor would be altered and an erroneous reading would result.

FIGURE 19.3. Wire gauge and measuring technique. (a) Wire gauge to measure AWG sizes from #0 to #36. Courtesy of L.S. Starrett Company, Athol, Massachusetts. (b) Perpendicular gauge and wire positioning for accurate measurement in gauge slot.

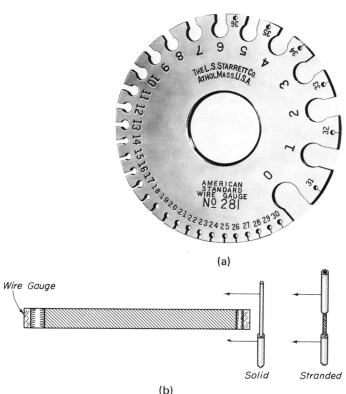

(a)

(b)

When measuring stranded wire, the wire gauge reading will be slightly larger than the corresponding AWG number for solid wire because of the slight increase in diameter caused by the spaces between strands. Therefore, when measuring stranded wire, the gauge slot into which the wire fits will be one AWG number *less* than the actual size and will also yield an *odd*-numbered AWG value. For example, a stranded AWG No. 20 wire will properly fit into the gauge slot marked AWG No. 19. Therefore, this rule of thumb indicates that the correct size of this wire is AWG No. 20. (It may be well to emphasize at this point that *odd* AWG number sizes are not commonly stocked and, for this reason, should not be specified.) An optional method of measuring stranded wire is to measure *one strand* in the gauge, count the number of strands in the total conductor, and refer to Table 19.1 to determine the equivalent AWG number.

To provide additional help in selecting wire, we will use the stereo amplifier to determine the most suitable type of wire plating, type (stranded vs solid), and size.

Example: Power amplifier requirements:

Current — 1.6 amperes dc maximum

Operating temperature range — 70°C (see Sec. 19.2)

Flexibility requirements — pc board is rigidly supported — no vibration problems — minimum flexibility required for assembly

Solution: Since flexibility is not a prime factor, solid wire is selected because it is the least expensive. The temperature requirements are minimal, thus allowing for the use of tin coating that will further reduce costs and ensure availability. The size of the wire will now be selected based on the current requirements of the circuit. Most of the interconnections are signal lines in which current demands are very low. The only wires that will be expected to handle 1.6 amperes are those originating from the power supply and feeding the amplifier boards. These wires are specifically the dc power lines V_{cc} (L) and V_{cc} (R), and the ground lines. Appendix XII indicates that AWG No. 20 wire will handle 2 amperes. All signal wires could be a great deal smaller than this; but for the sake of uniformity and simplicity, a realistic choice would be AWG No. 20 wire for *all* interconnections. Although the cost of wire is proportional to size, the savings realized by the choice of a smaller wire are negligible.

19.2 INSULATION MATERIAL

Hook-up wire used in making interconnections in electronic equipment must be provided with an insulation to prevent short circuits and to protect the conductor. The only exception to this rule involves the use of bus wire, which because of rigidity and short length, poses no insulation problems.

The selection of the most suitable insulation for a specific application is based on three criteria: *electrical*, *mechanical*, and *environmental*. Each of these criteria will be discussed.

Common wire insulations and insulation coatings are divided into three main categories: *rubbers*, *vinyls*, and *special high-temperature* or *special-application materials*. A comparative tabulation of various insulating materials is given in Table 19.2. This table compares the following mechanical and physical properties:

1. Voltage breakdown
2. Insulation resistance
3. Low dissipation factor
4. Abrasion resistance
5. Flexibility
6. Cut-through resistance
7. Water absorption resistance
8. Flame resistance
9. Acid resistance
10. Useful temperature range

The *voltage breakdown* and *insulation resistance* are the two factors that are of key importance in high-voltage applications. Insulations designated *Excellent* are the only ones that should be considered for this purpose. As seen from Table 19.2, silicone, polyethylene (PE), and Mylar possess high corona resistance and are commonly used for high-voltage cables. Neoprene, urethane, and nylon insulations would not be suitable for this application because of their poor rating. The amount of vibration that a wire is exposed to is directly related to its insulation voltage breakdown rating. Therefore, this factor needs to be considered in the choice of insulation.

Signal attenuation at high frequencies is a function of the insulation's *dissipation factor*. Generally, insulations with a low dissipation factor are preferable. Teflon has the lowest dissipation factor among those listed in Table 19.2.

Mechanical considerations, such as *abrasion resistance*, *flexibility*, and *cut-through resistance*, represent a compromise between cost versus installation ease. Abrasion and cut-through resistance are generally directly related — the higher the value of one, the higher the value of other. Of the insulations listed in Table 19.2, Kynar is the toughest, having excellent abrasion and cut-through resistance characteristics. The insulation flexibility becomes an extremely important consideration, especially in high-density packaging for ease of installation. The *rubber* family (silicon, neoprene, and butyl) exhibits the highest flexibility and the *vinyls* (PE and poly) the lowest.

Family	Material	Common Designation	Voltage Breakdown	Insulation Resistance	Low Dissipation Factor	Abrasion Resistance
Rubber	Silicone Rubber	Silicone	Exc.	Exc.	Good	Fair / Poor
	Polychloro- prene Rubber	Neoprene	Fair	Poor	Fair	Exc.
	Butyl Rubber	Butyl	Good	Good	Good	Good / Exc.
Polyo- lefin (Vinyl)	Polyvinyl chloride	PVC	Good	Good	Fair	Good
	Polyethy- lene	PE	Exc.	Exc.	Good	Fair / Good
	Polypro- pylene	Poly	Good / Exc.	Exc.	Good	Good / Exc.
Fluorocarbon	Polytetra- fluoro- ethylene	Teflon† (TFE)	Good	Good	Exc.	Exc.
	Monochlo- rotrifluo- roethylene	Kel-F*	Good	Exc.	Fair	Good / Exc.
	Polyvinyl- idine fluoride	Kynar**	Exc.	Fair	Good	Exc.
	Polyure- thane	Urethane	Good	Poor	Fair	Exc.
	Polyamide	Nylon	Poor	Poor	Fair	Exc.
	Polyester Film	Mylar†	Exc.	Exc.	Poor	Good

Ratings are in decreasing order: Excellent (Exc.), Good, Fair, Poor.

†   Trademark of DuPont
*   Trademark of Minnesota Mining & Manufacturing Company
**   Trademark of Pennsalt Chemical Corp.

Water-absorption resistance is an environmental consideration. Rubber and urethane materials are more susceptible to moisture than the vinyl materials. Teflon materials are the least susceptible to water absorption.

Flame-resistant materials find application in electronic equipment in which explosions or flammable conditions may be encountered. Teflon and Kel-F are the materials specified for this application, since they do not support combustion. All the other insulation materials listed in Table 19.2 will support combustion in varying degrees.

Insulations that are highly *acid resistant*, such as butyl rubber, polypropylene, and Teflon, find wide application in the space industry. Urethane, nylon, and silicone rubber are inferior in this property.

The useful temperature range for PVC, nylon, and PE is narrow, whereas for Teflon and Kel-F it is wide. In considering this property, it is

TABLE 19.2
Comparative Ratings for Wire Insulation

Flexibility	Cut-through Resistance	Water Absorption Resistance	Flame Resistance	Acid Resistance	Useful Temperature Range (°C)
Exc.	Poor	Fair	Fair	Fair	- 60 to 200
Exc.	Good	Good	Good	Good	- 30 to 80
Exc.	Fair	Good	Poor	Good/Exc.	- 40 to 85
Good	Good	Good	Good/Exc.	Good/Exc.	- 55 to 105
Fair	Good	Exc.	Poor	Good/Exc.	- 60 to 80
Poor	Good	Exc.	Fair/Poor	Exc.	- 20 to 125
Good	Fair	Exc.	Exc.	Exc.	- 70 to 250
Good	Good	Exc.	Exc.	Exc.	- 80 to 200
Good	Exc.	Good	Good	Good	- 65 to 125
Exc.	Good	Poor	Poor	Fair	- 55 to 80
Poor	Good	Fair/Poor	Good/Fair	Fair	- 40 to 120
Fair	Exc.	Good	Fair/Poor	Good	- 65 to 120

important to use good judgment in terms of cost. For example, it would not be advisable to choose Teflon (upper limit 250°C) based on its superior temperature characteristics if the temperature range of PE (upper limit 80°C), as well as its other characteristics, will satisfy the design requirements because Teflon is considerably more expensive.

From the wide range of materials available, it is evident that the selection of the most appropriate wire insulation for a specific application is a compromise between specifications and cost.

To illustrate the use of Table 19.2, the following problem is considered:

Circuit — stereo power amplifier
Electrical specifications —
    Voltage 36 volts dc maximum
    Current 1.6 amperes dc maximum
    Frequency 20 to 20 kHz

Mechanical specifications
    Stationary
    No expected high-level vibrations
    Low-density packaging
    All leads pass through grommets

Environmental specifications—temperature range +50 to 90°F (+10 to 32°C)
Humidity 40% typical

These specifications appear to present no critical circuit requirements insofar as wire insulation selection is concerned. Since voltage and current requirements are low, any insulation with a *voltage breakdown* and *insulation resistance* comparison rating of *good* or *excellent* is acceptable. This stipulation eliminates neoprene, Kynar, urethane, and nylon from further consideration.

With a frequency range specified at 20 Hz to 20 kHz, any insulation with a dissipation factor rating of *good* or *excellent* will be suitable. Low-dissipation-loss factors become critical only at higher frequencies.

Because all wires will be passed through rubber grommets, the possibility of abrasion is nonexistent and the quality of *abrasion resistance* need not be considered in this evaluation even if a *fair* or *poor* rating is assigned a specific material. Therefore, for this consideration, any of the wire insulations listed are suitable.

Flexibility also is not critical because of the low-density packaging of the circuit and the resultant uncomplicated wiring. Therefore, no insulation material need be rejected because of its flexibility rating.

To reduce assembly time, `especially if large-scale production is considered, an insulation with a comparison rating for *cut-through resistance* of *good* or *excellent* should be considered. This eliminates silicone, butyl, and Teflon from further consideration.

Since the use of this type of power amplifier is usually in the home or other location wherein temperature and humidity are controlled at levels that exclude critical environmental problems, any insulation with a *water-absorption resistance* rating of *good* or *excellent* is acceptable.

Flame and *acid resistance* is, of course, of no consideration in the intended environment. Therefore, no insulation, even with a *fair* rating for either of these categories, is omitted from consideration.

Finally, the expected ambient temperature range of the unit is approximately +10 to 32°C, with a factor of 2 used in derating on each end of the temperature range to allow for extreme conditions (the inside case temperature will increase under full power output). The range of derated temperatures now considered becomes 5 to 64°C. Table 19.2

shows that all the insulations listed have *useful temperature ranges* that are well within these specifications. As a result, no insulation material will be eliminated because of temperature limitations.

An evaluation chart is provided in Table 19.3 to show more clearly the procedure for eliminating insulation materials that are below minimum ratings. The *Analysis Results* column of Table 19.3 indicates that either polyethylene (PE) or polypropylene (poly) plastic vinyl insulation is suitable for the example problem. The final solution between these two materials now becomes a matter of cost and availability.

For purposes of identification, hook-up wire insulation is provided with a color code. This provides for rapid identification of wires, which is an indispensible aid when working with cables and harnesses. Color-coding is extremely useful in both assembly and in troubleshooting. Military and industrial standards have been established for color-coding insulation. These standards are listed in Table 19.4. Similar to the resistance color code adopted from EIA (Electronic Industries Association) standards, the solid color of the insulation is designated by numbers 0 through 9 and represents the first digit of the color-code number. With 10 basic colors, 10 numerical possibilities are available to identify individual wires. For complex wiring involving more than 10 leads that must be identified, insulation is also available with a broad tracer band that is a different color from the body color and spirals the length of the wire. These tracer colors now extend the available different color combinations to 100. When more than 100 wires are involved, as in the case of complex harnesses, insulation is available with a *second* tracer that is narrower than the first. Both tracers are closely spaced, running parallel with each other, and spiral the entire length of the wire. The use of the second tracer extends the numeric coding capabilities to 910 possible combinations. To illustrate the numbering system, a wire with a *red* body color, broad *green* tracer, and narrow *orange* tracer (red/green/orange) can be classified by the three-digit number *253.*

Standards also have been established that do not employ a numbering system but rather designate wire color with respect to circuit function. Two such standards are the Government Standard MIL-STD-122 color code for chassis wiring and the EIA color code for power transformers. These classifications are shown in Table 19.5.

19.3 SPECIAL WIRE CONFIGURATIONS

There are several readily available wire configurations that are used for special applications. *Shielded wire*, shown in Fig. 19.4a, is a variation of hook-up wire previously discussed. It consists of an insulated length of hook-up wire enclosed in a conductive envelope in the form of a braided wire shield. The shield is formed from fine strands of tinned, annealed copper wire. The strands are woven into a braid that provides approxi-

Material	Voltage Breakdown	Insulation Resistance	Low Dissipation Factor	Abrasion Resistance	Flexibility
Silicone	√	√	√	√	√
Neoprene	*	*	*	√	√
Butyl	√	√	√	√	√
PVC	√	√	*	√	√
PE	√	√	√	√	√
Poly	√	√	√	√	√
Teflon (TFE)	√	√	√	√	√
Kel-F	√	√	*	√	√
Kynar	√	*	√	√	√
Urethane	√	*	*	√	√
Nylon	*	*	*	√	√
Mylar	√	√	*	√	√

†   Total number of insulation characteristics which do not meet minimum required ratings
√   Insulation characteristic meets minimum required rating
*   Insulation characteristic below example problem minimum required rating

mately 85% coverage of the underlying insulated conductor. Shielded wire is also available with an outer insulation covering to prevent shorts when running the wire between termination points. Another variation of shielded wire has *two* inner conductors twisted around each other. This twist improves unwanted signal cancellation.

Shielded wire is employed in audio circuits at the input stage where the signal level is low and undesired noise (typically 60-hertz hum) or feedback from the output stage of a high-gain amplifier could produce unwanted signals. The shield is either connected to the chassis at the circuit input or it may be connected to the chassis at both ends to provide the most effective method of minimizing undesired signal pick up.

Coaxial cable (coax), shown in Fig. 19.4b, is used exclusively in RF circuits. It is similar in construction to shielded wire, with the outer insulating jacket isolating the shield from ground. The major difference between coaxial cable and shielded wire is in their electrical characteristics. Coax is designed specifically to transmit RF energy, from one point to another, with minimum loss (attenuation). Insulation material and thickness is controlled with extreme accuracy during manufacture to produce cables possessing 50, 75, or 95 ohms characteristic impedance for proper matching. Shielded wire, on the other hand, is intended for low-frequency applications in which its impedance is not critical.

Losses from *skin effect* at high frequencies must be minimized when using coaxial cable. Skin effect is the term used to describe the type of

TABLE 19.3

Sample Evaluation Chart for the Selection of Wire Insulation

Cut-through Resistance	Water Absorption Resistance	Flame Resistance	Acid Resistance	Useful Temperature Range	Analysis[†] Results
*	*	✓	✓	✓	2
✓	✓	✓	✓	✓	3
*	✓	✓	✓	✓	1
✓	✓	✓	✓	✓	1
✓	✓	✓	✓	✓	0-Suitable
✓	✓	✓	✓	✓	0-Suitable
*	✓	✓	✓	✓	1
✓	✓	✓	✓	✓	1
✓	✓	✓	✓	✓	1
✓	*	✓	✓	✓	3
✓	✓	✓	✓	✓	3
✓	✓	✓	✓	✓	1

TABLE 19.4

Color-Coding Standards for Wire Insulation

Color Code		Numeric Wire Code	
		Numerical Possibilities	Description
0	Black	10	Solid color insulation
1	Brown	90	Solid color with one broad trace
2	Red		
3	Orange	810	Solid color with one broad trace and one narrow trace
4	Yellow		
5	Green		(910 total numeric combinations)
6	Blue		
7	Violet		
8	Gray		
9	White		

TABLE 19.5

Wire Color Identification Related to Circuit Function

Functional Color Coding for Chassis Wiring		
Color	Code Number	Function
Black	0	Grounds, grounded elements and returns
Brown	1	Heaters, or filaments not connected to ground
Red	2	Power Supply, +Vcc or B+
Orange	3	Screen grid
Yellow	4	Transistor emitters or cathodes
Green	5	Transistor bases and control grids
Blue	6	Transistor collectors, Anodes for semiconductor elements and tube plates
Violet	7	Power supply, minus -Vcc or B-
Gray	8	AC power lines
White	9	Miscellaneous, above ground returns

Color Code for Power Transformers		
Winding		Color Code
Primary	Untapped	Black
	Tapped	Common-Black Tap Black/yellow
High-voltage Plate		Red
High voltage center-tap		Red /yellow
Rectifier Filament		Yellow
Rectifier Filament Center-tap		Yellow/blue
Filament (# 1)		Green
Filament (# 1) Center-tap		Green/yellow
Filament (# 2)		Brown
Filament (#2) Center-tap		Brown/yellow
Filament (# 3)		Gray
Filament (# 3) Center-tap		Gray/yellow

FIGURE 19.4. (a) Shielded wire. (b) Coaxial cable.

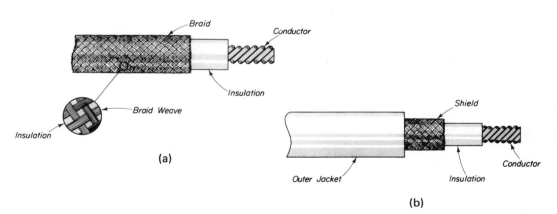

(a)

(b)

loss or attenuation of the signal caused by the current traveling along the surface (skin) of the conductor and partly into the adjacent insulation. Because of its outstanding insulating characteristics at high frequencies, nylon is the type of insulation most often used in the manufacture of coaxial cables.

Another special wire configuration is *Litzendraht* (Litz) wire, which is used in the manufacture of tuning coils. At short-wave frequencies, this type of wire has low skin effect, owing to its configuration. Litz wire is a stranded conductor containing 25 or more extremely fine wire strands. Each strand is independently insulated with cotton insulation and wound from the center to the outer surface of the overall conductor. This technique of winding tends to equalize the skin effect problem.

19.4 TOOLS FOR WIRE PREPARATION AND ASSEMBLY

To form a sound electrical connection between the hook-up wire used for interconnections and the terminal to which it is to be soldered, *all* insulation must be completely removed from the end of the wire to expose the conductor. In addition, it may become necessary to remove contaminants, such as dirt, finger oils, or oxidation, from the bare wire. (Oxidation will occur if the bare conductor has been exposed to the atmosphere for a prolonged time before soldering.) It is vitally important when preparing to make a solder connection that all surfaces to be soldered (alloyed) are absolutely free of all contaminants.

Mechanical wire strippers, such as the three styles shown in Fig. 19.5, are among the most common tools used to remove wire insulation. Each has its advantages and disadvantages with regard to resulting quality and ease of operation.

The *fully adjustable* wire stripper, shown in Fig. 19.5a and in its proper use in Fig. 19.6a, will accommodate most of the wire sizes used in the electronic industry. This stripper is adjusted for a particular gauge wire by first closing the cutting jaws around a portion of *uninsulated* wire until the jaws are *almost* in contact about the circumference of the wire. The slide-adjusting button is then moved along the radius groove until the jaws are set to the desired opening. Stripping is accomplished by inserting the wire with the insulation to be removed into the cutting jaws, which are positioned perpendicular to the axis of the wire (Fig. 19.6a). The handles are then squeezed together, causing the cutting jaws to sever the insulation about the conductor. While holding the wire securely with one hand, the severed length of insulation is then pulled off the conductor with the stripper. Two points of caution should be observed when using this stripper. First, never cock or twist the cutting jaws from their perpendicular orientation, since this may nick the conductor. Second, when working with insulations having high cut-through resistance ratings (see Table 19.2), it may be necessary to squeeze and rotate the cutting jaws back and forth around the insulation prior to removal. If, for any reason, solid wire is

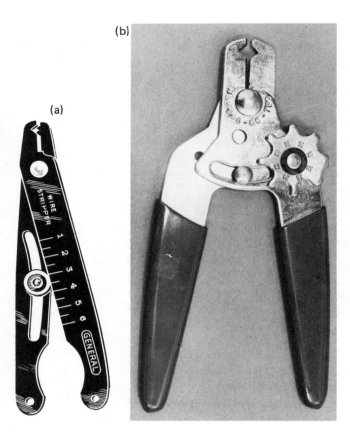

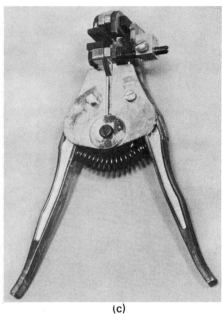

FIGURE 19.5. Mechanical wire strippers. (a) Fully-adjustable. Courtesy of General Tool Company. (b) Step-adjustable. (c) Adjustable.

nicked or individual strands of stranded wire are broken during the stripping operation, the resulting stripped end should be rejected. Either of these deficiencies will reduce the effective diameter of the wire, thereby reducing its current-carrying capability. One of the disadvantages of this stripper is the difficulty in avoiding damage to the conductor.

The *step-adjustable* stripper, shown in Fig. 19.5b, is used in the same way as the fully adjustable stripper but can be set quicker and more accurately for a particular wire gauge. All that is required to set the jaws is to rotate the *selector step wheel* until the desired gauge number is oriented with the *stop pin.* Although this type of stripper greatly reduces the possibility of conductor damage, its obvious disadvantage is that it is capable of stripping only eight gauge wire sizes (AWG No. 12 to AWG No. 26). To minimize the possibility of conductor damage when using this stripper, it should be held perpendicular to the wire when pulling off the insulation.

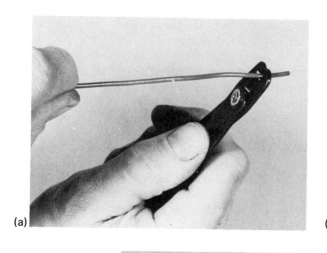

(a)

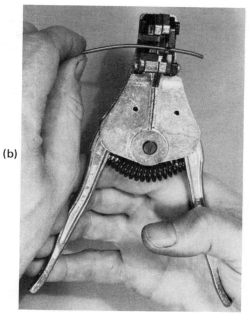

(b)

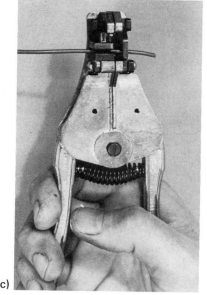

(c)

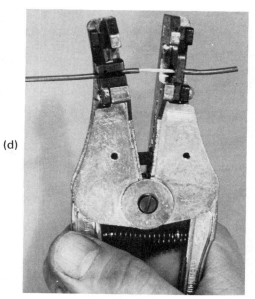

(d)

FIGURE 19.6. Insulation removal with mechanical wire strippers. (a) Proper removal of insulation. (b) Initial wire position. (c) Clamping and cutting operation. (d) Insulation removal.

481

The third and perhaps most popular mechanical stripper is the *automatic* style shown in Fig. 19.5c and being used in Fig. 19.6b, c, and d. This stripper has six marked cutting positions on the jaws to accommodate various gauge sizes. The wire to be stripped is placed into the appropriate cutting jaw position and between the gripping jaws, as shown in Fig. 19.6b. With the wire held in position, the handles are gently squeezed together. This causes the movable upper cutting jaws to first press the insulation down into the lower stationary cutting jaws. Additional squeezing of the handles causes the insulation to be severed about the conductor while the upper movable gripping jaw firmly grasps the wire with the lower stationary gripping jaw (Fig. 19.6c). Further squeezing of the handle causes both sets of jaws, while fully engaged, to separate, thereby removing the insulation from the end of the wire (Fig. 19.6d). As pressure on the handles is removed, both sets of jaws disengage, thus releasing the stripped wire. These strippers are also available with a stop that, when set to the desired length of strip, will consistently remove the same length of insulation from wires.

Thermal wire strippers, although not popular in prototype construction, are another means of removing the insulation from wires. They consist of a heating element and two pivoted handle grips having electrodes that serve as the stripping jaws. The wire to be stripped is positioned between the electrodes, which are then clamped together around the wire. Squeezing the insulated handles activates a switch that causes the electrodes to heat the insulation to its melting point. With the electrodes energized, the wire is rotated so that the insulation melts uniformly. When the insulation has melted, the pressure on the handles is released, the wire is removed, and the insulation is pulled off of the conductor. Thermal strippers will not nick or cut conductors. They are more expensive than mechanical strippers, however, and may also cause irritating fumes as the insulation is melted.

Whichever method of wire stripping is employed, more wire than is necessary to make the connection should be stripped. The reasons for this are (1) in the case of stranded wire where tinning (discussed later in this section) of the strands is required, solder buildup usually occurs at the conductor end and can be removed simply by cutting; and (2) if each wire is cut and stripped to its exact required dimension, this will not provide the necessary "play" that the technician may need to adjust wire positions during wiring operations.

As mentioned previously, it may be necessary to further clean the exposed wire prior to soldering. This can be quickly accomplished with the use of the lead cleaner shown in Fig. 18.32. Designed primarily for use on solid wire (including component leads), this eraser-style cleaner will remove contaminants such as dirt and oxides from the surface of the wire. A component lead being cleaned is shown in Fig. 19.7a.

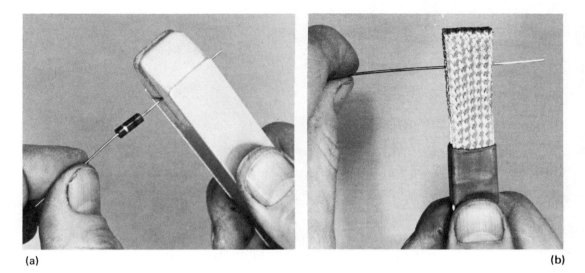

(a)   (b)

FIGURE 19.7. Lead and wire cleaning. (a) Mild abrasion. (b) Coarse abrasion.

To strip enamel or any type of film insulation from solid wire, an *abrasive* lead cleaner, such as that shown in Fig. 19.7b, may be used. It is made of braided wire formed about a piece of aluminum. Details for constructing this type of cleaner are given in Problem 20.1.

The final step prior to interconnection is the *tinning* of stranded wire. After stripping the insulation from stranded wire, the individual strands must be firmly twisted together and soldered (tinned). This step is necessary to allow stranded wire to form around a terminal in a manner that will result in as neat a joint as is obtainable with solid wire. Stranded wire is best tinned using the technique shown in Fig. 19.8. The insulation of the wire is gently held between the jaws of a bench vise with the conductor to be tinned extended. The soldering iron is positioned in contact with the bottom of the wire. The solder is applied from the top directly onto the wire strands. (Notice that the work to be soldered is always placed *between* the solder and soldering iron tip.) Heat from the iron will melt the solder and draw it through the wire strands (melting solder follows the direction of heat flow). Drawing the iron, and the solder, along the stripped wire from the end to the insulation and back again will ensure solid, uniform bonding of the individual strands. Correct tinning is achieved when sufficient solder has wet all conductor surfaces, forming a solid alloy (bond) between each strand. Excessive solder should be avoided. The contour of each strand should be plainly visible.

A problem associated with tinning or making soldered connections with stranded wire is a condition termed *wicking*, which causes solder to

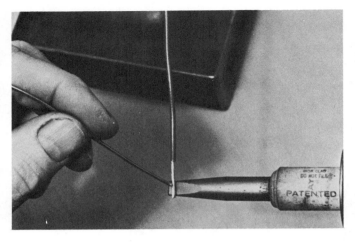

FIGURE 19.8. Stranded wire tinning.

flow by capillary action *under* the insulation. As a result, the wire becomes rigid where the most flexibility is required, thus defeating the purpose of using stranded wire. To minimize this problem, *anti-wicking tools* (conductor heat sinks) such as the type shown in Fig. 19.9a are used. The handles are first squeezed to open the jaws to accept the conductor and insulation (Fig. 19.9b). The tool is then moved so that each jaw nose rests firmly against the conductor adjacent to the insulation (Fig. 19.9c). The anti-wicking tool protects the insulation from the heat of the soldering iron by shunting the flow of heat from the conductor as it passes into the insulation, thereby reducing wicking.

After the tinning operation, the excess wire length along with the accumulation of solder that usually builds up at the wire end can be removed with a pair of *diagonal cutters*. Selecting the most suitable type of diagonal cutters (dykes) depends primarily on the size of the wire being used as well as accessibility during assembly. Because of the variety of cutting situations encountered, many styles of diagonal cutters are available. A selection of common styles used for electronic wiring is shown in Fig. 19.10. The size of these diagonal cutters ranges from 6 inches long for heavy-gauge wire (AWG No. 14 maximum) to 4 inches long for use with smaller wire such as AWG No. 24 and smaller.

The cutting edges of the diagonal cutters can be classified as *regular*, *semiflush*, or *full-flush*. Regular cutters, used for general-purpose applications, leave a "point" at the end of the wire after cutting. Semiflush cutters leave a less pronounced point. Full-flush cutters, intended for use only on annealed copper wire, leave no point or chamfer but rather a smooth, flat end perpendicular to the wire axis. Since the regular and semiflush cutters distort the end of a conductor, this could create a problem when attempting to insert the wire into a close tolerance hole such as those associated with pot-type terminals, small eyelets, and holes in pc board terminal pads. In addition, the sharp points these cutters leave are undesirable when protruding from terminal connections. For these reasons, the full-flush cutter is recommended.

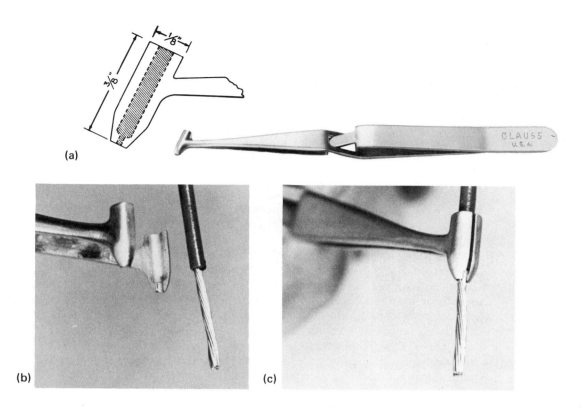

FIGURE 19.9. Use of anti-wicking tool. (a) Anti-wicking tool. (b) Wire insertion. (c) Proper tool position for soldering.

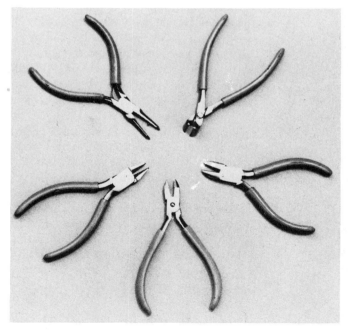

FIGURE 19.10. Various styles of wire cutting tools.

The jaws of most diagonal cutters produce a slight "shock" as they come together when cutting a wire. When used on hook-up wire, this presents no problem. However, sensitive semiconductor devices could be damaged by this shock. For this reason, *shear-type* cutters are recommended for delicate work. The jaws of these cutters pass each other similar to a pair of scissors, thereby minimizing the cutting shock.

Diagonal cutters are available in *rounded*, *tapered*, and *offset* noses to fulfill the requirements of a variety of cutting needs. The most appropriate cutting jaws, shapes, and edge types are selected after the work to be performed is evaluated. When heavier-gauge wires are to be cut and close cutting tolerances and accessibility are not critical, less delicate cutters may be employed. Intricate wiring using finer-gauge wire will require more delicate cutters.

Rubber jaw inserts are available on some styles of cutters, which grip the end of the wire to be removed. These inserts prevent the cut end from flying out of the jaw when the wire is cut. There is a serious safety hazard involved in using diagonal cutters not equipped with rubber inserts. It is always a good practice to point the conductor end *downward* to prevent the flying end from causing injury.

Cutters with plastic grips are preferable to cutters with no grips because they not only provide more positive gripping, but also reduce hand fatigue.

19.5 WIRE CONNECTIONS

One of the most common types of terminal used to make interconnections between pc boards and chassis-mounted components is the *swaged turret terminal* discussed in Chapter 14. To obtain a good electrical connection, it is vitally important that a solid mechanical connection first be made. The wire to be soldered to a turret-type terminal should be formed as tightly as possible around the turret without scraping or nicking the conductor surface or deforming the wire contour. With the insulation positioned no more than $1/8$ inch away from the terminal, the conductor is first tightly formed 180 degrees around the turret with long-nose pliers (Fig. 19.11a). The excess length of wire is then cut at point A, which leaves a sufficient length to continue forming a complete 360-degree *wrap-around* connection. This wrap being formed with long-nose pliers is shown in Fig. 19.11b. Although 180-degree wraps are often used in electronic wiring, the 360-degree wrap provides the most secure mechanical connection. Long-nose pliers with smooth jaws should be used when making mechanical connections to any terminal configuration to avoid nicking and deforming the conductor surface.

Soldering to a turret terminal is performed using the soldering techniques discussed in Chapter 15 for soldering leads to pc board terminal pads. A small amount of solder is first applied to the soldering iron tip (solder bridge). The tip is next brought into contact with the wire and

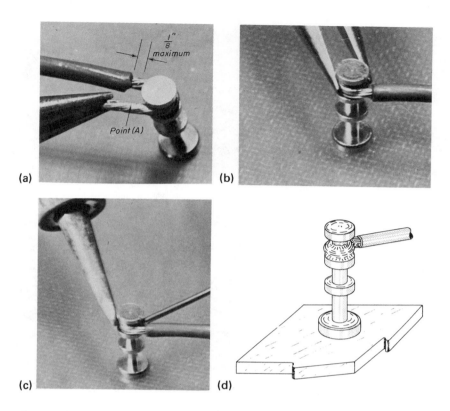

FIGURE 19.11. Wiring to a turret type terminal. (a) Initial 180°
wrap. (b) Forming the complete 360° wrap. (c) Correct iron tip and
solder position. (d) Properly soldered terminal.

shoulder of the terminal, as shown in Fig. 19.11c. Positioning the tip
directly opposite from the insulation will minimize heat damage to plastic
and rubber insulation. Solder is applied immediately when the connection
has reached the proper temperature (i.e., when the solder of the bridge
or the solder used in tinning stranded wire begins to flow). Only a *small*
amount of solder is necessary to make a sound electrical connection. If
properly soldered, the contour of both the wire and the terminal will be
plainly visible. A properly wired and soldered turret terminal connection
is shown in Fig. 19.11d.

During the soldering operation, the anti-wicking tool, shown in Fig.
19.9, or a lead heat sink, shown in Fig. 14.24b, may be used to prevent
wicking or excessive melting or burning of the insulation.

Wrap-around connections also find wide application when wiring to
any eyelet-type terminal, such as those associated with *terminal strips*,
switches, *potentiometers*, and *pc board connectors*. These terminal types
are shown in Fig. 19.12a. Wire forming about an eyelet-type terminal lug
begins by positioning the insulation no more than ⅛ inch from the

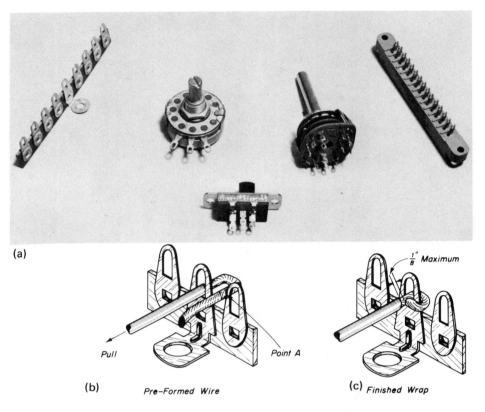

FIGURE 19.12. Wire forming for eyelet terminals. (a) Components and hardware with eyelet type terminals. (b) Preformed wire. (c) Finished wrap.

eyelet opening. The wire wrap is then formed in a similar manner to the 360-degree wrap for turret terminals, but since the cross section of the eyelet is a very thin rectangular shape, the wrap is more difficult to form. It is recommended that the first two 90-degree contours, as shown in Fig. 19.12b, be preformed before hooking the conductor through the eyelet. Once positioned into the eyelet, the excess lead length is cut at point A and the final 90-degree bend is made to complete the wrap (Fig. 19.12c). It is not good practice to make more than three wrap-around connections to a single eyelet nor to solder them until all have been mechanically formed.

When mechanical strength requirements are minimal, wires can be soldered more easily into small-diameter holes, such as those found on solder lugs, by means of a *feed-through* connection, shown in Fig. 19.13. For this type of connection, the diameter of the wire should be approximately equal to that of the hole in the lug. As with other types of connections, the insulation is positioned at a maximum of $\frac{1}{8}$ inch from the

FIGURE 19.13. Feed-through connection.

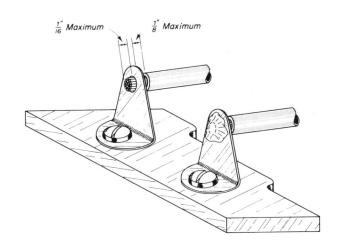

terminal, with the prepared wire end passing through the lug hole. The excess wire is trimmed to within approximately $\frac{1}{16}$ inch from the terminal. Because the diameter of the wire approximates that of the hole, soldering will properly alloy both circumferences. The resulting connection is mechanically satisfactory and electrically sound, although feed-through connections should be avoided in equipment that will be subjected to severe shock and vibration.

Wires may be *dressed* between connections by either *point-to-point* or *square-corner* wiring techniques. Point-to-point wiring, as the name implies, is a wiring technique used primarily in high-frequency circuits or any economical wiring application. Wiring interconnections are made with regard to the shortest routing path possible to minimize lead length, which is critical at high frequencies. This wiring technique is quicker and uses less wire than the other methods discussed here. Point-to-point wire dress usually does not add to the neatness of the package. However, with careful wiring positioning and neat soldered connections, the resulting wiring will be reliable and will display quality workmanship.

Square-corner wiring is a wire dress technique in which all wires installed in the package run either parallel or perpendicular to one another. When any wire path direction changes, the change is made with a 90-degree bend. Although not used in high-frequency work, this technique imparts a very neat appearance to the package. The only disadvantage to square-corner wiring is the additional amount of wire and time necessary to complete a package.

19.6 SPECIAL WIRE ASSEMBLIES AND TECHNIQUES

Many connectors made for terminating the stranded conductors used in cables are provided with *solder pot* terminals, such as those shown in Fig. 19.14a. Soldering a wire into this type of terminal requires that the pot itself be initially prepared. The solder pot must first be heated and filled

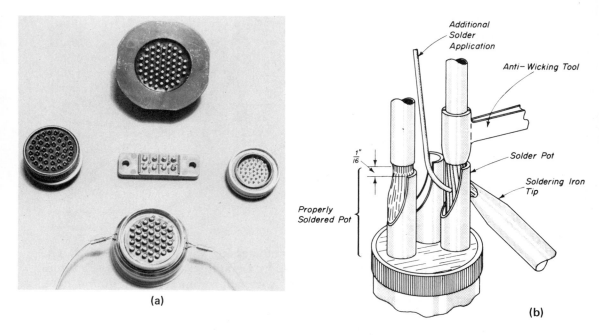

FIGURE 19.14. Technique for wiring pot type terminals. (a) Typical pot type terminals. (b) Soldering wire into pot type terminal.

at least one-half way with solder. The solder should not be allowed to flow out of the pot onto the outside surface. No more solder is applied after the pot is approximately half full, but the iron tip is allowed to remain in contact with the pot for several more seconds to allow any trapped flux or gases generated by the volatilized rosin inside the terminal to escape. (It is advisable to wear safety glasses when soldering, especially when working with solder pots, since the generated gases can cause solder to splatter out of the pot.) A properly tinned wire is then held with an anti-wicking tool, as shown in Fig. 19.14b, and inserted into the pot after the solder has been reheated with the iron. It will be difficult to insert the conductor into the solder pot if it has not been reheated long enough for the solder to completely melt to the bottom or if too much solder was applied to the wire when it was tinned. In a properly soldered pot connection, the wire insulation should be no further than $^1/_{16}$ inch from the top of the pot. This length can be predetermined on an unsoldered pot. There should be a minimum amount of solder used so that the contour of the wire strands between the insulation and the pot is plainly visible. In addition, no excess solder should be allowed to flow onto the outside surface of the pot. Excess solder puddling at the base of adjacent pots can cause short circuits. A properly formed solder pot connection is shown completed in Fig. 19.14b.

A common yet difficult wiring connection involves the preparation and installation of shielded wire to phono plugs and jacks. The function of the braid of the wire is to cover and thereby isolate the inner (signal) conductor. There should be minimum exposure of the signal wire insulation after the shield and signal wire connections are made to the connectors. If the shielded wire has an outer insulating jacket, approximately $1\frac{1}{2}$ inches of the jacket should be removed. The shield is then pushed back from the end so as to "loosen" it. A pick (scriber) is then carefully worked into the braid without damaging any individual strands to form a hole large enough through which the insulated signal lead can be pulled. This technique is shown in Fig. 19.15a. The pick is then passed into the hole between the braid and the signal wire insulation so that it protrudes from the other side. Doubling the shielded wire back at the breakout point and lifting the pick will remove the signal wire from the hole in the braid. This technique is shown in Fig. 19.15b. The end of the braid is then tinned and will serve as the ground lead. The prepared shield and signal wire ready for soldering to a connector are shown in Fig. 19.16a. Soldering the signal wire to a plug requires preparation similar to that for solder pot terminals. The hollow center prong of the plug is first filled with solder. This step can best be done by gently gripping the outer shell of the plug in a bench vise. The correct soldering iron tip position and application of solder is shown in Fig. 19.16b. When solder begins to appear at the tip of the prong, the iron is removed. Solder must not be allowed to flow in the outer surface of the prong or it may prove to be impossible to plug it into a jack. To install the wires onto the plug, at least $\frac{3}{4}$ inch of signal wire is stripped and tinned, leaving only $\frac{3}{16}$ inch of insulation between the stripped signal conductor and the braid. The braid is also tinned. The center prong of the plug is reheated and the signal wire is inserted into the top of the plug until the braid is even with the top of

FIGURE 19.15. Technique for preparing shielded wire for soldering. (a) Scribe used to provide access hole in shield. (b) Removing signal lead from shield.

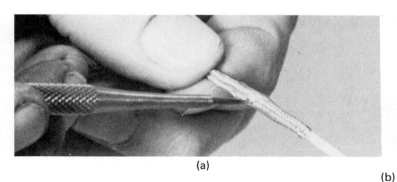

(a)

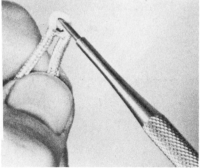

(b)

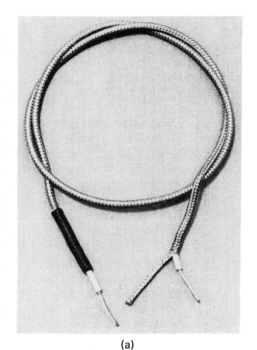

(a)

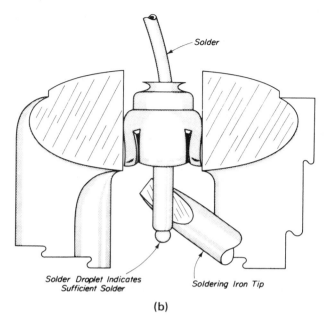

Solder

Solder Droplet Indicates
Sufficient Solder

Soldering Iron Tip

(b)

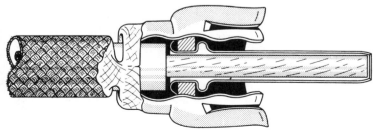

Proper Position of Iron Tip and Solder

FIGURE 19.16. Preparation and soldering of shielded wire to plugs and jacks. (a) Shielded wire prepared for soldering. (b) Preparation of prong with solder. (c) Proper position of iron tip and solder. (d) Correctly prepared jack.

Cutaway View Showing Correct Assembly of Wire and Shield

(c)

FIGURE 19.16. (Cont.)

(d)

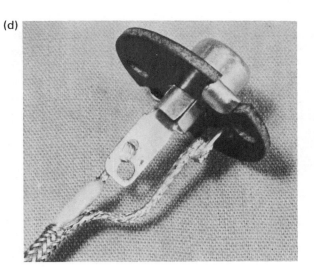

the ferrule. Some of the signal conductor may protrude from the end of the prong. This is removed flush with diagonal cutters. The braid is next wrapped 360 degrees around the ferrule and then soldered, using the technique shown in Fig. 19.16c. Proper installation of shielded wire to a phono jack is shown in Fig. 19.16d.

19.7 SOLDERLESS WIRING TECHNIQUES

Solderless- or *crimp*-type terminals provide a strong metal-to-metal (terminal-to-wire) bond that is suitable for applications using either stranded or solid wire. These terminals are particularly desirable in some production assemblies because they eliminate soldering. Two common configurations of solderless-type (crimp) connectors are the *ring tongue* and *fork tongue* styles shown in Fig. 19.17.

Selecting the most suitable style depends on the application. If rapid connection and disconnection are not required but a higher degree of mechanical security is necessary, the ring tongue should be used. If, however, rapid connecting and disconnecting requirements are specified with no critical mechanical requirements, the fork tongue style is preferable.

FIGURE 19.17. Crimp terminals.

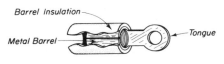

Ring Tongue

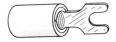

Fork Tongue

493

The *barrels* of crimp terminals are available in uninsulated or insulated forms. The insulator is usually polyvinyl chloride (PVC) and offers a degree of strain relief for the wire if strong vibration is expected. These insulators are commonly available in colors so that the range of AWG sizes that each barrel will accept can be quickly identified (see Table 19.6).

Solderless terminals in common use in electronics have barrels that are capable of accepting wire sizes from AWG No. 10 to AWG No. 22. A No. 10 terminal will accept both AWG Nos. 10 and 12 wire; a No. 14 terminal will accept AWG Nos. 14 and 16 wire; and a No. 18 terminal will accept AWG Nos. 18, 20, and 22 wire. The terminals are available to accept screw sizes from No. 4 to $3/8$ inch.

TABLE 19.6

Color Code for Insulated Crimp Connector Barrel Sizes

Terminal Number	Barrel Insulation Color	Accepts AWG Sizes
No. 18	Red	22 to 18
No. 14	Blue	16 to 14
No. 10	Yellow	12 to 10

Conductors are assembled into solderless terminals with a *crimping tool* such as the one shown in Fig. 19.18a. The wire insulation is stripped to a length equal to the metal barrel length. The conductor is then inserted into the terminal until the insulation seats against the end of the metal barrel. The barrel is then placed into the appropriate jaw crimping position of the tool. The split side of the barrel should be placed against the smooth portion of the jaw opposite the crimping tooth. When the handles are firmly squeezed together, as shown in Fig. 19.18b, the barrel will be crimped against the conductor and will make a sound mechanical and electrical connection.

Solderless terminals are most often used with barrier terminal strips such as in the example package of the automobile security alarm system shown in Fig. 21.16.

Another form of solderless connection employs *wire-wrap terminal posts*. Specifications for these terminals are given in Sec. 14.2 on pc hardware and assembly along with the recommended number of wraps for a given gauge wire. Mechanical and electrical connection for wire-wrap terminals is achieved by wrapping, under tension, solid wire around a post terminal that has at least two sharp edges. These edges, as well as the wire itself, are deformed when tightly wrapped. The average pressure between the wire and the sharp contact points of the post is in the order of 30,000 psi, which is more than sufficient to produce a sound mechanical and

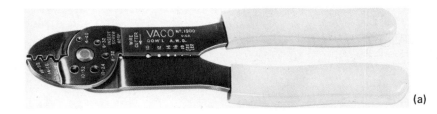
(a)

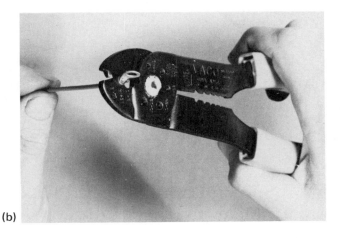
(b)

FIGURE 19.18. Technique for installing crimp terminals. (a) Crimping tool. (b) Position of tool and terminal for crimping.

electrical connection. Two properly formed types of wire-wrap connections are shown in Fig. 19.19.

Manually operated wire-wrap tools, such as the one shown in Fig. 19.20a, are available to quickly form wire-wrap connections. Pneumatic wire-wrap tools are also available, but are usually employed only in mass-production applications.

To use the tool shown in Fig. 19.20a, solid wire of the desired gauge is first stripped and then inserted into the outermost hole at the end of the wire wrapping tip (Fig. 19.20b). The tip is then positioned over the terminal and the trigger is quickly squeezed. The nose of the gun spins the wire around the terminal to form the desired connection. Application of the wire-wrap tool on the amplifier circuit is shown in Fig. 19.20c.

The number of turns of wire wrapped is determined by the specifications provided in Table 14.1. The amount of insulation that must be stripped from the end of the wire can be quickly determined by trial and error.

Should it be necessary to unwrap a connection that is not readily accessible (on high-density packaging, wrapped terminals may be within 0.1 inch away from each other) *de-wrapping* tools such as the one shown in Fig. 19.20d are available. To remove a wrapped connection, the center

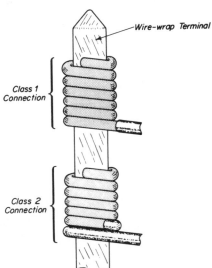

Wire-wrap Terminal

Class 1 Connection

Class 2 Connection

FIGURE 19.19. Two styles of wire-wrapping.

FIGURE 19.20. Wire-wrapping tools and methods. (a) Hand-operated wire-wrap tool. (b) Wire placement into nose of tool. (c) Correct position of tool over terminal. (d) De-wrapping tool.

(a)

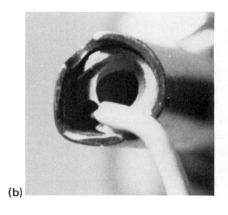

(b)

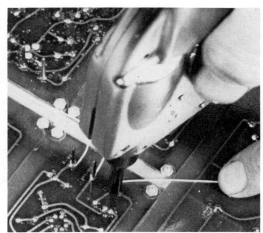

(c)

(d)

hole in the nose is positioned over the terminal. Downward pressure of the tool against the wire will quickly unwrap the wire from the post. The tool must be kept vertical during the entire operation to avoid bending the terminal post.

With the wiring completed, the finished amplifier system is shown in Fig. 19.21.

FIGURE 19.21. Completed stereo amplifier.

PROBLEM SET

19.1 Construct a pair of test leads such as those shown in Fig. 19.22. The test leads are to be made from 36-inch lengths of plastic insulated No. 18 AWG stranded wire. Strip ½ inch of insulation from all four lead ends and twist and tin the stranded conductors. Slip the rubber insulated grips onto the leads before soldering the alligator clips to the lead ends. After soldering the clips, slide the rubber insulators into position.

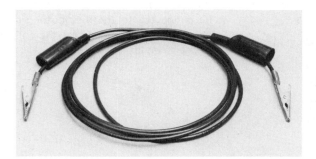

FIGURE 19.22.

19.2 Construct a set of interconnecting leads for a stereo system by first centering a 2-foot length of shrinkable tubing about two 3-foot lengths of insulated shielded conductor cables. Slide 2-inch lengths of shrinkable tubing over all four ends of the cables. Remove sufficient outer insulation from all cable ends to prepare the copper braid and the inner insulated conductors for soldered connections to RCA phono plugs. Once the phono plugs have been soldered, slip the shrinkable tubing onto the plugs until the end of the tubing is even with the leading edges of the outer portion of the connector. Finally, shrink the tubing about the cables and plugs. (Refer to Chapter 20.)

19.3 Construct four 6-foot speaker leads from plastic insulated No.16 AWG stranded wire. Slide $1\frac{1}{2}$-inch lengths of shrinkable tubing over all lead ends. Strip and tin all lead ends to a length of $\frac{1}{2}$ inch. Position the uninsulated barrels of solderless fork tongue connectors onto all prepared conductor ends and crimp. Slide the shrinkable tubing over the entire barrel length and shrink into place. (Refer to Chapter 20.)

19.4 Using the technique discussed in this chapter for straightening bus wire, prepare a $32\frac{1}{2}$-inch length of tinned No. 14 AWG bus wire to construct a UHF antenna. Form the wire into a loop with a $7\frac{3}{4}$-inch diameter. Adjust the loop so that both loop ends overlap equally. Crimp solderless spade lugs onto both loop ends and form two 90-degree bends on each end for attachment to terminal lugs on the rear of a television set. The first 90-degree bends are to be made 4 inches from the end of the spade lugs and perpendicular to the circumference of the loop. The remaining 90-degree bend is to be located 2 inches from the ends of the lugs for the most convenient attachment.

19.5 Install plastic insulated No. 20 AWG stranded wire between the chassis-mounted components and hardware and the pc board for the seven fixed-interval UJT timer designed, fabricated, and assembled in Problems 11.3, 13.5, 14.3, 15.3, and 16.5.

19.6 Install interconnection wires to the tachometer designed, fabricated, and assembled in Problems 11.5, 13.4, 14.5, 15.5, and 16.6. All leads will be plastic insulated No. 18 AWG stranded wire. The power lead insulation color is to be red, with lead length sufficient for connection to the battery side of the fuse block. The ground lead insulation color should be black. This lead is to be terminated with a solderless spade lug and attached to any convenient screw on the car chassis. Select green for the insulation color for the remaining lead to be connected to the distributor side of the coil. A solderless ring tongue connector of sufficient size for the coil connection should be crimped onto the end of the green lead.

20

Harness and Cable Fabrication

20.0 INTRODUCTION

Wire harnessing and cabling are alternate interconnection methods and extensions of the wiring procedures previously considered in Chapter 19.

Whenever two or more wires are mechanically secured or bundled together, a *harness* or *cable* is formed. The distinction between these two terms depends on the application. Harnesses are defined as wires mechanically bundled together to complete the electrical interconnections within a chassis assembly. They are characterized by leads leaving the main trunk and routed to components, terminals, or subassemblies such as pc boards. A cable is a bundle of individual wires each starting at one end of the bundle and terminating at the opposite end. The bundle is generally formed by passing wires through a protective insulated tubing. The assembly is completed by soldering connectors at both ends to provide for electrical interfaces. The ends of the cable may also be terminated with the individual leads stripped and tinned for soldering to terminal points. Cables are employed for external interconnections among separate units in a system.

This chapter provides information for the design and construction of harnesses and cables, including harness assembly drawings, termination and wire-coding techniques, harness jigs, lacing, cable ties, and cabling tools. Also considered are shrinkable and track closure tubing and various types of cable clamps for securing harnesses and cables.

20.1 CONSIDERATIONS FOR HARNESS DESIGN

The decision to construct a harness will depend on the mechanical and electrical requirements of the circuit. Although ribbon and flat printed circuit cable, such as those shown in Fig. 20.1, have in many instances replaced individual wiring, twisted pairs, coaxial cable, and standard round-wire harnesses, conventional cables and harnesses continue to be widely used because lead ends require no special connectors for termination and standard soldering techniques are employed.

Since harnessing is employed in prototype construction in anticipation of production, it must be considered as a means of interconnection. Not all circuits are suitable for harnessing, however. Circuits with few peripheral connections within a chassis are not usually harnessed and may best be wired by the routing techniques discussed in Chapter 19. Circuits with many interconnections may be harnessed if the electrical considerations will allow. For example, high-frequency circuits demand that lead lengths be as short as possible. This requirement necessitates point-to-point wiring and a harness should not be employed. In high-gain, low-

FIGURE 20.1. Ribbon and flat printed circuit cables facilitate interconnections. (a) Flat cable. Courtesy of Fortin Laminating Corp.

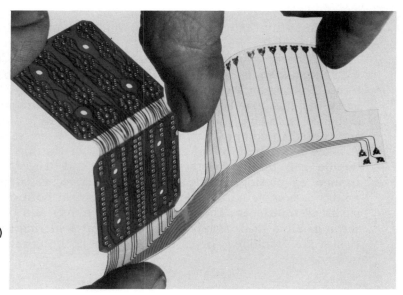

(a)

(b)

FIGURE 20.1. (cont.) (b) Ribbon. Courtesy of 3M Company.

frequency circuits, harnessing is not recommended, since interstage coupling between parallel conductors may cause signal interference, especially at the low-signal-level inputs. For most other types of circuits whose wiring serves such functions as dc power and ground leads, high-level signal leads, low-level signal leads employing shielded or coaxial cable, ac power leads (when formed into twisted pairs), as well as bypass and coupling leads, may be considered for harnessing. Harnesses are also used extensively when shock and vibration problems exist.

Harnessing lends itself readily to mass-production applications because the harness can be formed apart from the assembly for which it is intended by any of the methods discussed in Sec. 20.3. When the harness is properly bundled and formed, it is positioned into the chassis and interconnections are made among the various chassis-mounted components.

20.2 HARNESS ASSEMBLY DRAWINGS

To permit a better understanding of harness design and construction, some common terms used to identify the various parts of a harness, shown in Fig. 20.2, will first be defined.

Breakout — two or more unlaced leads leaving any portion of a harness assembly at the same breakout point.

Breakout point — space between lacing or cable ties where a lead or group of leads leave any portion of a harness.

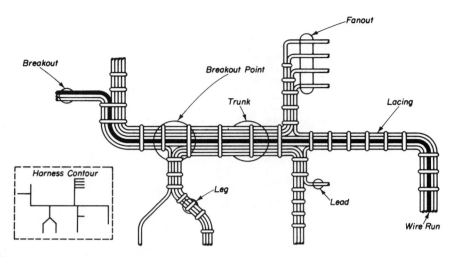

FIGURE 20.2. Harness nomenclature.

Fanout — individual leads leaving a leg at regular spaced intervals.

Harness contour — overall shape of the harness assembly.

Lacing — method of bundling individual leads to form a harness using waxed lacing cord or twine or nylon lacing tape.

Lead — individual insulated conductor.

Leg — minor run of two or more laced leads.

Trunk — major run of the harness assembly containing the largest number of laced leads.

Wire run — complete route of an individual lead in a harness assembly.

Harness design begins with a detailed *harness assembly drawing* such as that shown in Fig. 20.3 for the amplifier. This drawing will be used to construct both the harness jig and the actual harness. For this reason, it must be detailed to provide sufficient information for the harness contour, including all wire runs and configurations of leads, legs, breakouts, and fanouts. In addition, the type and number of terminations on each breakout or leg as well as a descriptive designation for each major termination on the harness should be provided.

The harness assembly drawing is produced from the chassis and component layout drawings in conjunction with the schematic diagram. The exact harness dimensions and contour are determined from the chassis and component layout drawings, whereas routing paths and originating and terminating points are obtained from these drawings and the schematic diagram. To be useful, the harness assembly drawing should be constructed using a 1:1 scale. The dimensional and positional informa-

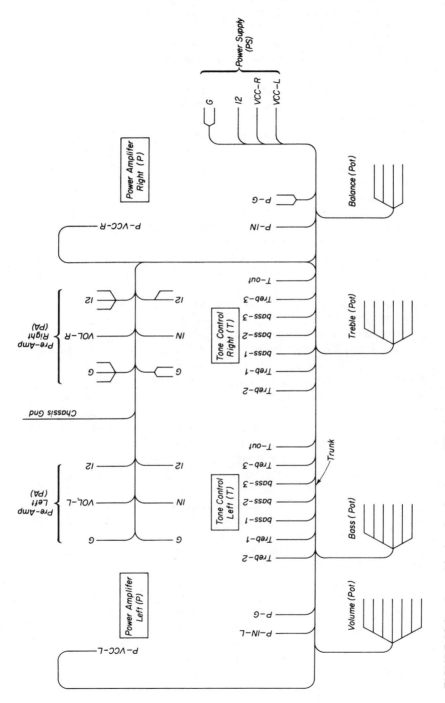

FIGURE 20.3. Assembly drawing for the amplifier harness.

tion can be most accurately shown if constructed on a grid system having 10 divisions per inch. For clarity, all harness segments (trunks and legs), regardless of how many individual leads they may contain, are indicated by single solid lines that form the harness contour. (Note that the harness assembly drawing is a *bottom* view of the harness, which rests against the chassis.) In addition, each lead of a fanout or breakout is also represented by an individual single line. Dimensioning of the drawing should be done in a random fashion for simplification if a one-to-one scale and grid system are not used.

The trunk of the harness shown in Fig. 20.3 is so indicated. Identifying the trunk allows rapid determination of the most complex portion of the harness to be constructed, since it contains the largest number of leads.

Since a harness may contain many leads for terminations, each should be literally described concerning its function to simplify initial harness fabrication and final installation. This information is tabulated in a *Wiring Table* such as the one shown in Table 20.1 for the harness of Fig. 20.3. As seen in this table, the letters *PS* are used to designate the terminations of the wire runs connecting the power supply pc board. Minor classifications within the major designations should also be descriptive. Within the major designation of *PS*, *VCC-L* is used to indicate the lead connection for the dc voltage of the left channel and *PA-12* designates the dc voltage for the preamplifier pc board.

The wiring table used in conjunction with the assembly drawing provides sufficient information to accurately construct a harness. Each entry in the wiring table represents an individual lead in the harness. Each wire run is listed separately with its complete descriptive information relative to its *To* and *From* locations, lacing termination distances, type of lead termination at each location, color of lead insulation, AWG number, type of insulation, type of wire, and length of each run.

An explanation of the first entry in Table 20.1 will serve to show the complete description of a wire run. A 19-inch lead made of No. 20 AWG solid copper wire with red plastic insulation having a white stripe (tracer) will connect to the (PS-VCC-L) power supply pc board at the left channel dc voltage terminal where the insulation is stripped ¾ inch from the end. Lacing will terminate ⅜ inch from this end. The other end of this lead will terminate at the (P-VCC-L) left channel dc voltage terminal of the power amplifier pc board where the insulation is stripped ¾ inch from the end. Lacing will terminate 9 inches from this end.

The wiring table may be modified for purposes of simplicity when applicable. For example, in low-volume production work, it may be economically advantageous to individually cut to size and strip each lead as it is placed into the harness, thus eliminating the need for the *lead length* column. Where all stranded or all solid wire with the same type of insulation is to be used, these two designations are also unnecessary.

Finally, it may be desirable from the standpoint of uniformity to select a common wire size for the entire harness that meets all electrical requirements of the circuit. This would further simplify the wiring table. In any event, the wiring table should contain the following minimum information: To-From locations, lacing termination distances, type of terminations, and insulation color code.

The harness assembly drawing and the wiring table are developed only for the fabrication of a harness. Once the harness has been installed into the system, there is no further use for these aids. Circuits with large complex harnesses should be provided with a *functional wiring layout* to aid the service technician in testing or troubleshooting. This type of layout for the amplifier circuit is shown in Fig. 20.4. Functional wiring layouts are best developed on a grid system having four divisions per inch.

FIGURE 20.4. Functional wiring layout for the amplifier circuit.

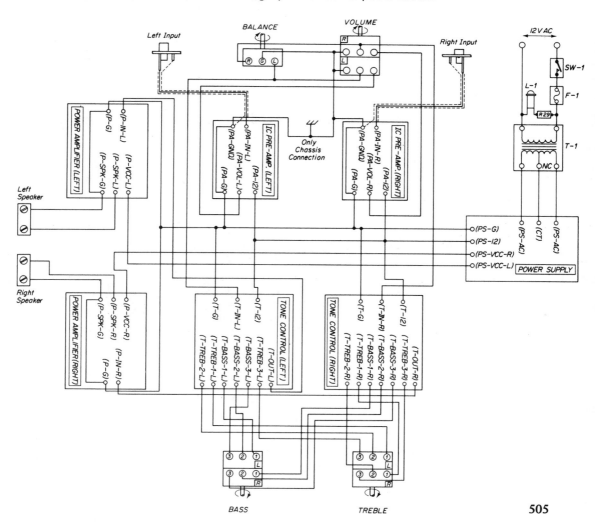

TABLE 20.1

Wiring Table for Amplifier Harness

From Location	Lacing Termination Distance	Lead Termination	Color of Insulation	AWG # (Copper)	Insulation	Type** of Wire	Lead* Length in Inches	To Location	Lacing Termination Distance	Lead Termination
PS-VCC-L	3/8"	3/4" stripped	Red/w	20	PE †	S	19	P-VCC-L	9"	3/4" stripped
PS-VCC-R	3/8"	3/4" stripped	Red	20	PE	S	14	P-VCC-R	5"	3/4" stripped
PS-12	3/8"	3/4" stripped	Red/yel	24	PE	S	9-1/2	PA-12-R	5/8"	1/2" stripped
PA-12-L	5/8"	1/2" stripped	Red/yel	24	PE	S	5-1/2	PA-12-R	5/8"	1/2" stripped
T-12-L	5/8"	1/2" stripped	Red/yel	24	PE	S	5-1/2	T-12-R	5/8"	1/2" stripped
T-12-R	5/8"	1/2" stripped	Red/yel	24	PE	S	3-3/8	PA-12-R	5/8"	1/2" stripped
PA-G-L	1"	1/2" stripped	Black/W	20	PE	S	5-1/2	PA-G-R	5/8"	1/2" stripped
T-G-L	1"	1/2" stripped	Black/W	20	PE	S	5-1/2	T-G-R	5/8"	1/2" stripped
T-G-R	5/8"	1/2" stripped	Black/W	20	PE	S	3-3/8	PA-G-R	5/8"	1/2" stripped
PA-G-R	5/8"	3/4" stripped	Black/W	20	PE	S	12-1/2	PS-G	1-1/2"	3/4" stripped
P-G-L	5/8"	1/2" stripped	Black	20	PE	S	9-1/2	P-G-R	5/8"	1/2" stripped
P-G-R	5/8"	1/2" stripped	Black	20	PE	S	6-1/2	PS-G	1-1/2"	3/4" stripped
P-In-L	5/8"	1/2" stripped	Green	24	PE	S	6-1/2	T-Out-L	5/8"	1/2" stripped
P-In-R	5/8"	1/2" stripped	Green/W	24	PE	S	4-1/2	T-Out-R	5/8"	1/2" stripped
T-Bass-1-R	5/8"	1/2" stripped	Blue	24	PE	S	7-1/2	B	2-1/4"	1" stripped
T-Bass-2-R	5/8"	1/2" stripped	Green	24	PE	S	7-3/4	A	2-1/4"	1" stripped

Label	Length*	Strip	Color	Gauge	Insul.†	Type**	No.	Destination	Length*	Strip
T-Bass-3-R	5/8"	1/2" stripped	Orange	24	PE	S	8	S	2-1/4"	1" stripped
T-Bass-1-L	5/8"	1/2" stripped	Blue/W	24	PE	S	5-1/2	P	2-1/4"	1" stripped
T-Bass-2-L	5/8"	1/2" stripped	Green/W	24	PE	S	5-3/4	O	2-1/4"	1" stripped
T-Bass-3-L	5/8"	1/2" stripped	Orange/W	24	PE	S	6	T	2-1/4"	1" stripped
T-Treb-1-R	5/8"	1/2" stripped	Brown	24	PE	S	5-1/4	T	2-1/4"	1" stripped
T-Treb-2-R	5/8"	1/2" stripped	Violet	24	PE	S	5-1/2	R E B	2-1/4"	1" stripped
T-Treb-3-R	5/8"	1/2" stripped	Gray	24	PE	S	6	L E	2-1/4"	1" stripped
T-Treb-1-L	5/8"	1/2" stripped	Brown/Grn	24	PE	S	7-3/4	P	2-1/4"	1" stripped
T-Treb-2L	5/8"	1/2" stripped	Violet/Grn	24	PE	S	8	O	2-1/4"	1" stripped
T-Treb-3-L	5/8"	1/2" stripped	Gray/Grn	24	PE	S	6-34	T	2-1/4"	1" stripped
Chassis GND	5"	1" stripped	Black/Red	22	PE	STD	15-1/2		2-1/4"	1" stripped
V	2-1/4"	1" stripped	White/Brn	22	PE	STD	12	Balance	2-1/4"	1" stripped
O	2-1/4"	1" stripped	White/Blk	22	PE	STD	12		2-1/4"	1" stripped
L U M E	2-1/4"	1" stripped	White/Red	24	PE	STD	12	Pot	2-1/4"	1" stripped
	2-1/4"	1" stripped	White/Org	24	PE	S	17	PA-Vol-L	5/8"	1/2" stripped
P	2-1/4"	1" stripped	White/Yel	24	PE	S	15-1/2	PA-Vol-R	5/8"	1/2" stripped
O	2-1/4"	1" stripped	White/Grn	24	PE	S	17	T-IN-L	5/8"	1/2" stripped
T	2-1/4"	1" stripped	White/Blue	24	PE	S	15-1/2	T-IN-R	5/8"	1/2" stripped

* Additional length included for tie off
** S – solid, STD– stranded
† PE – Polypropylene vinyl plastic

Major chassis-mounted components are represented along the edges of the paper. Examples of these components are pc boards, transformers, lamps, fuses, meters, switches, and potentiometers. These components are randomly oriented and in such a location as to simplify the wiring layout. No attempt should be made to position them in accordance with their location on the actual chassis. With all components randomly positioned on the drawing, their terminals are identified using the same coding system used on the wiring table. Wire runs are then drawn using the grid lines as guides for spacing. These wire runs are drawn horizontally and vertically with routing only at right angles between *To* and *From* positions established from the schematic diagram. The object of this type of drawing is to present a clear, easily read wire trace of the package. For this reason, closely spaced lines representing wire runs must be avoided. In addition, all wire runs should be clearly identified in terms of color code (obtained from the wiring table) for rapid identification in servicing.

20.3 HARNESS JIG

The simplest method of developing a jig for the construction of a harness is through the use of a plywood panel and wire brads. A grid system having 10 divisions to the inch is first taped securely to the plywood panel. From the harness assembly drawing, the harness contour is accurately reproduced on the grid system. An alternative method is simply to tape a copy of the original harness assembly drawing onto the panel. All lead terminations and identifications are accurately positioned. With the harness contour completed, 1-inch brads (serving as *jig pins*) are driven into the board to help position the wire runs exactly over the harness contour lines. Brads are positioned on both sides of the contour lines at a spacing that will accept the diameter of that portion of the harness that will be developed as the various wire runs are passed between them. Additional brads are used at each breakout point to guide the leads of various breakouts or legs into the correct position. Breakout points for individual leads or fanouts along the harness will require more closely spaced brads since they will accommodate only a single lead diameter between them. However, these brads should not be spaced so close as to interfere with the lacing process. Finally, brads are positioned beyond the lead termination point at each end of all wire runs to tie off each lead end. Each lead is tied off as it is placed into the jig to hold it into position and to maintain all leads taut as the harness is being formed. The brads inserted for this purpose are positioned beyond the termination points of the harness so that wire distortion caused in tying off lead ends will be removed when leads are cut to size as indicated on the harness assembly drawing. The lead length specified in Table 20.1 includes an additional length for tie-off purposes over and above the actual termination length necessary for the completed harness. An arbitrary additional length of approximately 3 inches per end is usually sufficient to tie off a wire run.

With all brads positioned, the individual wire runs are checked off from the wiring table as each is placed into the harness jig. When positioning the wire runs that will form a breakout from the trunk, it is important to orient these leads closest to the brads on the side of the harness from which the breakout will occur. This will minimize crossovers that tend to distort the harness contour. Continual checking of wire run positions during this phase of construction is absolutely essential to ensure optimum routing as well as to verify that all required wire runs are included before the leads are bundled. Once a harness has been completely laced, it is a very difficult and time-consuming task to add a wire run or to change the routing of a lead. A harness jig with the harness partially formed is shown in Fig. 20.5a.

An alternative, although more expensive, method of harness construction involves the use of commercially available harnessing aids such as *saddles*, *clamps*, and *fanout strips*. These aids are shown in Fig. 20.5b. They are available to accept a variety of harness diameters and are mounted to a plywood harness board with nails, screws, or with their own adhesive backing. Harness boards constructed of three layers of fine-mesh wire, each separated by a honeycombed spacer, have become popular because they permit more rapid jig construction. Once the harness layout drawing is taped to the board surface, the wire saddles, clamps, and fanout strips can be quickly secured to the board using nails requiring only finger pressure for penetration through the board (Fig. 20.5c). These aids make harness construction much easier than plywood panels and brads.

Whichever method of harness construction is employed, the major concern is to produce a harness that will exactly duplicate the contour and dimensions on the harness assembly drawing so that it will properly fit into the finished package.

When all the leads have been correctly placed in the harness jig, the bundling of the harness may begin. All lacing of the harness must be performed in place. If any section is lifted out from between the brads during lacing, dimensioning accuracy between breakout points may be lost. The two common methods of bundling a harness are *lacing* and employing *cable* ties. Both methods result in a rigid and secure harness.

20.4 LACING

Lacing involves the formation of a series of evenly spaced locking *hitches* or "knots" along the trunk and legs. This form of bundling requires the use of lacing *cord*, *twine*, or *tape*. Many types of lacing material are commercially available, such as *waxed linen*, *polyester*, *nylon*, and *Teflon*. Each of these is available with their fibers impregnated with substances such as *wax*, *silicone*, or *vinyl resin* to better grip the lock hitches. Flat lacing tapes are better than round cord or twine, since they provide a larger contact surface around the leads of the harness and are less likely to cut into the wire insulation when pulled tight. All these lacing materials

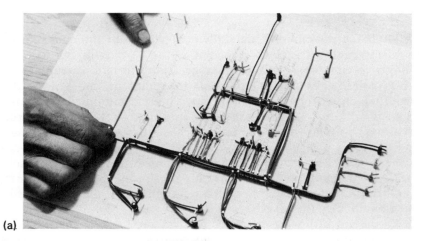

(a)

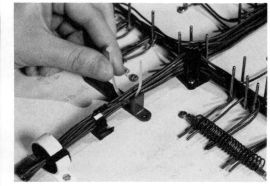

(b)

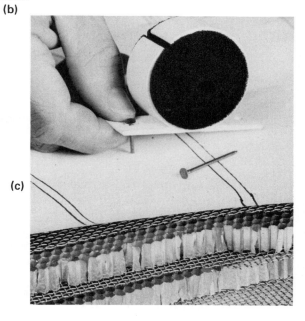

(c)

FIGURE 20.5. Harness construction aids. (a) Amplifier's harness being completed on jig. (b) Saddles, clamps, and fanouts. (c) Rapid installation of clamps onto honeycombed harness board. (b) Courtesy of Panduit Corp. and Thomas & Betts Co. (c) Courtesy of Thomas & Betts Co.

510

have a minimum breaking force in pounds, varying with the material and the size. A typical $\frac{1}{16}$-inch-wide tape has a minimum breaking force of approximately 25 pounds, although lacing materials are available with breaking forces as high as 200 and as low as 10 pounds.

Initial lacing of the harness begins at one end of the trunk. To aid in lacing, a *bobbin* or *shuttle*, as shown in Fig. 20.6, may be used. A sufficient length of lacing tape is wound on one of these lacing aids to complete the lacing of the entire trunk of the harness. This length should approximate four to five times that of the section to be laced to avoid splicing. Splicing of lacing materials is highly undesirable since a potential break is formed if the splice is weak and the splice makes an unsightly knot. The bobbin or shuttle is also used to help pull the lacing tape to tighten the locking hitches. If a great deal of lacing is to be performed, a lacing *glove*, shown in Fig. 20.7, should be worn to protect the hand and fingers from the abrasive action of the tape.

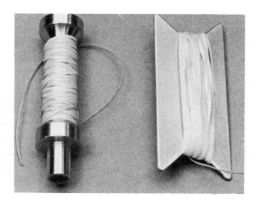

FIGURE 20.6. Lacing bobbin and shuttle.

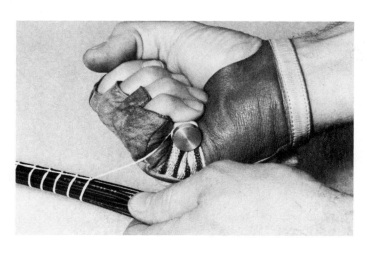

FIGURE 20.7. Lacing gloves are used for protection.

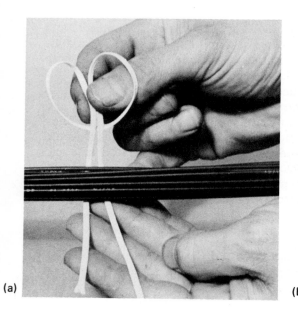

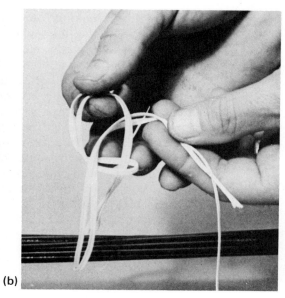

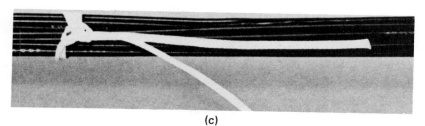

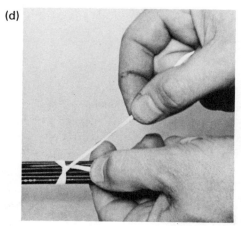

FIGURE 20.8. Sequential procedure for forming a girth hitch. (a) Forming double loop. (b) Both lacing ends are passed through double loop. (c) Partially tightened girth hitch. (d) Final dress.

The *girth hitch* is often used to initiate the lacing process. The sequential procedure for forming a girth hitch is shown in Fig. 20.8. In Fig. 20.8a, one end of the lacing tape is wrapped into a double loop. The free end of the tape and the bobbin are then passed through this loop (Fig. 20.8b). Tension is simultaneously applied to the free end of the tape and the bobbin to *dress* the girth hitch firmly against the leads. Completion of the girth hitch is shown in Fig. 20.8c and d. The trailing end of the lacing tape is then laid flat between two adjacent leads and a series of four tightly grouped locking hitches are formed against the girth hitch as well as over the free end of the tape to prevent the girth hitch from slipping out of position. This completes the originating point of the harness, which is shown in Fig. 20.9a. One of several types of these locking hitches is shown in Fig. 20.9b. After the initial four locking hitches are formed against the girth hitch, the *running hitches* may be begun. These are formed in the same way as locking hitches. The term *locking hitch* refers only to tightly grouped hitches whose function is to "lock" the girth hitch and breakout point positions. When this type of hitch is used solely to bundle the remaining portions of a harness, it is referred to as a *running hitch.*

FIGURE 20.9. Forming lock and running hitches. (a) Originating point of harness. (b) Formation of a running hitch. (c) Evenly spaced running hitches.

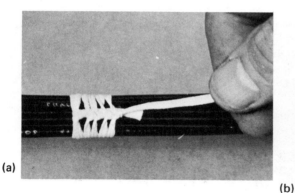

(a)

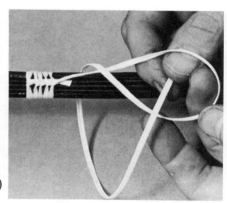

(b)

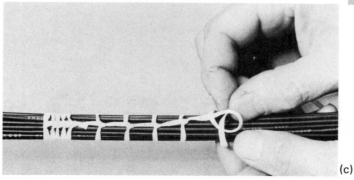

(c)

The type of hitch shown in Fig. 20.9 is preferable since it provides excellent security at each individual tie point. The running hitches should be uniformly spaced at a distance approximately equal to the diameter of the harness at any point but never less than ½ inch apart (Fig. 20.9c). The selected spacing distance should be consistent throughout the entire harness. Each hitch should be pulled firmly to ensure positive gripping but should not be so tight as to damage the insulation of the individual leads within the harness.

When the lacing termination point on a trunk or leg of the harness is reached, as specified on the wiring table, a special technique for termination is used, shown in Fig. 20.10. After the last running hitch is formed, a small loop of the tape is placed flat against the harness extending toward the termination point (Fig. 20.10a). Four tightly grouped lock hitches are then formed over the loop. This procedure is shown in Fig. 20.10b.

FIGURE 20.10. Terminating technique for lacing. (a) Loop used for termination. (b) Terminating lock hitches. (c) End of lacing tape feed-through loop. (d) Completed termination.

(a)
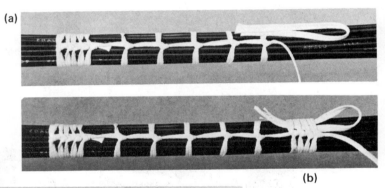
(b)

(c)
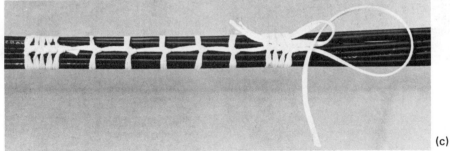

(d)

To complete the termination, a sufficient length of tape is cut from the bobbin and inserted through the exposed end of the loop extending from under the four previously formed lock hitches (Fig. 20.10c). The ends of the loop are then pulled back, drawing the end of the lacing tape under the tightly formed lock hitches. The completed termination with the end of the tape secured and trimmed is shown in Fig. 20.10d.

The technique of lacing just discussed is repeated on all legs of the harness. When the lacing is completed, all leads are cut to size, using the underlying harness assembly drawing as a guide. Finally, all leads are stripped and tinned in accordance with the specifications provided in the wiring table. The harness is then removed from the jig and is ready for assembly into the chassis.

A properly constructed harness possesses the following characteristics:

1. All individual leads within a trunk or leg are parallel to one another.
2. Leads leave any leg or breakout point at right angles to the run, with all leads leaving from one side of the run whenever possible.
3. All hitches are uniformly spaced and perpendicular to the run.
4. All running hitches appear between the same two leads for the length of a trunk or leg and do not spiral the harness.
5. Each hitch is sufficiently secure to keep itself from slipping along the run.
6. Tightly grouped lock hitches are formed when beginning or terminating the lacing.
7. Three tightly grouped lock hitches bound the breakout point for a leg, whereas two are sufficient about the breakout points for leads, breakouts, or fanouts.

The completed harness ready for chassis assembly is shown in Fig. 20.11. Notice that optimum appearance is achieved when the hitches are formed so that when the harness is installed, the knots are against the chassis and not visible.

20.5 CABLE TIES

Another method of forming a harness other than continuous lacing is with the use of commercially available *cable ties*. They consist of a nylon strap with a locking eyelet. Small closely spaced grooves are on the inside surface of the strap perpendicular to the length of the strap. These grooves will be in contact with the harness when assembled. The cable tie is first looped around the leg to be bundled. The tapered end of the strap is then fed through the eyelet and pulled tightly. The strap is held securely by the locking eyelet, which engages the grooves and prevents the strap from

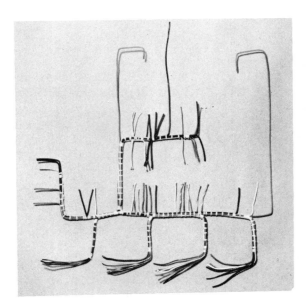

FIGURE 20.11. Completed harness formed with lacing tape.

slipping. Excess strap length is cut off with diagonal cutters close to the eyelet. The proper procedure for bundling leads with cable ties is shown in Fig. 20.12.

Special hand-operated cable tie assembly tools are available to facilitate the use of these bundling aids. One such tool is shown in Fig. 20.13. It consists of a gripping and cutting *head*, a *tension control*, and an *actuating trigger*. The tension control, roughly gauged from *loose* to *tight*, is set for the desired tension. The specific setting is determined by "feel," using the trial-and-error method. With the cable tie looped around the harness, the tapered end of the strap is fed through the eyelet into the gripping head of the assembly tool. Each time the trigger is actuated, the head grips the strap and pulls it for a short distance through the eye-

FIGURE 20.12. Cable ties are used to bundle leads. (a) Insertion of tab through locking eyelet. (b) Trimming excess tab length.

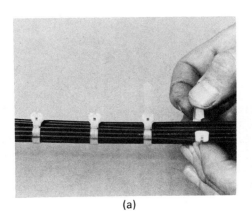

(a)

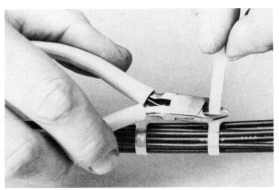

(b)

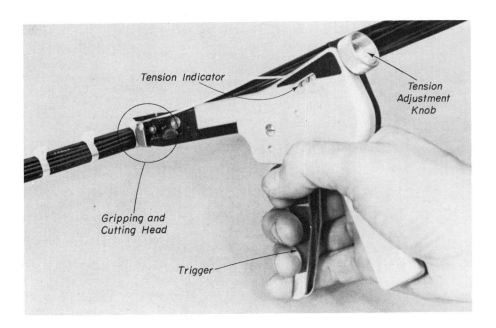

FIGURE 20.13. Hand-operated cable tie assembly tool.

let. It may be necessary to actuate the trigger several times to reach the amount of tension that will firmly grip the leads. When the preset tension setting is reached, the cutting head shears the strap close to the eyelet. The use of this tool allows for rapid and uniformly controlled cable tie assembly.

Cable ties are positioned at regularly spaced intervals along the harness and at each breakout point. Two tying techniques can be employed at breakout points. The first involves the use of three ties around the breakout. This technique is shown in Fig. 20.14a. An alternative method is to loop a single cable tie in a *crossed* fashion around the breakout (Fig. 20.14b). A completed harness bundled with cable ties is shown in Fig. 20.15.

FIGURE 20.14. Techniques for securing breakout points. (a) Three-tie method. (b) Single crossed-tie method.

(a)

(b)

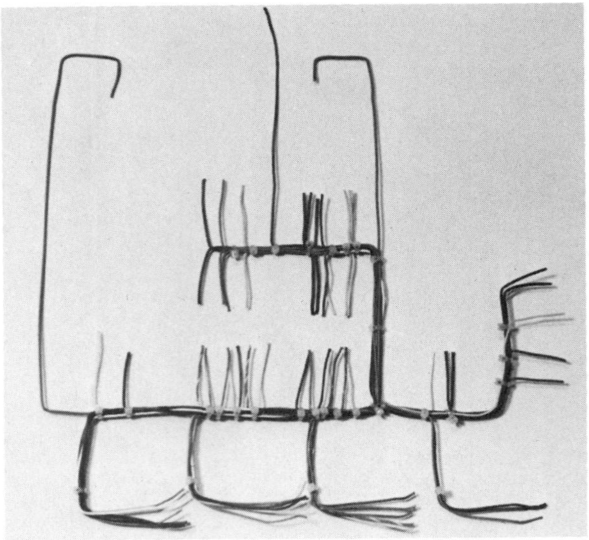

FIGURE 20.15. Harness bundled with cable ties.

20.6 SHRINKABLE TUBING

Continuous lacing or cable ties for harness construction are the two most popular methods used. Another technique of harnessing employs plastic and elastomer materials classified as *heat-shrinkable tubing*. This tubing is available in at least 12 different materials, each having its unique properties. These materials together with their characteristics and applications are shown in Appendix XIII.

Manufacturers of shrinkable tubing employ special mechanical and thermal means to expand the molecules of the tubing and "fix" them in a strained state. When heated, the strains established during the manufacturing process are relieved, thus returning the tubing diameter to its original size. In addition, the wall thickness, which is reduced during manufacture, returns to its original thickness when heated. *Shrink temperatures* for some of the common tubing materials range from 175 to over 600°F (79 to 316°C).

In its expanded state, shrinkable tubing is easily slipped over the leads to be bundled after being cut to the required length (Fig. 20.16a). Shrink temperature is achieved with the use of a portable electric *heat gun*, shown in Fig. 20.16b, which concentrates heated air completely around the tubing, thus shrinking it and forming a tight fit. The effects of heat on shrinkable tubing are shown in Fig. 20.16c. This process requires only one or two practice trials to determine the optimum length of time for the application of heat to shrink the tubing to the desired size.

Shrinkable tubing exhibits several desirable characteristics that are not attainable with continuous lacing and cable tie methods. Shrinkable tubing provides a dust-free, waterproof, and an abrasive and oil-resistant

(a)

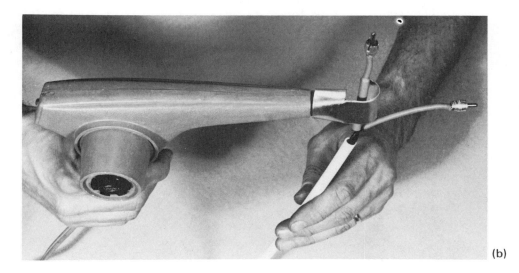

(b)

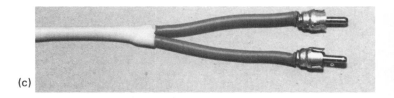

(c)

FIGURE 20.16. Installation of shrink tubing. (a) Initial placement of tubing sections. (b) Air deflector allows for uniform heating. (c) Finished wire assembly.

covering over the harness as well as the ends of the connectors. Short lengths of shrinkable tubing may also be used on a laced or cable-tied harness where abrasion at a particular point could be a problem. The one disadvantage of this tubing is that it can only be used where accessibility will permit.

20.7 CABLE CLAMPS

When the harness has been completed, it is initially installed into the system. If connectors have been preassembled onto the harness, they are first mounted. All leads requiring soldered connections are then assembled onto the subassemblies. The orientation of the installed harness must be such as to protect the wire insulation from abrasion. In addition, the harness must not contact high-temperature components.

The harness must finally be rigidly fastened to the chassis to prevent movement or undue strain on terminal connections. Commercially available *cable clamps*, such as those shown in Fig. 20.17, are used for this purpose. These clamps are constructed of nylon or Teflon and are provided with either a clearance hole for typical machine screw sizes or with an adhesive backing for securing the clamp to the chassis. They are available in sizes to accept harness diameters from $\frac{1}{16}$ inch to several inches and provide an excellent means of securing a harness rigidly to chassis elements.

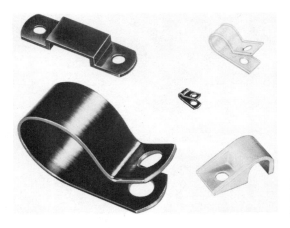

FIGURE 20.17. Typical cable clamps. Courtesy of Weckesser Company, Inc.

20.8 CABLE DESIGN AND FABRICATION

Cable design is less involved than that for harnesses since all leads run the full length of the cable with no breakout points. Therefore, the basic information necessary for the fabrication of a cable is (1) overall length; (2) termination hardware and connectors; (3) type, size and insulation of the wire; (4) type of protective tubing; (5) number of leads; and (6) *to* and *from* terminal connections for each lead. This information may be listed in a table similar to the wiring table of Table 20.1, with appropriate modifications to include only the specific information listed above. Similar to harnesses, the addition of leads or terminations once a cable has been completed is a tedious and time-consuming task. For this reason, the wiring table must be developed with care and the cable fabricated in exact accordance with the table.

To fabricate a cable such as that shown in Fig. 20.18a, the leads must first be run through flexible tubular insulation. With the ends of the tubing secured in the ends of the connectors, the leads become completely enclosed, thus reducing undesirable environmental effects. The two basic styles of tubing are the solid and split wall (track closure) configurations. The split-wall type is shown in Fig. 20.18b. When solid-wall tubing is used, a *wire snake* may be necessary. One end of each lead is secured to the wire snake and all are pulled through the tubing simultaneously. When cables are extremely long, this process can become very difficult. For this

FIGURE 20.18. Flexible tubular cable insulation. (a) Solid wall tubing. (b) Split wall (zipper) tubing.

(a)

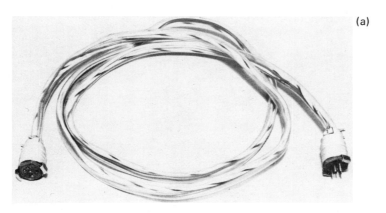

(b)

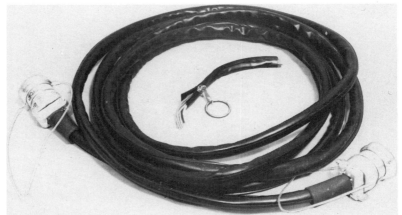

application, split-wall tubing is preferred, since it can be opened along its entire length. The edges (track) of this tubing are designed to interlock when forced together through a zipper action, thereby completely enclosing the leads that are laid into it. If the cable is to be bent to a small radius when installed, a sealer, such as ZT Sealer,* is spread onto the mating edges of the track before they are interlocked. This sealer causes the track to chemically fuse, preventing the tubing from opening when bent.

After the leads have been enclosed in the tubing, one connector is soldered on one end to secure both the tubing and all leads at that position. The flexible tubing is then forced back from the opposite lead end to allow accessibility for soldering leads to the remaining connector. It is important to point out that all hardware associated with the installation of the connectors (locking collars and seals) are slid over the tubing *before* the lead ends are soldered. After the leads are soldered, the tubing is smoothed into position and secured with the locking collar of the connector. When an additional measure of sealing is desired at the point where the leads enter the connector, a length of shrinkable tubing may be positioned to extend from the tubing onto the rear portion of the connector.

*Product of the Zippertubing Company.

PROBLEM SET

20.1 Fabricate a lead cleaner such as that shown in Fig. 20.19. Use a ⅝- by 10-inch strip of 16-gauge aluminum, two 5-inch lengths of ⅝-inch-wide copper braid, and two 1-inch lengths of shrinkable tubing.

FIGURE 20.19.

20.2 Determine the spacing between running hitches when bundling the trunk of a harness that consists of 23 leads. Each lead is No. 18 AWG stranded conductor with an insulation thickness of 1/64 inch.

20.3 Determine the suitable expanded inside diameter of shrinkable tubing to secure a bundle of leads such as described in Problem 20.2.

20.4 Construct a lacing shuttle such as is shown in Fig. 20.6. The shuttle is to be fabricated from a 1- by 2½-inch blank of 16-gauge aluminum. A 90-degree notch is to be sheared into each end to a depth of 7/16 inch. Break all sharp points and edges with an appropriate file.

20.5 Construct two wooden frames from 1- by 2-inch No. 1 pine having inside dimensions of 18 by 24 inches. Using heavy-duty staples, aluminum window screen, corrugated cardboard, and the frames, fabricate a harnessing board. First staple one 19¾- by 25¾-inch piece of screen to one of the frames. Shear enough ¾- by 18-inch strips of corrugated cardboard to fill the inside of the frame over the screen when the strips are stood on edge. Once the honeycombed center section is completed, staple another 19¾- by 25¾-inch screen to the frame to secure the cardboard sections in place. The second frame is used as a stand to provide clearance below the harness panel when the harness pins pass through the lower screen. Masking tape should be run over all edges of the screens on both sides of the frame to eliminate any sharp pieces of screen wire from interfering with the use of the board.

20.6 Using the harnessing board constructed in Problem 20.5 and Table 20.1, fabricate the harness shown in Fig. 20.3. Draw the harness contour on a piece of 17- by 22-inch grid paper. Secure the contour drawing onto the board with masking tape. Press 2½-inch (8d) finish nails through the board about the contour where necessary to position and tie off all wire runs given in the table. Once all wire runs are in place, form the harness with nylon lacing tape.

21

Projects

21.0 INTRODUCTION

This chapter offers projects for the student to construct that will allow him or her to develop further the skills discussed in this text. The projects presented illustrate varying degrees of construction complexity. Each of the circuits has been tested and performs as described. The projects were selected to ensure familiarity with the application of as many devices as possible. The expense of each project also was considered. Although they vary in cost of construction, many of the projects can be completed at a nominal price.

The authors are aware of the necessity to substitute parts and devices because of availability considerations. Many devices and components are not critical, and where indicated, equivalencies may be substituted. For this reason, and also to further familiarize the student with the selected circuit, it is strongly recommended that the circuit be breadboarded and tested before fabrication. This approach also provides an opportunity for the advanced student to modify a project.

Each project begins with an introduction to acquaint the student with its purpose and function. The circuit operation is explained in general. Finally, construction hints are provided to avoid difficulties.

The finished packages shown pictorially are to be considered as typical. The student, of course, is not restricted to the design shown and should feel free to employ ingenuity in fabricating a package that will reflect his or her own requirements and desires.

It is hoped that these projects will not only serve a useful purpose but will also make students confident of their ability to apply all of their skills in constructing circuits of which they can be proud.

The following projects are presented in this chapter:

Portable Multimeter

Variable-Voltage-Regulated Power Supply

Transistor Curve Tracer

Automotive Accessory Alarm

Automotive Safety Flasher

Automobile Security Alarm System

Digital Automotive Clock

400-Watt Lamp Dimmer

To aid the student in construction, terminal configurations for all of the devices used in these projects are provided in Fig. 21.1.

FIGURE 21.1. Device configurations for projects.

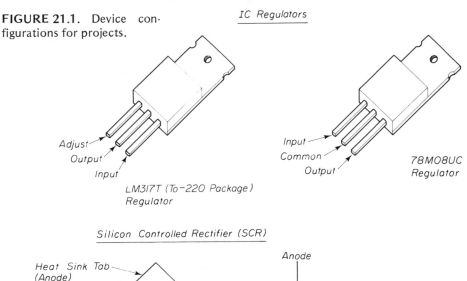

IC Regulators

Adjust
Output
Input

LM3I7T (To-220 Package)
Regulator

Input
Common
Output

78M08UC
Regulator

Silicon Controlled Rectifier (SCR)

Heat Sink Tab
(Anode)

Cathode
Anode
Gate

CIO6 B1
CIO6 Y1

Anode

Gate

Cathode

525

FND 357 Seven−Segment Display

Top View

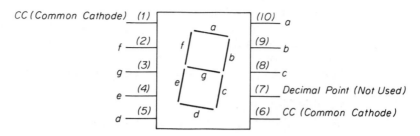

CC (Common Cathode) (1) ——— (10) a
f (2) ——— (9) b
g (3) ——— (8) c
e (4) ——— (7) Decimal Point (Not Used)
d (5) ——— (6) CC (Common Cathode)

Triac

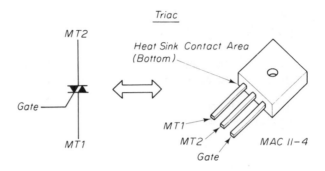

MT2

Gate

MT1

Heat Sink Contact Area
(Bottom)

MT1
MT2
Gate

MAC 11−4

Bilateral Trigger Diode

Style 2
1) Cathode
2) Anode

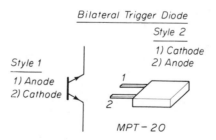

Style 1
1) Anode
2) Cathode

1
2

MPT−20

Leads are Interchangeable

FIGURE 21.1. (cont.)

526

Diodes

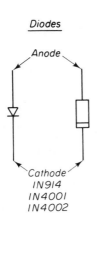

Anode

Cathode
IN9I4
IN400I
IN4002

Zener Diode

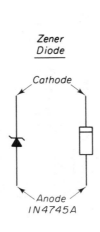

Cathode

Anode
IN4745A

Light Emitting Diode (LED)

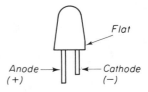

Flat

Anode
(+)

Cathode
(−)

Transistors

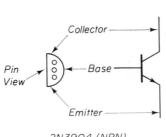

Collector

Pin View

Base

Emitter

2N3904 (NPN)
2N3906 (PNP)

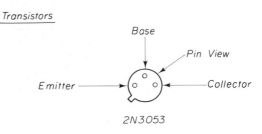

Base

Pin View

Emitter

Collector

2N3053

FIGURE 21.1. (cont.)

527

555 Timer

Top View

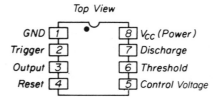

GND 1	8 V_{CC} (Power)
Trigger 2	7 Discharge
Output 3	6 Threshold
Reset 4	5 Control Voltage

National Oscillator / Timer

Top View

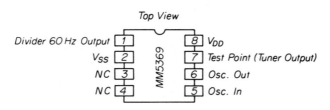

MM5369

Divider 60 Hz Output 1	8 V_{DD}
V_{SS} 2	7 Test Point (Tuner Output)
NC 3	6 Osc. Out
NC 4	5 Osc. In

National Clock Chip

Top View

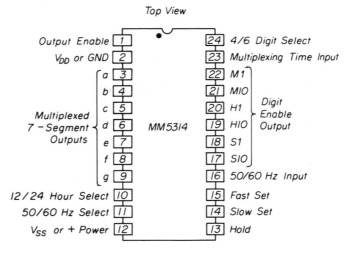

MM5314

Output Enable 1	24 4/6 Digit Select
V_{DD} or GND 2	23 Multiplexing Time Input
a 3	22 M1
b 4	21 MIO
c 5	20 H1
d 6	19 HIO
e 7	18 S1
f 8	17 SIO
g 9	16 50/60 Hz Input
12/24 Hour Select 10	15 Fast Set
50/60 Hz Select 11	14 Slow Set
V_{SS} or + Power 12	13 Hold

Multiplexed 7 – Segment Outputs (pins 3–9)

Digit Enable Output (pins 17–22)

FIGURE 21.1. (cont.)

21.1 PORTABLE MULTIMETER

One of the more useful pieces of test equipment to the electronic technician is a basic multimeter. A versatile multimeter should be able to measure reasonable values of resistance, dc currents, and dc voltages in the ranges commonly encountered in electronics. To extend the versatility of a multimeter, it is also advantageous to have the ability to measure ac voltages.

A basic portable multimeter circuit, shown in Fig. 21.2, meets all of the foregoing specifications with the following ranges:

Ohmmeter:	1 kΩ (center scale)
DC current ranges:	0 to 500 mA
	0 to 50 mA
	0 to 5 mA
	0 to 0.5 mA
DC voltage ranges:	0 to 500 volts
	0 to 50 volts
	0 to 5 volts
AC voltage ranges:	0 to 500 volts (rms)
	0 to 50 volts (rms)
	0 to 5 volts (rms)

Circuit Operation. The portable multimeter project is constructed around the basic movement M_1 with the adjustment and protection circuit consisting of R_2, R_3, R_4, R_5, D_1, and D_2.

To improve the accuracy of the basic meter movement M_1, which is a standard 50-μA full-scale unit ($\pm$3%) with an internal resistance of approximately 1 kΩ, two fixed resistors and two trim pots are added as shown in Fig. 21.2. This circuitry can be adjusted to produce a very accurate 75-μA, 2.00 kΩ internal resistance "equivalent meter" between points a and b in Fig. 21.2. The series combination of R_2 and R_3 is placed in parallel with M_1. R_2 is adjusted to shunt an additional current ($\approx$25 μA) around the meter so that exactly 75 μA (full-scale) of "equivalent meter" current is flowing into point b when M_1 reads full scale at 50 μA. In this way, any error in current required to produce full-scale deflection of M_1 (i.e., current value slightly above or below 50 μA) is eliminated and a precise 75-μA equivalent meter is attained. Resistors R_4 and R_5 are used to set the internal resistance of the equivalent meter to exactly 2.00 kΩ. Since an exact 75 μA must flow through the equivalent meter to produce full-scale deflection on M_1, R_5 is adjusted until the voltage between points a and b is exactly 0.150 volts dc when full-scale current (I_{fs}) = 75μA. In equation form:

Equivalent meter voltage $V_{ab} = I_{fs} R_{ab}$

$$= 75 \, \mu A \times 2 \text{ k}\Omega = 0.150 \text{ volts dc}$$

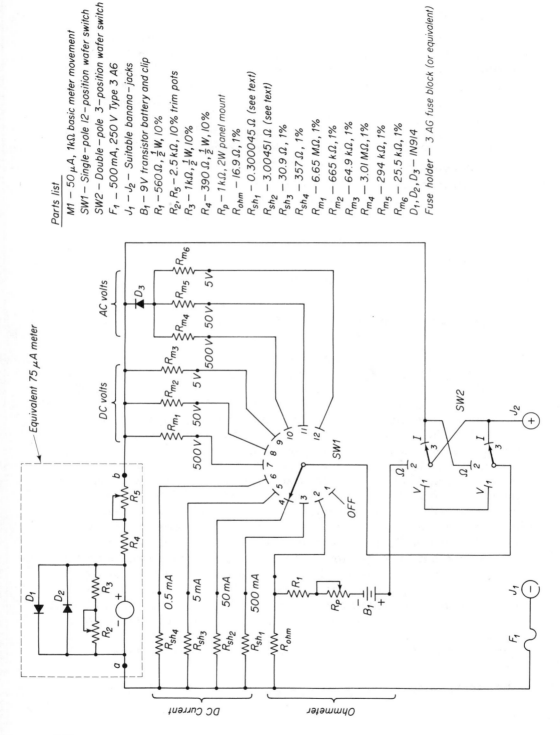

Parts list

M1 – 50 μA, 1kΩ basic meter movement
SW1 – Single–pole 12–position wafer switch
SW2 – Double–pole 3–position wafer switch
F_1 – 500 mA, 250 V Type 3 A6
$J_1 - J_2$ – Suitable banana–jacks
B_1 – 9V transistor battery and clip
R_1 – 560 Ω, $\frac{1}{2}$ W, 10%
R_2, R_5 – 2.5 kΩ, 10% trim pots
R_3 – 1kΩ, $\frac{1}{2}$ W, 10%
R_4 – 390 Ω, $\frac{1}{2}$ W, 10%
R_p – 1 kΩ, 2W panel mount
R_{ohm} – 16.9 Ω, 1%
R_{sh_1} – 0.300045 Ω (see text)
R_{sh_2} – 3.00451 Ω (see text)
R_{sh_3} – 30.9 Ω, 1%
R_{sh_4} – 357 Ω, 1%
R_{m_1} – 6.65 MΩ, 1%
R_{m_2} – 665 kΩ, 1%
R_{m_3} – 64.9 kΩ, 1%
R_{m_4} – 3.01 MΩ, 1%
R_{m_5} – 294 kΩ, 1%
R_{m_6} – 25.5 kΩ, 1%
D_1, D_2, D_3 – IN914
Fuse holder – 3 AG fuse block (or equivalent)

FIGURE 21.2. Basic multimeter circuit.

A simple circuit to adjust the equivalent meter is shown in Fig. 21.3. Two silicon diodes, D_1 and D_2, are placed directly across the meter movement to protect it from overvoltage, which could cause it damage. Under normal operation of the meter, both diodes are effectively open circuits and only conduct current (around M_1) under fault conditions.

To realize a reasonable degree of accuracy, 1% tolerance resistors are specified where necessary.

The basic 50-μA meter movement and designed equivalent 75-μA meter circuit were selected for two reasons. First, all voltage and current ranges are multiples of 5 which require no special scales to be constructed for interpreting pointer positions. The resistance measurement scale requires only one simple reference chart to be constructed. Second, this multimeter provides a sensitivity of 15,000 ohms per volt.

Fuse F_1 provides protection for the entire multimeter circuit.

Ohmmeter Circuitry. The "equivalent meter" circuitry, together with R_{ohm}, R_1, R_p, and B_1, constitute the ohmmeter function of the multimeter. When the *function selector* switch Sw_2 is set to ohms (Ω) in position 2 and the *range selector* switch Sw_1 is set to position 2, the circuit acts as an ohmmeter with a center scale value of 1 kΩ. To adjust the ohmmeter to 0 Ω, jacks J_1 and J_2 are shorted and R_p is adjusted to achieve full-scale deflection (0 Ω value). A scale can then be constructed by measuring known resistor values and recording the equivalent meter reading from the meter pointer position.

FIGURE 21.3. Circuit to adjust the "equivalent meter." Adjust the power supply until I_1 reads exactly 75 μA. Vary R_2 until M_2 reads full scale exactly. Next adjust R_5 until V_1 reads 0.15V.

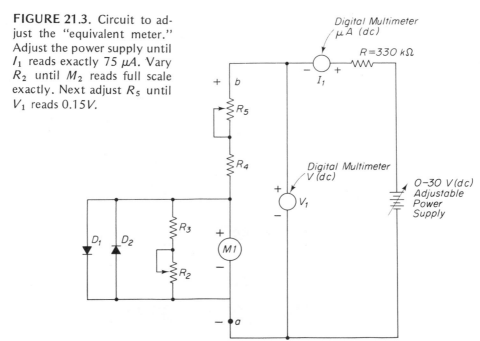

DC Current Circuitry. With the function selector set to current (I), position 3, range selector positions 3, 4, 5, and 6 will measure dc currents. Shunt resistors R_{sh1}, R_{sh2}, R_{sh3}, and R_{sh4}, together with the equivalent meter circuit, provide the ranges shown in Fig. 21.2. R_{sh1} and R_{sh2} are standard 1% resistors. R_{sh3} and R_{sh4} are not available in the standard 1% range but can be constructed using series and parallel combinations of available values and measured to obtain the required shunt values.

DC Voltage Ranges. With the function selector switch set to voltage (V), position 1, voltage measurements are made using range selector positions 7, 8, and 9. Three standard 1% resistors, R_{m1}, R_{m2}, and R_{m3}, are used as multiplier resistors.

AC Voltage Ranges. With the function selector switch set to voltage (V), position 1, ac voltages are measured using range selector positions 10, 11, and 12. Standard 1% resistors are used for R_{m4}, R_{m5}, and R_{m6}. Diode D_3 blocks the negative half-cycle of the incoming voltage to obtain an up-scale rms reading. As with all meters of this type, only rms values of sine waves of voltage can be accurately measured.

Construction Hints. There are no special problems encountered in the construction of this multimeter. A typical package is shown in Fig. 21.4. Resistor mounting is accomplished between two 9-pin terminal strips mounted in parallel and wired with standard hook-up wire. The equivalent meter components shown enclosed in dashed lines in Fig. 21.2 are mounted onto a small piece of perforated insulation board, wired, and then secured onto the rear of meter M_1.

FIGURE 21.4. Typical meter package.

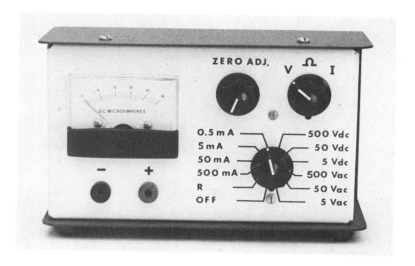

21.2 VARIABLE-VOLTAGE-REGULATED POWER SUPPLY

One of the most useful projects the experimenter can construct is a laboratory-type dc power supply. Applications of this type of circuit are almost unlimited. However, for a power supply to be most useful, it should have the following features: (1) adjustable output-voltage capability over a wide range, (2) high-load-current capability, (3) excellent voltage regulation at all load currents, (4) current limiting with short-circuit protection to prevent damage to the supply when accidental overloads or shorts are applied to the output terminals, and (5) metering to monitor both output current and voltage.

The variable-voltage-regulated power supply circuit, shown in Fig. 21.5, will meet all of the foregoing requirements with the following specifications:

Output voltage: adjustable from 1.2 to 20 volts dc

Output current: 0 to 1 ampere

Voltage regulation: 0.1%

Including: Load-current limiting

Thermal-overload protection

Safe-area-device protection

The requirements listed for a basic laboratory-type power supply are easily met with the use of an integrated regulator chip having only three external terminals: input (in), output (out), and adjustment (adj). These are shown in Fig. 21.5. The resulting circuit is (1) low in overall parts count, (2) inexpensive, and (3) simpler to package than a comparable-type discrete power supply.

Circuit Operation. For purposes of analysis, the power supply circuit schematic diagram, shown in Fig. 21.5, can be divided into two basic sections: an unregulated power supply and a voltage regulator section. The unregulated power supply consists of transformer T_1, diodes D_1 to D_4 in a full-wave bridge configuration, and the filter capacitor C_1.

The basic regulator section consists of the LM317T regulator chip and support components R_3, R_4, R_5, C_2, and C_3.

The LM317T regulator chip requires that the output of the unregulated supply, across filter capacitor C_1, be limited to a maximum of 40 volts dc. This specification is easily met by T_1, since its theoretical peak output voltage under no-load current conditions is approximately 34.5 volts dc. Under full-load current conditions of 1 ampere, the dc output voltage of the unregulated supply is approximately 27 volts dc.

All of the electronics necessary to produce an adjustable, positive regulated power supply are contained in one three-terminal integrated

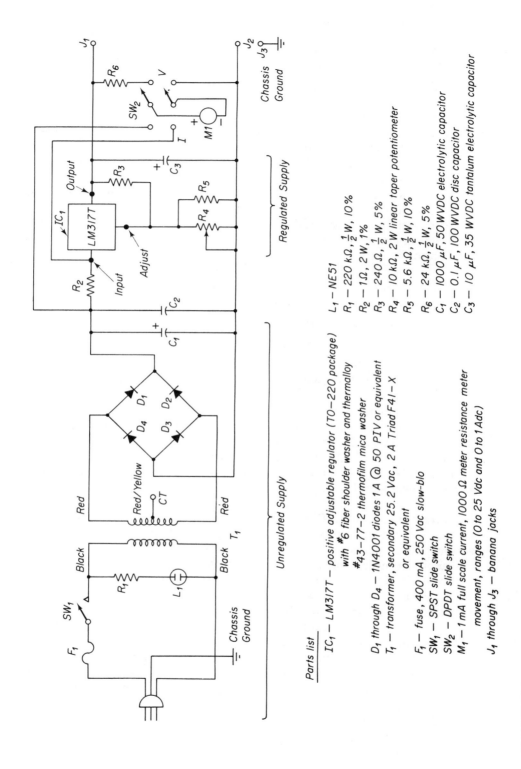

Parts list

IC_1 — LM317T — positive adjustable regulator (TO–220 package)
with #6 fiber shoulder washer and thermalloy
#43–77–2 thermofilm mica washer

D_1 through D_4 — 1N4001 diodes 1 A @ 50 PIV or equivalent

T_1 — transformer, secondary 25.2 Vac, 2 A Triad F41–X
or equivalent

F_1 — fuse, 400 mA, 250 Vac slow-blo

SW_1 — SPST slide switch

SW_2 — DPDT slide switch

M_1 — 1 mA full scale current, 1000 Ω meter resistance meter
movement, ranges (0 to 25 Vdc and 0 to 1 Adc)

J_1 through J_3 — banana jacks

L_1 — NE51

R_1 — 220 kΩ, $\frac{1}{2}$ W, 10%

R_2 — 1 Ω, 2 W, 1%

R_3 — 240 Ω, $\frac{1}{2}$ W, 5%

R_4 — 10 kΩ, 2 W linear taper potentiometer

R_5 — 5.6 kΩ, $\frac{1}{2}$ W, 10%

R_6 — 24 kΩ, $\frac{1}{2}$ W, 5%

C_1 — 1000 μF, 50 WVDC electrolytic capacitor

C_2 — 0.1 μF, 100 WVDC disc capacitor

C_3 — 10 μF, 35 WVDC tantalum electrolytic capacitor

FIGURE 21.5. Schematic diagram of variable-voltage regulated power supply.

circuit package, available in several standard package styles. The TO-220 plastic case was selected for this project. An internal reference voltage of 1.2 volts (constant) is established between the output (out) and the adjustment (adj) terminals of the LM317T regulator. This reference voltage is applied directly across the 240-ohm resistor (R_3) establishing a constant reference current of 5 mA flowing through R_3. This reference current also flows through the potentiometer (R_4) and parallel padding resistor R_5. An increase in the resistance of R_4 results in larger output voltages while a decrease in R_4 results in lower voltages. The minimum dc output voltage can never be lower than the reference voltage of 1.2 volts.

An ac ripple voltage is present on the input terminal of the regulator chip. At full-load currents of 1 ampere, this ripple can be as high as 5 volts peak to peak. On the output terminal of the regulator, there is essentially no ripple voltage (5 mV or less). The ripple voltage that is present at the input terminal is absorbed by the regulator.

Capacitor C_2 is a 0.1-μF disc required if the lead length from the unregulated supply terminal is excessive (2 inches or longer). It is, however, good practice to always include this capacitor.

Capacitor C_3 (10-μF tantalum) is connected from the output terminal of the regulator to common to improve the transient response of the regulator.

Switch SW_2 is used to switch meter M_1 into the circuit for current or voltage measurements. When SW_2 is switched to position "I," a 1-ohm shunt resistor (R_2) appears across the meter movement to allow for a maximum reading of 1 ampere.

With SW_2 switched to position "V," a 24-kΩ resistor (R_6) is placed in series with the meter movement to allow for a full-scale reading of 25 volts.

Construction Hints. For optimum performance, the power supply circuit should be packaged in a well-ventilated enclosure. The LM317T regulator package will require heat sinking to achieve the required specifications. A metal chassis element can be used as a suitable heat sink provided that it consists of at least 25 square inches of $\frac{1}{16}$-inch-thick

FIGURE 21.6. Assembly diagram of LM317T regulator package to chassis.

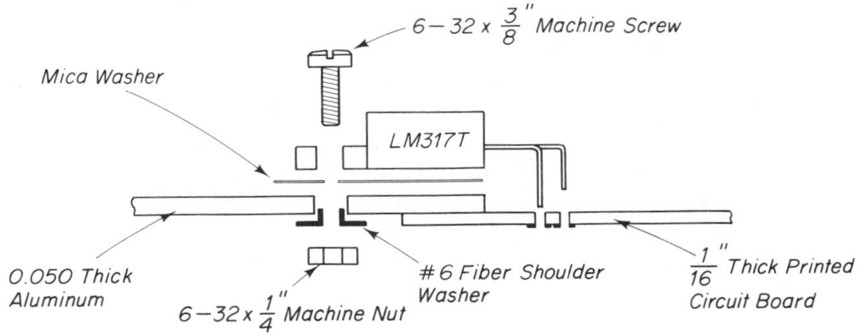

6 – 32 x $\frac{3}{8}$" Machine Screw

Mica Washer

LM317T

0.050 Thick Aluminum

6 – 32 x $\frac{1}{4}$" Machine Nut

#6 Fiber Shoulder Washer

$\frac{1}{16}$" Thick Printed Circuit Board

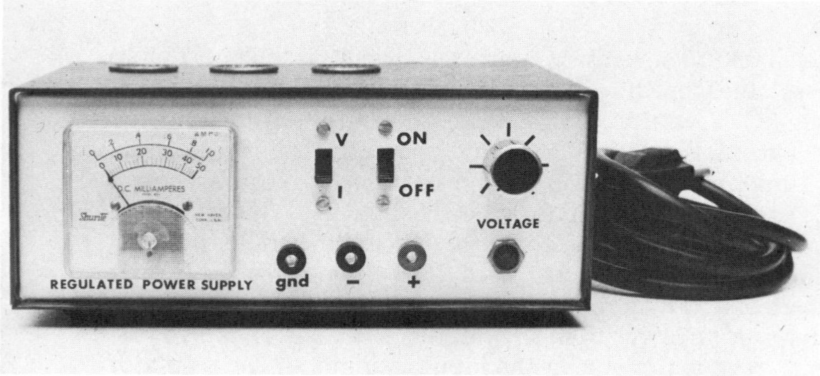

FIGURE 21.7. Basic package configuration for power supply project.

aluminum. The regulator chip *must* be electrically insulated from the chassis. This is done with the use of a silicon-greased mica washer and insulating shoulder washer (Fig. 21.6). Note that the leads are bent at 90 degrees onto a pc board. The pc board is used to assemble all the small, fragile circuit components.

A three-pronged plug is essential for electrical grounding of the chassis.

The basic power supply package is shown in Fig. 21.7.

21.3 TRANSISTOR CURVE TRACER

Often it becomes necessary to obtain exact transistor collector characteristics. A curve tracer, which is expensive and often not readily available, would normally be required for this purpose. The inexpensive curve tracer circuit of Figs. 21.8 and 21.9 can instead be used for most cases. This simple circuit used in conjunction with a standard oscilloscope will yield a family of collector curves for both NPN- and PNP-type silicon transistors. With the use of an oscilloscope mounted camera using several exposures on one frame, a family of characteristic curves for six values of base current can be obtained. A family of curves obtained in this manner for a 40406 PNP transistor is shown in Fig. 21.9.

Circuit Operation. With switch SW_1 closed, 12.6 volts ac appear across the potentiometer R_2. Diode D_1 acts as a half-wave rectifier to provide positive collector voltage for NPN transistors (switch SW_2 at NPN position) and negative collector voltage for PNP transistors (switch SW_2 at PNP position). Collector-to-emitter voltage is variable across the test transistor from 0 to approximately 18 volts peak by adjusting potentiometer R_2.

The oscilloscope vertical amplifier is connected across J_1 and J_2 with the gain set at 0.01 volt per centimeter. Since R_3 is chosen to be 10 ohms, this automatically sets the vertical scale for a collector current sensitivity of 1 mA per centimeter. The horizontal input is connected at J_3 and the gain set at 1 volt per centimeter.

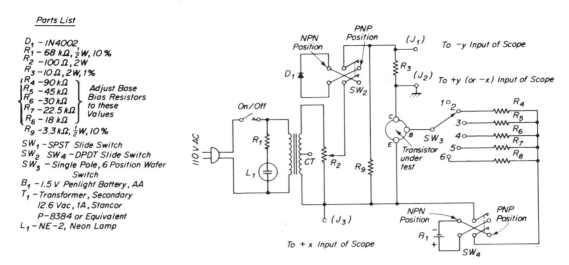

Parts List

$D_1 - 1N4002$
$R_1 - 68\ k\Omega, \frac{1}{2}W, 10\%$
$R_2 - 100\ \Omega, 2W$
$R_3 - 10\ \Omega, 2W, 1\%$
$\left\{\begin{array}{l} R_4 - 90\ k\Omega \\ R_5 - 45\ k\Omega \\ R_6 - 30\ k\Omega \\ R_7 - 22.5\ k\Omega \\ R_8 - 18\ k\Omega \end{array}\right\}$ Adjust Base Bias Resistors to these Values
$R_9 - 3.3\ k\Omega; \frac{1}{2}W, 10\%$
$SW_1 -$ SPST Slide Switch
$SW_2\ SW_4 -$ DPDT Slide Switch
$SW_3 -$ Single Pole, 6 Position Wafer Switch
$B_1 - 1.5\ V$ Penlight Battery, AA
$T_1 -$ Transformer, Secondary 12.6 Vac, 1A, Stancor P-8384 or Equivalent
$L_1 - NE-2$, Neon Lamp

FIGURE 21.8. Transistor curve tracer.

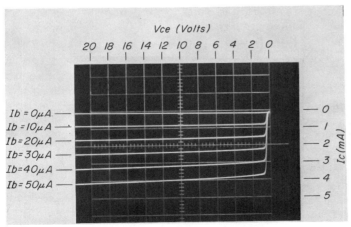

(a)

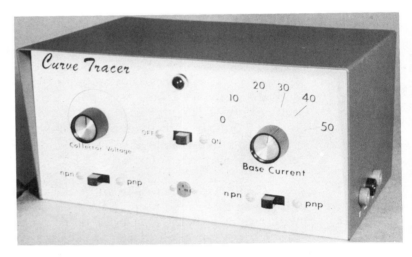

(b)

FIGURE 21.9. (a) Transistor curve tracer package. (b) Collector characteristics for 40406 transistor.

Base bias current is controlled through switch SW_4, battery B_1, and one of the six positions of switch SW_3, which introduces a specific value of base resistance to set the base current from 0 μA at position 1 to a maximum of 50 μA at position 6. Resistors R_4, R_5, R_6, R_7, and R_8 should be chosen as close as possible to those values given in the parts list for accurate base current calibration.

To obtain a collector characteristic curve for any silicon NPN or PNP transistor, both polarity switches (SW_2 and SW_4) are moved to the appropriate transistor-type position. With SW_3 in position 1, R_2 is varied to obtain a sweep of the $I_B = 0$ characteristic curve. Switch SW_3 is then moved to position 2 for the $I_B = 10$ μA curve. This procedure is continued to position 6 to complete the family.

Construction Hints. There are no special construction problems with this circuit. The front-panel controls (SW_1 through SW_4 and R_2) should be assembled to avoid crowding. In addition, jacks J_1, J_2, and J_3 should be positioned for ready access.

21.4 AUTOMOTIVE ACCESSORY ALARM

Installing electronic equipment such as a tape deck, transmitter/receiver, or FM converter in an automobile introduces the problem of security from theft. These accessories are normally mounted under the dashboard and are easily accessible to a potential thief. If the automobile is not equipped with an intrusion alarm system, expensive equipment can be removed quickly with the use of a screwdriver and a pair of wire-cutting pliers. This automobile accessory alarm project is simple in design yet provides some degree of security.

Circuit Operation. The automobile accessory alarm shown in Figs. 21.10 and 21.11 requires only three electrical connections for installation. With arming switch SW_1 closed, dc power is applied to the alarm circuit through the horn coil at jack J_1. With detector wire (J_3) connected to chassis ground, transistor Q_1 is in *cutoff*, holding SCR_1 *off*. When the detector wire is removed from ground (cut or pulled off the accessory chassis), current flows through R_1 into the base of Q_1. The value of R_1 is chosen to saturate Q_1, thus causing a current of approximately 1 mA to flow through the emitter resistor R_3. This current provides a voltage drop to the gate of SCR_1 sufficient for it to fire, thus providing a virtual ground on the horn coil and activating the horn as the alarm. Once triggered, SCR_1 will continue to conduct even if an attempt is made to replace the detector wire connection. The horn can now be disabled only by switching SW_1 to the "off" position.

Construction Hints. Constructing the automotive accessory alarm requires no special precautions. A pc board is recommended but is not

FIGURE 21.10. Automobile accessory alarm.

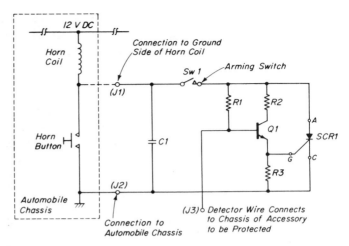

Parts List

$Q_1 - 2N3904$
$SCR_1 - C106Y1$
$R_1 - 27\,k\Omega, \frac{1}{2}\,W, 10\%$
$R_2 - 2.2\,k\Omega, \frac{1}{2}\,W, 10\%$

$R_3 - 1.2\,k\Omega, \frac{1}{2}\,W, 10\%$
$C_1 - 0.01\,\mu F, 100\,WVdc\,Disc$
$SW_1\,SPST, 2A, Slideswitch$

Chassis $-$ LMB Type MOO
or Equivalent

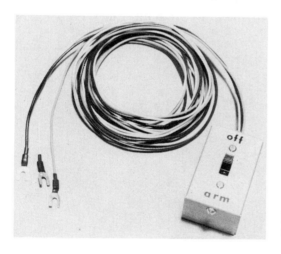

FIGURE 21.11. Package for automobile accessory alarm.

necessary. Care should be taken if the alarm chassis is connected to the car chassis that the metal tab of SCR_1 is electrically insulated from any ground connections. The alarm circuit should be mounted inside the car and well hidden from view, but the arming switch must be accessible. The wire from the ground of the horn coil (J_1) as well as the wire making the circuit ground (J_2) should be at least AWG No. 18 to handle the horn current. The detector wire should be AWG No. 26 so that it may easily be broken. In addition, the insulation color of this wire should be black, to appear as a ground wire, and it should be mounted directly under the head of one of the mounting screws of the accessory to be protected. All external electrical connecting wires may be provided with suitable spade lugs to simplify installation.

21.5 AUTOMOTIVE SAFETY FLASHER

Any driver who has been forced to stop on a highway to make emergency repairs or wait for road service realizes the dangers involved, especially at night. Other drivers approaching at high speeds may have little time to detect the disabled vehicle and avoid an accident. For this reason the automotive industry now provides emergency flasher controls for both the front and rear parking lights on all cars. For those cars not equipped with factory-installed flasher units or to provide an added degree of safety for those cars having them, the automotive safety flasher circuit, shown in Fig. 21.12, can be a real asset. This safety flasher can be positioned as much as 50 feet away from the disabled vehicle and emits a brilliant flashing light that can be seen at a considerable distance, particularly at night.

Circuit Operation. The Automotive Safety Flasher shown in Fig. 21.12 consists of a 555 integrated circuit timer chip operated in the

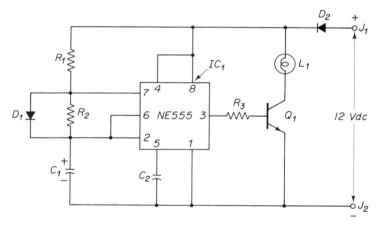

FIGURE 21.12. Circuit schematic for safety flasher.

Parts list

IC_1 — NE555 timer
Q_1 — 2N3053 transistor (TO−5 package)
R_1 and R_2 — 680 kΩ, $\frac{1}{2}$ W, 10%
R_3 — 470 Ω, $\frac{1}{2}$ W 10%
C_1 — 1μF, 15 WVDC electrolytic
C_2 — 0.01μF, 100 WVDC disc
L_1 — GE 93
D_1 and D_2 — 1N4001 diodes, 1A @ 50 PIV or equivalent

astable (free-running) multivibrator mode. Output pulses from the timer are used to switch the driver transistor Q_1 on and off and, as a result, the lamp L_1.

With 12 V dc applied to jacks J_1 and J_2, the output terminal pin 3 of the 555 timer will be high (approximately 12 volts), keeping the transistor on while simultaneously charging capacitor C_1 through resistors R_1 and R_2. Once C_1 charges up to approximately $\frac{2}{3} V_{cc}$, a discharge path is established from pin 7 to pin 1 of the 555 timer through an internal transistor connection between these two pins. The total discharge path for C_1 is comprised of R_2 and the path from pin 7 to pin 1 through the IC. Pin 3 of the timer will be at a very low voltage (approaching 0 volts) when capacitor C_1 is being discharged through resistor R_2. With the component values given, square-wave pulses occur at the output terminal of the timer, high voltage (approximately 12 volts dc) for about 1 second and low voltage (approximately 0 volts) for about $\frac{1}{2}$ second.

The base of the drive transistor Q_1 is connected, through the current-limiting resistor R_3, to the output terminal of the timer. Q_1, connected in the common-emitter mode, is gated on and off by the square-wave pulses at the output of the timer, causing lamp L_1 to be switched on for approximately 1 second and off for approximately $\frac{1}{2}$ second.

Equal on-and-off times can be achieved by adding diode D_1 across the resistor R_2. Other time intervals are possible by varying the values of either R_1 or R_2 or both. Diode D_2 may also be added to prevent the circuit from being subjected to a reversal of polarity in the event that the power plug is wired incorrectly or the battery terminals are reversed when alligator clips are used.

Construction Hints. Packaging this circuit should present no special construction problems. The flasher unit can be assembled into any suitable size chassis that will house the components without crowding. With an 8-pin IC package used, a pc board will allow easier installation of the IC, resulting in a neater package. This is not absolutely essential, however. Because of the small number of parts involved, assembly and interconnecting wiring can be easily produced with perforated board to mount the components. Whichever component assembly and wiring method is used, care must be taken not to connect any device case or lamp socket to the chassis. Electrical insulation of these components from the chassis is absolutely essential.

The power cord should be at least AWG No. 16 stranded wire no longer than 50 feet. The flasher circuit can be connected directly to the 12-volt battery terminals with the use of alligator clips or with a plug, as shown in Fig. 21.13. This plug connects directly into the cigarette lighter socket. Rubber feet should be attached onto the chassis base to prevent the surface from damage and to provide stability when the unit is placed on a smooth surface such as the hood, roof, or trunk of the vehicle.

FIGURE 21.13. A typical safety flasher package.

21.6 AUTOMOBILE SECURITY ALARM SYSTEM

Although no alarm system is foolproof, they do serve as a deterrent to automobile thefts, especially those thefts committed by the amateur. The most positive types of automobile alarm systems are those which employ mechanical door, hood, and trunk switches. Although this type of system is more difficult to install than the voltage- or current-sensing type of alarm systems, it is less susceptible to false triggering due to climate changes, extreme variations in temperature, or outside electrical interference. The most effective alarm systems are those that are completely concealed inside the automobile, thus giving potential thieves a false sense of security. The automobile security alarm system, shown in Fig. 21.14, is a reliable, relatively inexpensive system that will provide a high degree of security for an automobile.

Circuit Operation. An automobile security alarm system is usually installed in the trunk of the automobile. All the required electrical connections are then made from this location.

Switch SW_1 is a DPDT slide or toggle switch that serves two functions: it "arms" or "disarms" the alarm system. In the "armed" position, the switch acts to connect the distributor side of the ignition points to ground, thus preventing the car from being started. This arm/disarm switch is usually mounted under the dashboard of the car, hidden from view but accessible to the driver.

The alarm system includes (1) an exit delay, (2) an entrance delay, (3) alarm signal and recycle, and (4) an "instant on" protection feature for both the hood and trunk.

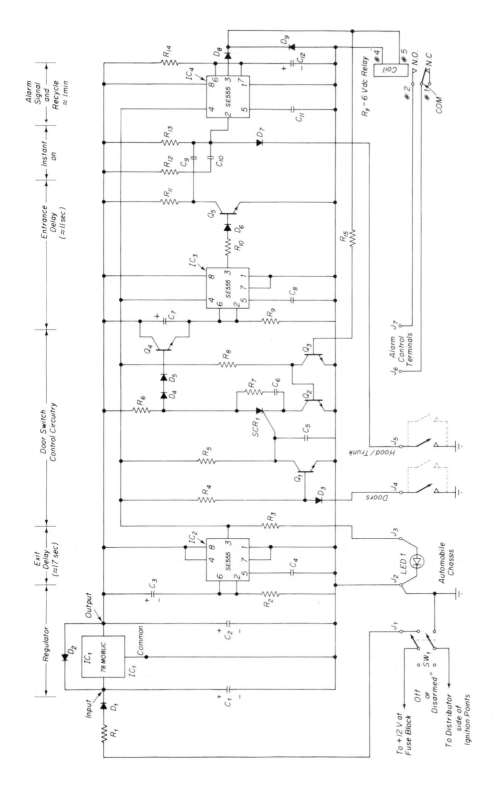

FIGURE 21.14. Circuit schematic for the automobile security alarm system.

543

FIGURE 21.14. (cont.)

Parts list

IC_1 – 78M08UC positive 8 Vdc IC regulator

IC_2, IC_3, IC_4 – SE555 timer

Ry – 6 Vdc relay (Guardian series 1345 pc mount
                          SPDT – NO)

SCR_1 – C106B1

Q_1, Q_2, Q_3, Q_4, Q_5 – 2N3904 NPN transistor

D_1, D_3, D_7, D_8, D_9 – 1N4001 diodes (I_o = 1A PIV = 50 V)

D_2, D_4, D_5, D_6 – 1N914 diodes (I_o = 10 mA PIV = 75 V)

C_1, C_2, C_3, C_7, C_{12} – 4.7 μF @ 15 WVDC tantalum
                          capacitors

C_5, C_6, C_9 – 0.1 μF disc capacitors

C_4, C_8, C_{11} – 0.01 μF disc capacitors

C_{10} – 0.001 μF disc capacitor

R_1 – 15 Ω, $\frac{1}{2}$ W, 10 %

R_2 – 3.3 MΩ, $\frac{1}{2}$ W, 10 %

R_3 – 220 Ω $\frac{1}{2}$ W, 10 %

R_4 – 100 KΩ, $\frac{1}{2}$ W, 10 %

R_5, R_{10} – 10 KΩ, $\frac{1}{2}$ W, 10 %

R_6 – 180 Ω, $\frac{1}{2}$ W, 10 %

R_7 – 10 Ω, $\frac{1}{2}$ W, 10 %

R_8, R_{11}, R_{12}, R_{15} – 1KΩ, $\frac{1}{2}$ W, 10 %

R_9 – 2.2 MΩ, $\frac{1}{2}$ W, 10 %

R_{13} – 22 KΩ, $\frac{1}{2}$ W, 10 %

R_{14} – 10 MΩ, $\frac{1}{2}$ W, 10 %

SW_1 – DPDT switch

LED_1 – Light emitting diode

Seven terminal barrier strip and
assorted hardware

To activate the system, the arm/disarm switch SW_1 is moved from the "off" or "disarmed" position to the "armed" position. After the switch has been activated, the operator has a preset exit delay time of approximately 17 seconds to leave the car and lock the doors. (If closing the doors takes longer than 17 seconds, the horn will sound.) The exit delay time interval is set by the timing capacitor C_3 and resistor R_2. At the end of the exit delay cycle, the output of IC_2, at pin 3, goes from approximately 0 volts to 6 volts, applying power to the reset terminals of IC_3 and IC_4 at pin 4. The system is now armed and ready to detect an intruder entering any protected door. If desired, LED-1 can be connected between jacks J_2 and J_3 as shown in Fig. 21.14. LED-1 will light at the end of the 17-second exit delay interval, indicating that the system is armed. This light can be mounted on the dashboard of the car.

Once the circuit is armed, the door switch control circuitry, consisting of R_4, R_5, Q_1, D_3, SCR_1, R_6, D_4, D_5, Q_4, and Q_2, is also powered. As long as all door switches are open (door switches are open when car doors are closed since the door position depresses the switch plunger), transistor Q_4 remains on and acts as a short circuit across C_7, preventing IC_3 from operating. An intruder opening a door causes the switch plunger to release and closes the switch, which causes the collector of Q_1 to go high, firing the gate of SCR_1. The base of Q_4 will go to approximately 0 volts, causing its collector-to-emitter to act as an open circuit across C_7. The timing capacitor C_7 (associated with IC_3) begins to charge. The intruder now has 11 seconds to find the arm/disarm switch and return it to the "off" position. If the switch is not thrown within the preset entrance delay time of 11 seconds, the charge developed across C_7 (through resistor R_9) will cause the output of IC_3 (pin 3) to go high, thus triggering IC_4. With IC_4 triggered, its pin 3 goes high, turning on diode D_8, which places 6 volts across the coil of the relay, Ry. The relay switches from its normally open condition to cause a closed condition at points J_6 and J_7. As can be seen in Fig. 21.15a, b, or c, when J_6 and J_7 are shorted together by the switching action of relay Ry, they complete the alarm circuit. This causes the alarm siren or automobile horn to sound for a duration of approximately 1 minute, this time interval being set by R_{14} and C_{12}. After the alarm sounds for 1 minute, the system will recycle. If, for example, during the 1-minute sound interval, the intruder is frightened off and closes the door which he initially opened, the system will recycle. Even though the door has been returned to its closed position, the sound will complete its 1-minute cycle, after which the system is set and ready to detect another door intrusion. If, on the other hand, the intruder leaves without closing the door which he initially opened, the alarm will sound for 1 minute, shut down for an 11-second interval, and then sound again for 1 minute. This 1 minute on/11 seconds off cycle will continue until the opened door is finally closed.

Since it is not advisable to have any delay on either the hood or trunk switches, the "instant-on" feature accomplished by C_{10}, R_{12}, and D_7, is included. Opening either the hood or trunk after the system has been armed will instantly sound the alarm for the same preset 1-minute interval as established by R_{14} and C_{12}.

Diode D_1 has been added to prevent damage to any of the components in the system should the correct polarity be reversed when wiring the system into the automobile. In addition, an on-board voltage regulator circuit has been added. This circuit consists of IC_1, C_1, C_2, and D_2, and its purpose is to minimize any voltage spikes from the car's electrical system being transferred to the IC timers, causing damage or false triggering.

Construction Hints. Because of the complexity of the circuit, it is strongly recommended that all components be assembled onto a pc board. The pc board should be mounted onto a chassis and protected by a sturdy

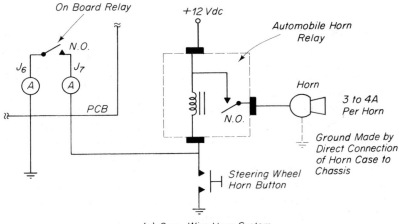

(a) One—Wire Horn System

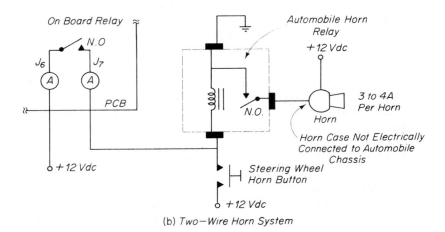

(b) Two—Wire Horn System

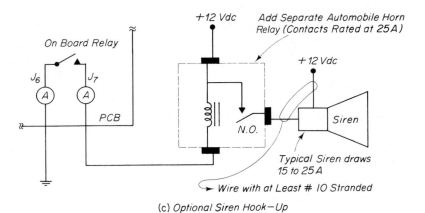

(c) Optional Siren Hook—Up

FIGURE 21.15. Horn and siren wiring diagrams.

enclosure. This unit is best secured in a safe position in the trunk.

Jack J_2 should be wired to the automobile chassis frame close to where the circuit enclosure is mounted. At least AWG No. 18 wire should be used and the connection tested to ensure that a good electrical contact has been made.

It is generally unnecessary to buy switches for the door, trunk, or hood, since these are often standard equipment on many cars. All existing switches should be checked for perfect working order. If a switch must be installed at any of the locations, the plunger type is most suitable for this application. A newly installed switch must be wired in parallel with either those using exit and entrance delay (door switches) or instant-on (trunk or hood). For example, hood and trunk switches should be wired in parallel with each other but not in parallel with the door switches (Fig. 21.14). Since diode D_3 isolates the alarm circuit from the 12-volt source and all door switches are in parallel, only a single wire connection is necessary between J_4 and the terminal connection on one door switch. In like manner, only one wire is necessary between J_5 and either the trunk or hood switch.

Caution. Never attempt to switch any automobile horn directly with the pc on-board relay. These horns draw in excess of 3 amperes each and should be controlled only from the car's horn relay. Depending upon the type of automobile horn system, wiring according to Fig. 21.15a or b is recommended. As seen in Fig. 21.14, J_6 and J_7 have been left uncommitted for controlled switching of any particular type of horn circuit. This allows for convenient wiring connections to be made between J_6 and J_7 and any of the circuits shown in Fig. 21.15a, b, or c. If the car has no horn relay or if a siren is desired, an automobile horn relay must be installed. Sirens for alarm systems for use on a 12-volt dc source can draw in excess of 20 amperes. A horn relay capable of handling at least this much current must be selected. The siren, +12 volts, and ground connections should be made with a minimum of AWG No. 10 wire.

The finished security alarm package is shown in Fig. 21.16.

FIGURE 21.16. Alarm package.

21.7 DIGITAL AUTOMOTIVE CLOCK

A six-digit crystal-controlled clock that can be used in an automobile or in the home is a popular and useful product. This clock project can be powered directly from any 12-volt dc negative-ground automobile battery or from household power with the use of an inexpensive transformer, such as that used to charge hand-held calculators.

This clock features a crystal-controlled oscillator circuit for accurate time keeping, a 6-digit display (hours, minutes, and seconds) whose output feeds inexpensive seven-segment LED readouts, a display enable/disable circuit to conserve power yet maintain accurate time, and zener diode protection circuitry to guard against the severe electrical environment in an automobile.

With the display disabled (darkened) the digital clock circuit, shown in Fig. 21.17, draws less than 10 mA of current to perform the time-keeping function. At this rate of drawing current, it would take about 6 months to discharge a normal car battery, assuming that the vehicle is not operated during that period. As a practical matter, there need be no concern about the possibility of running down an automobile battery with the use of this clock.

Circuit Operation. IC_2 is a CMOS integrated circuit (National Semiconductor Clock Chip MM5369) with 17 binary divider stages and is used with a crystal oscillator network to produce an accurate 60-Hz frequency at its output terminal (pin 1). The crystal oscillator network consists of C_5, C_6, C_7, R_{12}, and a quartz crystal having an oscillation frequency of 3.579545 MHz. This network, together with IC_2, forms the crystal-controlled oscillator. The use of the MM5369 chip as IC_2 allows the use of this inexpensive color-TV oscillator crystal. R_{12} is required for biasing and the trimmer capacitor C_5 allows adjustment of the crystal's frequency to establish the precise 60 Hz at the output of IC_2.

The 60-Hz output of IC_2 is fed directly to the 50/60-hertz input terminal of IC_1 at pin 16. Because the input frequency is 60 hertz (American operation), the 50/60-hertz select terminal at pin 11 of IC_1 must be connected to ground at pin 2 (V_{DD}) of IC_2 to obtain 60-hertz operation. Pin 10 of IC_1 provides both 12- and 24-hour formats. For a 12-hour format, pin 10 is also connected to pin 2 of IC_2, as shown in Fig. 21.17. To utilize a 24-hour format, the 12/24-hour select input at pin 10 is left disconnected.

Since this project is a six-digit clock, the 4/6-digit select input at pin 24 (IC_1) must be connected to ground at pin 2 (IC_2). If pin 24 remains disconnected, the *seconds* digits are disabled and only the *hours* and *minutes* readouts will be displayed.

The clock circuitry of IC_1 allows the display to be wired for multiplexer operation. This reduces the wiring from 42 leads to 13 leads for a six-digit clock. Pins 3 through 9 of IC_1 are labeled with the letters *a*

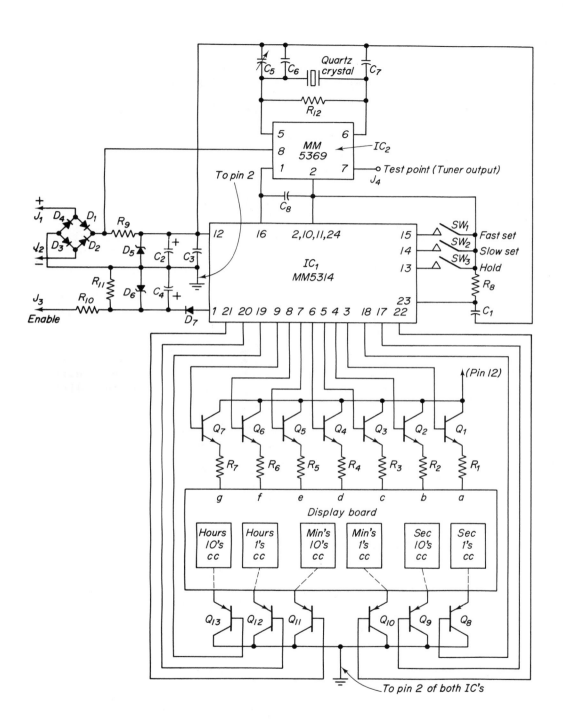

FIGURE 21.17. Digital clock diagrams.

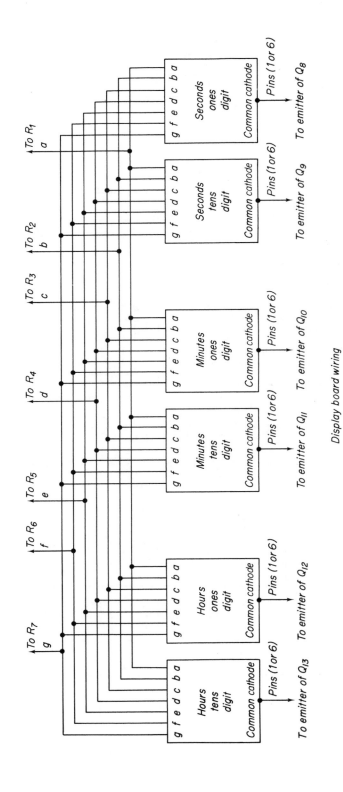

FIGURE 21.17. (cont.)

Parts list
IC₁ – MM5314 national clock chip
IC₂ – MM5369 national oscillator/timer

Parts list

IC$_1$ – MM5314 national clock chip
IC$_2$ – MM5369 national oscillator/timer
Seven – segment displays (FND – 357 or equivalent)
Q$_1$ through Q$_7$ – 2N3904 NPN transistors
Q$_8$ through Q$_{13}$ – 2N3906 PNP transistors
Quartz crystal – f$_0$ = 3.579545 MHz
D$_1$, D$_2$, D$_3$, D$_4$, D$_7$ – 1N4001 Diodes (I$_0$= 1A PIV = 50V)
D$_5$, D$_6$ – 1N4745A (16 V, 1A Zener diodes)
C$_1$, C$_3$, C$_8$ – 0.01 disc capacitors
C$_2$ – 220 μF @ 15WVDC electrolytic capacitor
C$_4$ – 10μF @ 15 WVDC electrolytic capacitor
C$_5$ – 7 to 25 pF trimmer capacitor
C$_6$ – 10 pF disc capacitors
C$_7$ – 20 pF disc capacitors
R$_1$ through R$_7$ – 15 Ω, $\frac{1}{4}$ W, 10%
R$_8$ – 220 KΩ, $\frac{1}{4}$ W, 10%
R$_9$ – 15 Ω, $\frac{1}{2}$ W, 10%
R$_{10}$ – 2.2 KΩ, $\frac{1}{4}$ W, 10%
R$_{11}$ – 22 KΩ, $\frac{1}{4}$ W, 10%
R$_{12}$ – 22 MΩ, $\frac{1}{4}$ W, 10%
SPST Mini dip switch (4 section for pc mount–fast slow and hold)
8 – Pin IC socket–DIP
24 – Pin IC socket–DIP
Suitable enclosure and assorted hardware

through g and are the seven-segment outputs of the clock chip. These pins are connected directly to the bases of the segment driver transistors Q_1 through Q_7. The emitters of these transistors are connected to the appropriate segments on the display board through the current limiting resistors R_1 through R_7 (Fig. 21.17).

Transistors Q_8 through Q_{13} are digit driver transistors connecting the common cathode of each seven-segment digit to the appropriate digit enable output terminals of IC_1 (pins 17 through 22). The frequency at which the displays are multiplexed is provided through an internal relaxation oscillator on IC_1 and the external RC circuit R_8 and C_1. The multiplexing time input (pin 23) of IC_1 is connected to this external RC circuit.

Dc power is supplied from the diode bridge D_1 through D_4 and R_9, C_2 and C_3. For automobile installation, J_1 is connected to any point in the car that is fused and is at +12 volts with the ignition off. J_2 is grounded to the automobile chassis. For household use, any transformer with a secondary voltage rating of 6.3 volts ac and a secondary current of at least 300 mA (ac) is suitable. The primary of the transformer is wired for 115 volts ac and is connected to regular house power. The secondary windings are connected to J_1 and J_2. For household applications, diodes D_1 through D_4 act as a full-wave bridge rectifier and C_2 as the filter capacitor.

The enable circuitry input at J_3 consists of R_{10}, R_{11}, and D_7. With J_3 disconnected, the display is disabled. Connecting J_3 to a positive voltage enables the display. For automobile installation, J_3 should be connected to the *accessory* wire from the ignition that connects +12 volts to the car radio. This will result in the display being enabled when the ignition is *on* and disabled when the ignition is *off*. For household use, J_3 is connected to pin 12 of IC_1 for constant display.

Protection against overvoltage is provided for both IC_1 and IC_2 by zener diodes D_5 and D_6.

Pins 13, 14, and 15 of IC_1 are used for setting the time. If the "fast set" input on pin 15 is switched to ground (V_{DD}), the *hours* reading will advance at the rate of 1 hour per second. Switching the "slow set" input at pin 14 to ground advances the *minutes* reading at the rate of 1 minute per second. Using the *fast* and *slow* set inputs to advance the time slightly beyond the correct time allows the use of the "hold" input at pin 13. This prevents further advance of the internal counter and displays. When the *hold* time displayed agrees with the correct time, the hold switch is opened and the time advances at a normal rate.

To adjust the accuracy of the clock, a frequency counter is connected across pin 7 (tuner output) of IC_2 and ground (pin 2 of IC_1). The output at pin 7 is provided so that the crystal frequency can be measured and adjusted without disturbing the crystal oscillator. If the frequency of the oscillator is found to be above or below 3.579540 MHz, the trimmer capacitor C_5 is slowly turned in the appropriate direction to obtain this frequency. Once installed in the automobile, clock accuracy can be monitored. If the clock is running slightly fast or slow, the trimmer capacitor can be adjusted. It is suggested that one-eighth of a turn *clockwise* be made on the capacitor and the clock readjusted to the correct time and monitored for at least 1 week. If the accuracy is worse, the capacitor is adjusted to *one-quarter* turn *counterclockwise*, and again the results monitored. With patience, accuracy to within 10 seconds per month can be realized. As with most automobile clocks, however, extreme temperatures will affect accuracy.

Construction Hints. The digital clock should be constructed using two separate printed circuit boards. The display board will require a double-sided pc board to interconnect the six seven-segment LED display circuitry shown in Fig. 21.17. A total of 13 external connections are required for interconnection to the mother board which will house all the remaining components (except the display) shown in Fig. 21.17. Seven terminals are required on the display board for the common *a* to *g* segments and six terminals for the digit drivers for the digits *seconds* (ones) through *hours* (tens). A single-sided pc board can be used to interconnect the components on the mother board. The display board is best assembled perpendicular to the mother board using short lengths of bus wire bent at 90 degrees. The assembly and soldering of IC_1 and IC_2

FIGURE 21.18.
Completed clock project.

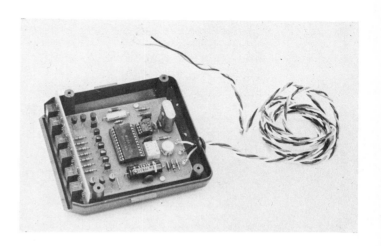

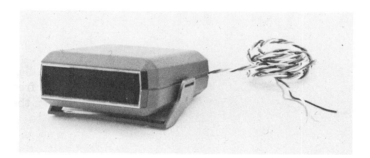

require precaution since they are in the CMOS family. They should remain in their protective foam pads until ready for assembly. During assembly, the ICs should be carefully removed from the pads, inserted into the pc board *without* touching the pins with the fingers, and finally soldered.

The completed clock project is shown in Fig. 21.18.

21.8 400-WATT LAMP DIMMER

An extremely simple but useful circuit is a lamp dimmer. Power control circuits, such as that shown in Figs. 21.19 and 21.20, are sold commercially for installation into a standard wall switch receptacle box and are used to control the intensity of overhead lights.

Circuit Operation. The lamp dimmer circuit shown in Fig. 21.19 is capable of controlling power to incandescent lamp loads of up to 400 watts. With SW_1 closed, phase control is provided for the triac Q_1 through the phase shift circuit of R_1, R_2, C_1, and C_2. During the positive half-cycle, C_2 charges quickly to the breakover potential of D_1, which is typi-

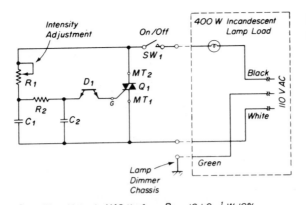

FIGURE 21.19. 400 watt lamp dimmer.

Q_1 — Triac, Motorola MAC 11-4
    or equivalent

D_1 — Bi-lateral trigger diode,
    Motorola MPT-20 or
    equivalent

R_1 — 250 kΩ potentiometer,
    Centralab # F1-250 k,
    or equivalent

R_2 — 10 kΩ, $\frac{1}{2}$ W, 10%

C_1 — 0.1 μF, 400 WVdc disc

C_2 — 0.01 μF, 400 WVdc disc

SW_1 — SPST, 5A, 125 Vac
    (Centralab # KR-1 or
    equivalent)

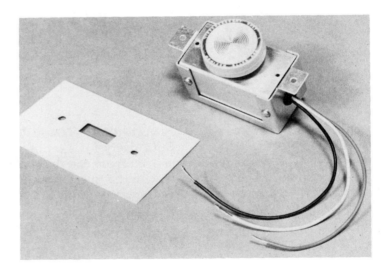

FIGURE 21.20. Lamp dimmer package.

cally 20 volts. When this breakover potential is reached, D_1 conducts current into the gate of Q_1, triggering it into conduction and providing a path for load current. A similar operation occurs during the negative half-cycle except that the direction of Q_1's gate current is reversed. The 250-kΩ potentiometer varies the time constant of the phase shift network, thus adjusting the amount of time during each half-cycle of the input voltage that Q_1 is in conduction.

Construction Hints. To provide for continuous duty, triac Q_1 should be mechanically heat sinked to the chassis. This, however, must be done with an insulating mica washer to prevent electrical connection of the triac to the circuit ground.

An optional feature of this circuit is its use as a variable-speed motor control. Small series-wound motors, such as those used in pistol-type drills, can be substituted for the 400-watt incandescent load as long as the load current does not exceed a maximum of 4 amperes. For this application, a receptacle for a three-prong plug should be installed in the chassis.

Appendices

Comparison of Aluminum and Steel Sheet Metal Gauges

Gauge Number	Aluminum (American or Brown & Sharpe)	Steel (United States Standard)
0000000	....	.500
000000	.5800	.46875
00000	.5165	.4375
0000	.4600	.40625
000	.4096	.375
00	.3648	.34375
0	.3249	.3125
1	.2893	.28125
2	.2576	.265625
3	.2294	.25
4	.2043	.234375
5	.1819	.21875
6	.1620	.203125
7	.1443	.1875
8	.1285	.171875
9	.1144	.15625
10	.1019	.140625
11	.09074	.125
12	.08081	.109375
13	.07196	.09375
14	.06408	.078125
15	.05707	.0703125
16	.05082	.0625
17	.04526	.05625
18	.04030	.05
19	.03589	.04375
20	.03196	.0375
21	.02846	.034375
22	.02535	.03125
23	.02257	.028125
24	.02010	.025
25	.01790	.021875
26	.01594	.01875
27	.01420	.0171875
28	.01264	.015625
29	.01126	.0140625
30	.01003	.0125
31	.008928	.0109375
32	.007950	.01015625
33	.007080	.009375
34	.006305	.00859375
35	.005615	.0078125
36	.005000	.00703125
37	.004453	.006640625
38	.003965	.00625
39	.003531	....
40	.003145	....

All decimals in inches

Stancor Transformer Type P-8388 Template for
Horizontal Mounting through Chassis

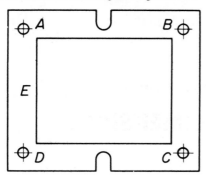

APPENDIX II
Hole Size

A	—	$\frac{3}{16}''$
B	—	$\frac{3}{16}''$
C	—	$\frac{3}{16}''$
D	—	$\frac{3}{16}''$
E	—	$1\frac{11}{16}'' \times 2\frac{1}{16}''$

Courtesy of Motorola Semiconductor Products, Inc.

DIMENSIONS ARE IN INCHES UNLESS OTHERWISE NOTED.

CASE OUTLINE DIMENSIONS

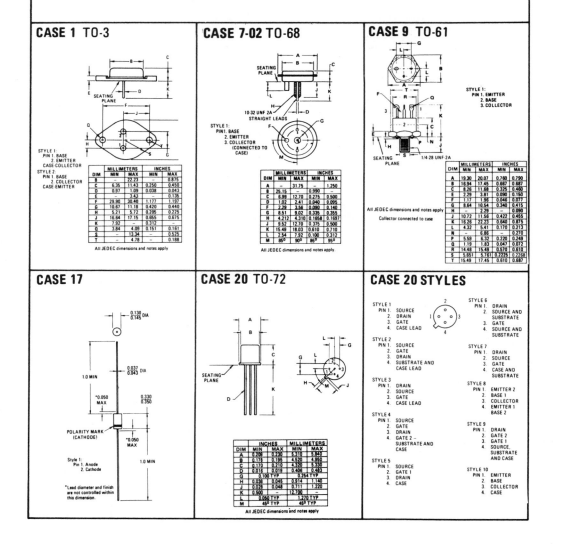

CASE 1 TO-3

STYLE 1:
PIN 1. BASE
2. EMITTER
CASE-COLLECTOR
STYLE 2:
PIN 1. BASE
2. COLLECTOR
CASE-EMITTER

DIM	MILLIMETERS		INCHES	
	MIN	MAX	MIN	MAX
B	–	22.23	–	0.875
C	6.35	11.43	0.250	0.450
D	0.97	1.09	0.038	0.043
E	–	3.43	–	0.135
F	29.90	30.40	1.177	1.197
G	10.67	11.18	0.420	0.440
H	5.21	5.72	0.205	0.225
J	16.64	17.15	0.655	0.675
K	7.92	–	0.312	–
Q	3.84	4.09	0.151	0.161
S	–	13.34	–	0.525
T	–	4.78	–	0.188

All JEDEC dimensions and notes apply

CASE 7-02 TO-68

10-32 UNF 2A
STRAIGHT LEADS

STYLE 1:
PIN 1. BASE
2. EMITTER
3. COLLECTOR
(CONNECTED TO
CASE)

DIM	MILLIMETERS		INCHES	
	MIN	MAX	MIN	MAX
A	–	31.75	–	1.250
B	25.15	–	0.990	–
C	6.99	12.70	0.275	0.500
D	1.02	2.41	0.040	0.095
F	2.29	3.56	0.090	0.140
G	8.51	9.02	0.335	0.355
H	4.212	4.310	0.1658	0.1697
J	9.52	12.70	0.375	0.500
K	15.48	18.03	0.610	0.710
L	2.54	7.92	0.100	0.312
M	85°	90°	85°	95°

All JEDEC dimensions and notes apply.

CASE 9 TO-61

STYLE 1:
PIN 1. EMITTER
2. BASE
3. COLLECTOR

1/4-28 UNF-2A

All JEDEC dimensions and notes apply

Collector connected to case

DIM	MILLIMETERS		INCHES	
	MIN	MAX	MIN	MAX
A	19.30	20.07	0.760	0.790
B	16.94	17.45	0.667	0.687
C	8.26	11.68	0.325	0.460
E	2.29	3.81	0.090	0.150
F	1.17	1.96	0.046	0.077
G	8.64	10.54	0.340	0.415
H	–	2.29	–	0.090
J	10.72	11.56	0.422	0.465
K	16.26	22.23	0.640	0.875
L	4.32	5.41	0.170	0.213
N	–	6.86	–	0.270
P	5.59	6.32	0.220	0.249
Q	1.19	1.83	0.047	0.072
R	14.48	15.49	0.570	0.610
S	5.651	5.761	0.2225	0.2268
T	15.49	17.45	0.610	0.687

CASE 17

0.130 / 0.145 DIA

0.037 / 0.043 DIA

1.0 MIN

*0.050 MAX

0.330 / 0.350

POLARITY MARK
(CATHODE)

*0.050 MAX

1.0 MIN

Style 1:
Pin 1. Anode
2. Cathode

*Lead diameter and finish
are not controlled within
this dimension.

CASE 20 TO-72

SEATING
PLANE

DIM	INCHES		MILLIMETERS	
	MIN	MAX	MIN	MAX
A	0.209	0.230	5.310	5.840
B	0.178	0.195	4.520	4.950
C	0.170	0.210	4.320	5.330
D	0.016	0.019	0.406	0.483
G	0.100 TYP		0.254 TYP	
H	0.036	0.045	0.914	1.140
J	0.028	0.048	0.711	1.220
K	0.500	–	12.700	–
L	0.050 TYP		1.270 TYP	
M	45° TYP		45° TYP	

All JEDEC dimensions and notes apply

CASE 20 STYLES

STYLE 1
PIN 1. SOURCE
2. DRAIN
3. GATE
4. CASE LEAD

STYLE 2
PIN 1. SOURCE
2. GATE
3. DRAIN
4. SUBSTRATE AND
CASE LEAD

STYLE 3
PIN 1. DRAIN
2. SOURCE
3. GATE
4. CASE LEAD

STYLE 4
PIN 1. SOURCE
2. GATE
3. DRAIN
4. GATE 2 –
SUBSTRATE AND
CASE

STYLE 5
PIN 1. SOURCE
2. GATE 1
3. DRAIN
4. CASE

STYLE 6
PIN 1. DRAIN
2. SOURCE AND
SUBSTRATE
3. GATE
4. SOURCE AND
SUBSTRATE

STYLE 7
PIN 1. DRAIN
2. SOURCE
3. GATE
4. CASE AND
SUBSTRATE

STYLE 8
PIN 1. EMITTER 2
2. BASE 1
3. COLLECTOR
4. EMITTER 1
BASE 2

STYLE 9
PIN 1. DRAIN
2. GATE 2
3. GATE 1
4. SOURCE,
SUBSTRATE
AND CASE

STYLE 10
PIN 1. EMITTER
2. BASE
3. COLLECTOR
4. CASE

CASE 21-02 TO-17

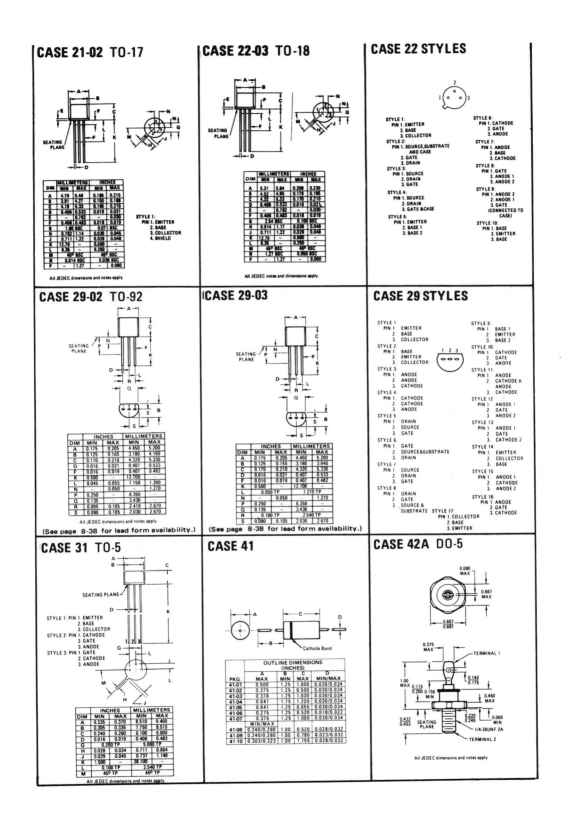

DIM	MILLIMETERS MIN	MAX	INCHES MIN	MAX
A	4.70	5.46	0.185	0.215
B	3.81	4.27	0.150	0.168
C	4.19	5.33	0.165	0.210
D	0.406	0.533	0.016	0.021
E	–	0.762	–	0.030
F	0.406	0.483	0.016	0.019
G	1.80 BSC		0.071 BSC	
H	0.762	1.14	0.030	0.045
J	0.711	1.22	0.028	0.048
K	12.70	–	0.500	–
L	6.35	–	0.250	–
M	45° BSC		45° BSC	
N	0.914 BSC		0.036 BSC	
P	–	1.27	–	0.050

STYLE 1:
PIN 1. EMITTER
2. BASE
3. COLLECTOR
4. SHIELD

All JEDEC dimensions and notes apply

CASE 22-03 TO-18

DIM	MILLIMETERS MIN	MAX	INCHES MIN	MAX
A	5.31	5.84	0.209	0.230
B	4.52	4.95	0.178	0.195
C	4.32	5.33	0.170	0.210
D	0.406	0.533	0.016	0.021
E	–	0.762	–	0.030
F	0.406	0.483	0.016	0.019
G	2.54 BSC		0.100 BSC	
H	0.914	1.17	0.036	0.046
J	0.711	1.22	0.028	0.048
K	12.70	–	0.500	–
L	6.35	–	0.250	–
M	45° BSC		45° BSC	
N	1.27 BSC		0.050 BSC	
P	–	1.27	–	0.050

All JEDEC notes and dimensions apply.

CASE 22 STYLES

STYLE 1:
PIN 1. EMITTER
2. BASE
3. COLLECTOR
STYLE 2:
PIN 1. SOURCE,SUBSTRATE AND CASE
2. GATE
3. DRAIN
STYLE 3:
PIN 1. SOURCE
2. DRAIN
3. GATE
STYLE 4:
PIN 1. SOURCE
2. DRAIN
3. GATE &CASE
STYLE 5:
PIN 1. EMITTER
2. BASE 1
3. BASE 2

STYLE 6:
PIN 1. CATHODE
2. GATE
3. ANODE
STYLE 7:
PIN 1. ANODE
2. BASE
3. CATHODE
STYLE 8:
PIN 1. GATE
2. ANODE 1
3. ANODE 2
STYLE 9:
PIN 1. ANODE 2
2. ANODE 1
3. GATE (CONNECTED TO CASE)
STYLE 10:
PIN 1. EMITTER
2. EMITTER
3. BASE

CASE 29-02 TO-92

DIM	INCHES MIN	MAX	MILLIMETERS MIN	MAX
A	0.175	0.205	4.450	5.200
B	0.125	0.165	3.180	4.190
C	0.170	0.210	4.320	5.330
D	0.016	0.021	0.407	0.533
F	0.016	0.019	0.407	0.482
K	0.500	–	12.700	–
L	0.045	0.055	1.150	1.390
N	–	0.050	–	1.270
P	0.250	–	6.350	–
Q	0.135	–	3.430	–
R	0.095	0.105	2.410	2.670
S	0.080	0.105	2.030	2.670

All JEDEC dimensions and notes apply
(See page 8-38 for lead form availability.)

CASE 29-03

DIM	INCHES MIN	MAX	MILLIMETERS MIN	MAX
A	0.175	0.205	4.450	5.200
B	0.125	0.155	3.180	3.940
C	0.170	0.210	4.320	5.330
D	0.016	0.021	0.407	0.533
F	0.016	0.019	0.407	0.482
K	0.500	–	12.700	–
N	–	0.050	–	1.270
P	0.250	–	6.350	–
Q	0.135	–	3.430	–
R	0.100 TP		2.540 TP	
S	0.080	0.105	2.030	2.670

All JEDEC dimensions and notes apply
(See page 8-38 for lead form availability.)

CASE 29 STYLES

STYLE 1
PIN 1. EMITTER
2. BASE
3. COLLECTOR
STYLE 2:
PIN 1. BASE
2. EMITTER
3. COLLECTOR
STYLE 3:
PIN 1. ANODE
2. ANODE
3. CATHODE
STYLE 4:
PIN 1. CATHODE
2. CATHODE
3. ANODE
STYLE 5
PIN 1. DRAIN
2. SOURCE
3. GATE
STYLE 6
PIN 1. GATE
2. SOURCE&SUBSTRATE
3. DRAIN
STYLE 7
PIN 1. SOURCE
2. DRAIN
3. GATE
STYLE 8
PIN 1. DRAIN
2. GATE
3. SOURCE & SUBSTRATE

STYLE 9
PIN 1. BASE 1
2. EMITTER
3. BASE 2
STYLE 10
PIN 1. CATHODE
2. GATE
3. ANODE
STYLE 11
PIN 1. ANODE
2. CATHODE & ANODE
3. CATHODE
STYLE 12
PIN 1. ANODE 1
2. GATE
3. ANODE 2
STYLE 13
PIN 1. ANODE 1
2. GATE
3. CATHODE 2
STYLE 14
PIN 1. EMITTER
2. COLLECTOR
3. BASE
STYLE 15
PIN 1. ANODE 1
2. CATHODE
3. ANODE 2
STYLE 16
PIN 1. ANODE
2. GATE
3. CATHODE
STYLE 17:
PIN 1. COLLECTOR
2. BASE
3. EMITTER

CASE 31 TO-5

STYLE 1: PIN 1. EMITTER
2. BASE
3. COLLECTOR
STYLE 2: PIN 1. CATHODE
2. GATE
3. ANODE
STYLE 3: PIN 1. GATE
2. CATHODE
3. ANODE

DIM	INCHES MIN	MAX	MILLIMETERS MIN	MAX
A	0.335	0.370	8.510	9.400
B	0.305	0.335	7.750	8.510
C	0.240	0.260	6.100	6.600
D	0.016	0.019	0.406	0.483
G	0.200 TP		5.080 TP	
H	0.028	0.034	0.711	0.864
J	0.029	0.045	0.737	1.140
K	1.500	–	38.100	–
L	0.100 TP		2.540 TP	
M	45° TP		45° TP	

All JEDEC dimensions and notes apply

CASE 41

Cathode Band

	OUTLINE DIMENSIONS (INCHES)			
PKG.	A MAX	B MIN	C MAX	D MIN/MAX
41-01	0.500	1.25	1.000	0.030/0.034
41-02	0.375	1.25	0.500	0.030/0.034
41-03	0.378	1.25	1.030	0.030/0.034
41-04	0.641	1.75	1.220	0.030/0.034
41-05	0.641	1.75	0.655	0.030/0.034
41-06	0.275	1.25	0.520	0.018/0.022
41-07	0.375	1.25	1.000	0.030/0.034
	MIN/MAX			
41-08	0.240/0.260	1.00	0.520	0.028/0.032
41-09	0.240/0.260	1.00	0.780	0.023/0.032
41-10	0.303/0.323	1.00	1.155	0.028/0.032

CASE 42A DO-5

0.080 MAX
0.667 MAX
0.667 0.687
0.375 MAX
1.00 MAX
0.115
0.200 0.156 MIN
0.140 0.175
0.450 MAX
TERMINAL 1
0.422 0.453 SEATING PLANE
0.220 0.249
0.060 MIN
1/4-28UNF 2A
TERMINAL 2

All JEDEC dimensions and notes apply

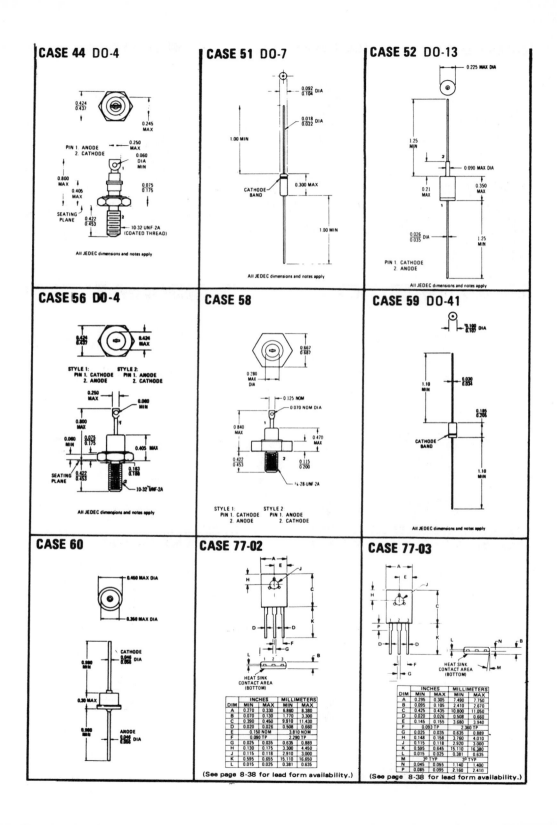

CASE 44 DO-4

PIN 1. ANODE
2. CATHODE

All JEDEC dimensions and notes apply

CASE 51 DO-7

CATHODE BAND

All JEDEC dimensions and notes apply

CASE 52 DO-13

PIN 1. CATHODE
2. ANODE

All JEDEC dimensions and notes apply

CASE 56 DO-4

STYLE 1:
PIN 1. CATHODE
2. ANODE

STYLE 2:
PIN 1. ANODE
2. CATHODE

SEATING PLANE

All JEDEC dimensions and notes apply

CASE 58

STYLE 1:
PIN 1. CATHODE
2. ANODE

STYLE 2:
PIN 1. ANODE
2. CATHODE

CASE 59 DO-41

CATHODE BAND

All JEDEC dimensions and notes apply

CASE 60

CATHODE

ANODE

CASE 77-02

HEAT SINK
CONTACT AREA
(BOTTOM)

DIM	INCHES		MILLIMETERS	
	MIN	MAX	MIN	MAX
A	0.270	0.330	6.860	8.380
B	0.070	0.130	1.770	3.300
C	0.390	0.450	9.910	11.430
D	0.020	0.026	0.508	0.660
E	0.150 NOM		3.810 NOM	
F	0.090 TP		2.290 TP	
G	0.025	0.035	0.635	0.889
H	0.130	0.175	3.300	4.450
J	0.115	0.118	2.910	3.000
K	0.595	0.655	15.110	16.650
L	0.015	0.025	0.381	0.635

(See page 8-38 for lead form availability.)

CASE 77-03

HEAT SINK
CONTACT AREA
(BOTTOM)

DIM	INCHES		MILLIMETERS	
	MIN	MAX	MIN	MAX
A	0.295	0.305	7.490	7.750
B	0.095	0.105	2.410	2.670
C	0.425	0.435	10.800	11.050
D	0.020	0.026	0.508	0.660
E	0.145	0.155	3.680	3.940
F	0.093 TP		2.360 TP	
G	0.025	0.035	0.635	0.889
H	0.148	0.158	3.760	4.010
J	0.115	0.118	2.920	3.000
K	0.595	0.645	15.110	16.380
L	0.015	0.025	0.381	0.635
M	3° TYP		3° TYP	
N	0.045	0.055	1.140	1.400
P	0.085	0.095	2.160	2.410

(See page 8-38 for lead form availability.)

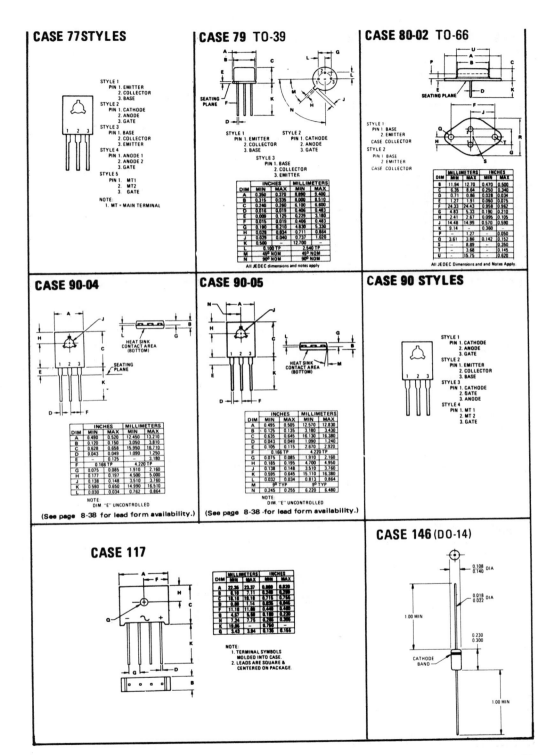

CASE 77 STYLES

STYLE 1
PIN 1. EMITTER
2. COLLECTOR
3. BASE
STYLE 2
PIN 1. CATHODE
2. ANODE
3. GATE
STYLE 3
PIN 1. BASE
2. COLLECTOR
3. EMITTER
STYLE 4
PIN 1. ANODE 1
2. ANODE 2
3. GATE
STYLE 5
PIN 1. MT1
2. MT2
3. GATE

NOTE:
1. MT = MAIN TERMINAL

CASE 79 TO-39

STYLE 1
PIN 1. EMITTER
2. COLLECTOR
3. BASE
STYLE 2
PIN 1. CATHODE
2. ANODE
3. GATE
STYLE 3
PIN 1. BASE
2. COLLECTOR
3. EMITTER

DIM	INCHES MIN	INCHES MAX	MILLIMETERS MIN	MILLIMETERS MAX
A	0.350	0.370	8.890	9.400
B	0.315	0.335	8.000	8.510
C	0.240	0.260	6.100	6.600
D	0.016	0.019	0.406	0.483
E	0.009	0.125	0.229	3.180
F	0.015	0.019	0.406	0.483
G	0.190	0.210	4.830	5.330
H	0.029	0.034	0.711	0.864
J	0.029	0.040	0.737	1.020
K	0.500	—	12.700	—
L	0.100 TP		2.540 TP	
M	45° NOM		45° NOM	
N	90° NOM		90° NOM	

All JEDEC dimensions and notes apply

CASE 80-02 TO-66

STYLE 1
PIN 1 BASE
2 EMITTER
CASE COLLECTOR
STYLE 2
PIN 1 BASE
2 EMITTER
CASE COLLECTOR

DIM	MILLIMETERS MIN	MAX	INCHES MIN	MAX
B	11.94	12.70	0.470	0.500
C	6.35	8.64	0.250	0.340
D	0.71	0.86	0.028	0.034
E	1.27	1.91	0.050	0.075
F	24.33	24.43	0.958	0.962
G	4.83	5.33	0.190	0.210
H	2.41	2.67	0.095	0.105
J	14.48	14.99	0.570	0.590
K	9.14	—	0.360	—
P	—	1.27	—	0.050
Q	3.61	3.86	0.142	0.152
S	—	8.89	—	0.350
T	—	3.68	—	0.145
U	—	15.75	—	0.620

All JEDEC Dimensions and Notes Apply.

CASE 90-04

HEAT SINK CONTACT AREA (BOTTOM)
SEATING PLANE

DIM	INCHES MIN	MAX	MILLIMETERS MIN	MAX
A	0.490	0.520	12.450	13.210
B	0.120	0.150	3.050	3.810
C	0.628	0.658	15.950	16.710
D	0.043	0.049	1.090	1.250
E	—	0.125	—	3.180
F	0.166 TP		4.220 TP	
G	0.075	0.085	1.910	2.160
H	0.177	0.197	4.500	5.000
J	0.138	0.148	3.510	3.760
K	0.590	0.650	14.990	16.510
L	0.030	0.034	0.762	0.864

NOTE:
DIM "E" UNCONTROLLED

(See page 8-38 for lead form availability.)

CASE 90-05

HEAT SINK CONTACT AREA (BOTTOM)

DIM	INCHES MIN	MAX	MILLIMETERS MIN	MAX
A	0.495	0.505	12.570	12.830
B	0.125	0.135	3.180	3.430
C	0.635	0.645	16.130	16.380
D	0.043	0.049	1.090	1.240
E	0.105	0.115	2.670	2.920
F	0.166 TP		4.220 TP	
G	0.075	0.085	1.910	2.160
H	0.185	0.195	4.700	4.950
J	0.138	0.148	3.510	3.760
K	0.595	0.645	15.110	16.380
L	0.032	0.034	0.813	0.864
M	9° TYP		9° TYP	
N	0.245	0.255	6.220	6.480

NOTE:
DIM "E" UNCONTROLLED

(See page 8-38 for lead form availability.)

CASE 90 STYLES

STYLE 1
PIN 1. CATHODE
2. ANODE
3. GATE
STYLE 2
PIN 1. EMITTER
2. COLLECTOR
3. BASE
STYLE 3
PIN 1. CATHODE
2. GATE
3. ANODE
STYLE 4
PIN 1. MT 1
2. MT 2
3. GATE

CASE 117

DIM	MILLIMETERS MIN	MAX	INCHES MIN	MAX
A	22.35	23.37	0.880	0.920
B	6.10	7.11	0.240	0.280
C	18.16	19.18	0.715	0.755
D	0.89	1.14	0.035	0.045
F	11.18	11.88	0.440	0.480
G	4.57	5.59	0.180	0.220
H	7.24	7.75	0.285	0.305
K	19.05	—	0.750	—
Q	3.43	3.94	0.135	0.155

NOTE:
1. TERMINAL SYMBOLS MOLDED INTO CASE.
2. LEADS ARE SQUARE & CENTERED ON PACKAGE.

CASE 146 (DO-14)

0.108 / 0.140 DIA
0.018 / 0.022 DIA
1.00 MIN
0.230 / 0.300
CATHODE BAND
1.00 MIN

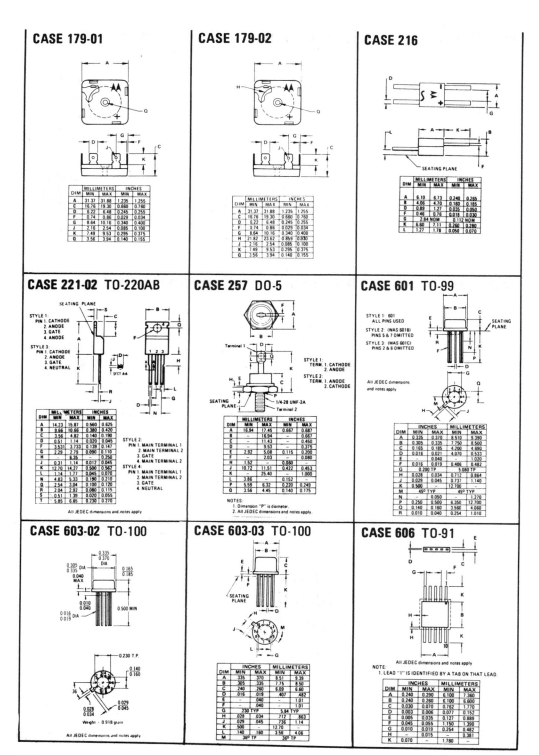

CASE 179-01

DIM	MILLIMETERS		INCHES	
	MIN	MAX	MIN	MAX
A	31.37	31.88	1.235	1.255
C	16.76	19.30	0.660	0.760
D	6.22	6.48	0.245	0.255
F	0.74	0.86	0.029	0.034
G	8.64	10.16	0.340	0.400
J	2.16	2.54	0.085	0.100
K	7.49	9.53	0.295	0.375
Q	3.56	3.94	0.140	0.155

CASE 179-02

DIM	MILLIMETERS		INCHES	
	MIN	MAX	MIN	MAX
A	31.37	31.88	1.235	1.255
C	16.76	19.30	0.660	0.760
D	6.22	6.48	0.245	0.255
F	0.74	0.86	0.029	0.034
G	8.64	10.16	0.340	0.400
H	21.82	23.62	0.859	0.930
J	2.16	2.54	0.085	0.100
K	7.49	9.53	0.295	0.375
Q	3.56	3.94	0.140	0.155

CASE 216

DIM	MILLIMETERS		INCHES	
	MIN	MAX	MIN	MAX
A	6.10	6.73	0.240	0.265
B	4.06	4.70	0.160	0.185
D	0.89	1.27	0.035	0.050
F	0.46	0.76	0.018	0.030
G	2.84 NOM		0.112 NOM	
K	6.60	7.11	0.260	0.280
L	1.27	1.78	0.050	0.070

CASE 221-02 TO-220AB

STYLE 1:
PIN 1. CATHODE
    2. ANODE
    3. GATE
    4. ANODE

STYLE 3:
PIN 1. CATHODE
    2. ANODE
    3. GATE
    4. NEUTRAL

SEATING PLANE

SECT A-A

DIM	MILLIMETERS		INCHES	
	MIN	MAX	MIN	MAX
A	14.23	15.87	0.560	0.625
B	9.66	10.66	0.380	0.420
C	3.56	4.82	0.140	0.190
D	0.51	1.14	0.020	0.045
F	3.531	3.733	0.139	0.147
G	2.29	2.79	0.090	0.110
H		6.35		0.250
J	0.31	1.14	0.012	0.045
K	12.70	14.27	0.500	0.562
L	1.14	1.77	0.045	0.070
N	4.83	5.33	0.190	0.210
Q	2.54	3.04	0.100	0.120
R	2.04	2.92	0.080	0.115
S	0.51	1.39	0.020	0.055
T	5.85	6.85	0.230	0.270

STYLE 2:
PIN 1. MAIN TERMINAL 1
    2. MAIN TERMINAL 2
    3. GATE
    4. MAIN TERMINAL 2

STYLE 4:
PIN 1. MAIN TERMINAL 1
    2. MAIN TERMINAL 2
    3. GATE
    4. NEUTRAL

All JEDEC dimensions and notes apply

CASE 257 DO-5

Terminal 1

1/4-28 UNF-2A

SEATING PLANE

Terminal 2

STYLE 1:
TERM. 1. CATHODE
      2. ANODE

STYLE 2:
TERM. 1. ANODE
      2. CATHODE

DIM	MILLIMETERS		INCHES	
	MIN	MAX	MIN	MAX
A	16.94	17.45	0.667	0.687
B	—	16.94	—	0.667
C	—	11.43	—	0.450
D	—	9.53	—	0.375
E	2.92	5.08	0.115	0.200
F	—	2.03	—	0.080
H	1.52	—	0.060	—
J	10.72	11.51	0.422	0.453
K		25.40		1.000
L	3.86		0.152	
P	5.59	6.32	0.220	0.249
Q	3.56	4.45	0.140	0.175

NOTES:
1. Dimension "P" is diameter.
2. All JEDEC dimensions and notes apply.

CASE 601 TO-99

SEATING PLANE

STYLE 1: 601
ALL PINS USED

STYLE 2: (WAS 601B)
PINS 5 & 7 OMITTED

STYLE 3: (WAS 601C)
PINS 2 & 6 OMITTED

All JEDEC dimensions
and notes apply

DIM	INCHES		MILLIMETERS	
	MIN	MAX	MIN	MAX
A	0.335	0.370	8.510	9.390
B	0.305	0.335	7.750	8.500
C	0.165	0.185	4.204	4.690
D	0.016	0.021	4.070	0.533
E		0.040		1.020
F	0.016	0.019	0.406	0.482
G	0.200 TP		5.080 TP	
H	0.028	0.034	0.712	0.864
J	0.029	0.045	0.737	1.140
K	0.500		12.700	
M	45° TYP		45° TYP	
N		0.050		1.270
P	0.250	0.500	6.350	12.700
Q	0.140	0.160	3.560	4.060
R	0.010	0.040	0.254	1.010

CASE 603-02 TO-100

0.335
0.370
DIA

0.305
0.335 DIA

0.165
0.185

0.040 MAX

0.010
0.040

0.016
0.019 DIA

0.500 MIN

0.230 T.P.

0.140
0.160

36°

0.028
0.034

0.029
0.045

Weight - 0.918 gram

All JEDEC dimensions and notes apply

CASE 603-03 TO-100

SEATING PLANE

DIM	INCHES		MILLIMETERS	
	MIN	MAX	MIN	MAX
A	.335	.370	8.51	9.39
B	.305	.335	7.75	8.50
C	.240	.260	6.09	6.60
D	.016	.019	.407	.482
E	.240			1.01
F		.040		1.01
G	.230 TYP		5.84 TYP	
H	.028	.034	.712	.863
J	.029	.045	.736	1.14
K	.500		12.70	
L	.140	.160	3.56	4.06
M	36° TP		36° TP	

CASE 606 TO-91

All JEDEC dimensions and notes apply

NOTE:
1. LEAD "1" IS IDENTIFIED BY A TAB ON THAT LEAD.

DIM	INCHES		MILLIMETERS	
	MIN	MAX	MIN	MAX
A	0.240	0.290	6.100	7.360
B	0.240	0.260	6.100	6.600
C	0.030	0.070	0.762	1.770
D	0.003	0.006	0.077	0.152
E	0.005	0.035	0.127	0.889
F	0.045	0.055	1.150	1.390
G	0.010	0.019	0.254	0.482
H	—	0.015	—	0.381
K	0.070		1.780	—

CASE 607 TO-86

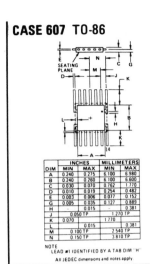

DIM	INCHES MIN	MAX	MILLIMETERS MIN	MAX
A	0.240	0.275	6.100	6.980
B	0.240	0.260	6.100	6.600
C	0.030	0.070	0.762	1.770
D	0.010	0.019	0.254	0.482
E	0.003	0.006	0.077	0.152
G	0.005	0.035	0.127	0.889
H		0.015		0.381
J	0.050 TP		1.270 TP	
K	0.070		1.770	
L		0.015		0.381
M	0.100 TP		2.540 TP	
N	0.150 TP		3.810 TP	

NOTE
LEAD #1 IDENTIFIED BY A TAB DIM "H"
All JEDEC dimensions and notes apply

CASE 608 TO-90

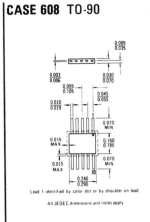

Lead 1 identified by color dot or by shoulder on lead

All JEDEC dimensions and notes apply

CASE 614

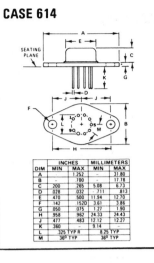

DIM	INCHES MIN	MAX	MILLIMETERS MIN	MAX
A	1.252	-	31.80	-
B	-	700	-	17.78
C	200	265	5.08	6.73
D	028	032	.711	.813
E	470	500	11.94	12.70
F	142	152D	3.61	3.86
G	050	075	1.27	1.90
H	958	962	24.33	24.43
J	477	483	12.12	12.27
K	360	-	9.14	-
L	325 TYP R		8.25 TYP	
M	36° TYP		36° TYP	

CASE 642 TO-76

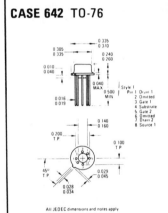

Style 1
Pin 1 Drain 1
2 Omitted
3 Gate 1
4 Substrate
5 Gate 2
6 Omitted
7 Drain 2
8 Source 1

All JEDEC dimensions and notes apply

CASE 623

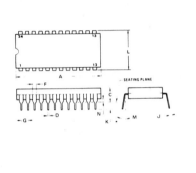

DIM	MILLIMETERS MIN	MAX	INCHES MIN	MAX
A	31.24	32.26	1.230	1.270
C	4.06	5.08	0.160	0.200
D	0.41	0.51	0.016	0.020
F	1.27	1.52	0.050	0.060
G	2.54 BSC		0.100 BSC	
J	0.20	0.30	0.008	0.012
K	2.92	3.43	0.115	0.135
L	15.37 BSC		0.605 BSC	
M	5°	15°	5°	15°
N	0.51	0.76	0.020	0.030

NOTES:
1. DIM "L" TO CENTER OF LEADS WHEN FORMED PARALLEL.
2. LEADS WITHIN 0.13 mm (0.005) RADIUS OF TRUE POSITION AT SEATING PLANE AT MAXIMUM MATERIAL CONDITION. (WHEN FORMED PARALLEL)

CASE 626

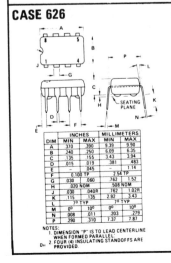

DIM	INCHES MIN	MAX	MILLIMETERS MIN	MAX
A	370	390	9.39	9.90
B	240	250	6.09	6.35
C	135	155	3.43	3.94
D	015	019	381	.483
E		045		1.14
F	0.100 TP		2.54 TP	
G	030	060	762	1.52
H	020 NOM		508 NOM	
J	030	040R	762	1.02R
K	115	135	2.92	3.43
L	7° TYP		7° TYP	
M	0°	10°	0°	10°
N	008	011	203	.279
P	290	310	7.37	7.87

NOTES:
1. DIMENSION "P" IS TO LEAD CENTERLINE WHEN FORMED PARALLEL.
2. FOUR (4) INSULATING STANDOFFS ARE PROVIDED.

CASE 632 TO-116

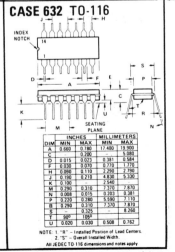

DIM	INCHES MIN	MAX	MILLIMETERS MIN	MAX
A	0.660	0.780	17.400	19.900
C		0.200		5.080
D	0.015	0.023	0.381	0.584
F	0.030	0.070	0.770	1.770
H	0.090	0.110	2.290	2.790
J	0.190	0.210	4.830	5.330
K	0.100		2.540	
M	0.290	0.310	7.370	7.870
N	0.008	0.015	0.203	0.381
P	0.220	0.280	5.590	7.110
R	0.290	0.310	7.370	7.870
S		0.325		8.260
T	90°	105°		
U	0.020	0.030	0.508	0.762

NOTE: 1. "R" – Installed Position of Lead Centers.
2. "S" – Overall Installed Width.
All JEDEC TO-116 dimensions and notes apply

CASE 635

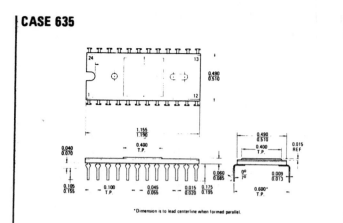

24 · 13
0.490 / 0.510
1 · 12

1.155 / 1.190
0.400 T.P.
0.040 / 0.070
0.060 / 0.085
0.105 / 0.155
0.100 T.P.
0.045 / 0.055
0.015 / 0.020
0.175 / 0.195

0.490 / 0.510
0.400 T.P.
0.015 REF
0° / 7°
0.009 / 0.013
0.600* T.P.

*Dimension is to lead centerline when formed parallel.

CASE 639

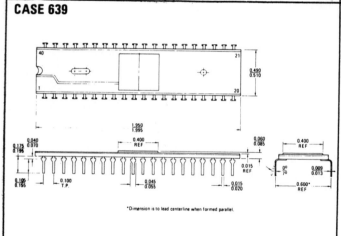

40 · 21
0.490 / 0.510
1 · 20

1.950 / 1.995
0.400 REF
0.175 / 0.195
0.040 / 0.070
0.060 / 0.085
0.400 REF
0.015 REF
0.105 / 0.155
0.100 T.P.
0.045 / 0.055
0.015 / 0.020

0° / 7°
0.009 / 0.013
0.600* REF

*Dimension is to lead centerline when formed parallel.

CASE 655 TO-71

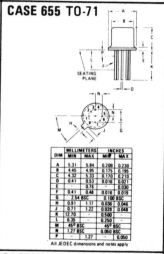

SEATING PLANE

DIM	MILLIMETERS		INCHES	
	MIN	MAX	MIN	MAX
A	5.31	5.84	0.209	0.230
B	4.45	4.95	0.175	0.195
C	4.32	5.33	0.170	0.210
D	0.41	0.53	0.016	0.021
E		0.76		0.030
F	0.41	0.48	0.016	0.019
G	2.54 BSC		0.100 BSC	
H	0.91	1.17	0.036	0.046
J	0.71	1.22	0.028	0.048
K	12.70		0.500	
L	6.35	–	0.250	–
M	45° BSC		45° BSC	
N	1.27 BSC		0.050 BSC	
P	–	1.27	–	0.050

All JEDEC dimensions and notes apply

CASE 683

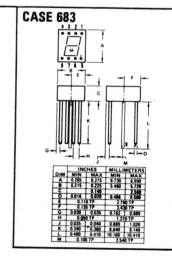

DIM	INCHES		MILLIMETERS	
	MIN	MAX	MIN	MAX
A	0.265	0.275	6.730	6.990
B	0.215	0.225	5.460	5.720
C		0.140		3.560
D	0.016	0.020	0.406	0.508
E	0.110 TP		2.790 TP	
F	0.135 TP		3.430 TP	
G	0.030	0.035	0.762	0.889
H	0.050 TP		1.270 TP	
J	0.035	0.040	0.889	1.020
K	0.340	0.360	8.640	9.140
L	0.400	0.410	10.160	10.410
M	0.100 TP		2.540 TP	

CASE 684

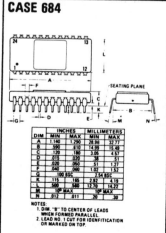

24 · 13

12

SEATING PLANE

DIM	INCHES		MILLIMETERS	
	MIN	MAX	MIN	MAX
A	1.140	1.290	28.96	32.77
B	.590	.610	14.99	15.49
C	.120	.180	3.05	4.57
D	.015	.020	.38	.51
E	.020	.050	.51	1.27
F	.040	.060	1.02	1.52
G	.100 BSC		2.54 BSC	
K	.115	.165	2.92	4.19
L	.500	.540	12.70	14.22
M	10° MAX		10° MAX	
N	.012	.011	.20	.30

NOTES:
1. DIM. "B" TO CENTER OF LEADS
   WHEN FORMED PARALLEL.
2. LEAD NO. 1 CUT FOR IDENTIFICATION
   OR MARKED ON TOP.

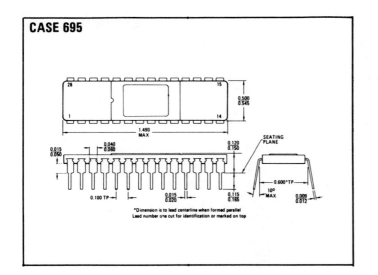

CASE 695

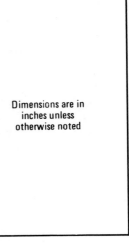

Dimensions are in inches unless otherwise noted

Twist Drill Sizes and Decimal Equivalents of Number (Wire Gauge), Letter, and Fraction Size Drills

Drill Size	Decimal Equivalent in Inches	Drill Size	Decimal Equivalent in Inches	Drill Size	Decimal Equivalent in Inches
80	.0135	40	.0980	2	.2210
79	.0145	39	.0995	1	.2280
1/64	.0156	38	.1015	A	.2340
78	.0160	37	.1040	15/64	.2344
77	.0180	36	.1065	B	.2380
76	.0200	7/64	.1094	C	.2420
75	.0210	35	.1100	D	.2460
74	.0225	34	.1110	E	.2500
73	.0240	33	.1130	1/4	.2500
72	.0250	32	.1160	F	.2570
71	.0260	31	.1200	G	.2610
70	.0280	1/8	.1250	17/64	.2656
69	.0292	30	.1285	H	.2660
68	.0310	29	.1360	I	.2720
1/32	.0312	28	.1405	J	.2770
67	.0320	9/64	.1406	K	.2810
66	.0330	27	.1440	9/32	.2812
65	.0350	26	.1470	L	.2900
64	.0360	25	.1495	M	.2950
63	.0370	24	.1520	19/64	.2969
62	.0380	23	.1540	N	.3020
61	.0390	5/32	.1562	5/16	.3125
60	.0400	22	.1570	O	.3160
59	.0410	21	.1590	P	.3230
58	.0420	20	.1610	21/64	.3281
57	.0430	19	.1660	Q	.3320
56	.0465	18	.1695	R	.3390
3/64	.0469	11/64	.1719	11/32	.3438
55	.0520	17	.1730	S	.3480
54	.0550	16	.1770	T	.3580
53	.0595	15	.1800	23/64	.3594
1/16	.0625	14	.1820	U	.3680
52	.0635	13	.1850	3/8	.3750
51	.0670	3/16	.1875	V	.3770
50	.0700	12	.1890	W	.3860
49	.0730	11	.1910	25/64	.3906
48	.0760	10	.1935	X	.3970
5/64	.0781	9	.1960	Y	.4040
47	.0785	8	.1990	13/32	.4062
46	.0810	7	.2010	Z	.4130
45	.0820	13/64	.2031	27/64	.4219
44	.0860	6	.2040	7/16	.4375
43	.0890	5	.2055	29/64	.4531
42	.0935	4	.2090	15/32	.4687
3/32	.0937	3	.2130	31/64	.4844
41	.0960	7/32	.2187	1/2	.5000

APPENDIX V Drilling-Speed Conversion Table

CUTTING SPEEDS IN FEET PER MINUTE

Revolutions per Minute

Drill Size	40	60	80	90	100	120	150	180	200	220	250
80	11317.	16976.	22635.	25464.	28294.	33953.	42441.	50929.	56588.	62247.	70736.
1/64	9794.	14691.	19588.	22036.	24485.	29382.	36728.	44073.	48971.	53868.	61213.
77	8488.	12732.	16976.	19098.	21220.	25464.	31831.	38197.	42441.	46685.	53052.
75	7275.	10913.	14551.	16370.	18189.	21827.	27283.	32740.	36378.	40016.	45473.
73	6366.	9549.	12732.	14324.	15915.	19098.	23873.	28648.	31831.	35014.	39789.
71	5876.	8814.	11753.	13222.	14691.	17629.	22036.	26444.	29382.	32320.	36728.
69	5232.	7848.	10465.	11773.	13081.	15697.	19621.	23546.	26162.	28778.	32703.
1/32	4897.	7345.	9794.	11018.	12242.	14691.	18364.	22036.	24485.	26934.	30606.
66	4629.	6944.	9259.	10417.	11574.	13889.	17362.	20834.	23149.	25464.	28937.
64	4244.	6366.	8488.	9549.	10610.	12732.	15915.	19098.	21220.	23342.	26526.
62	4020.	6031.	8041.	9046.	10051.	12062.	15077.	18093.	20103.	22114.	25129.
60	3819.	5729.	7639.	8594.	9549.	11459.	14324.	17188.	19098.	21008.	23873.
58	3637.	5456.	7275.	8185.	9094.	10913.	13641.	16370.	18189.	20008.	22736.
56	3285.	4928.	6571.	7393.	8214.	9857.	12321.	14786.	16429.	18071.	20536.
55	2938.	4407.	5876.	6611.	7345.	8814.	11018.	13222.	14691.	16160.	18364.
53	2567.	3851.	5135.	5777.	6419.	7703.	9629.	11555.	12839.	14123.	16049.
52	2406.	3609.	4812.	5413.	6015.	7218.	9023.	10827.	12030.	13233.	15038.
50	2182.	3274.	4365.	4911.	5456.	6548.	8185.	9822.	10913.	12004.	13641.
48	2010.	3015.	4020.	4523.	5025.	6031.	7538.	9046.	10051.	11057.	12564.
47	1946.	2919.	3892.	4379.	4865.	5839.	7298.	8758.	9731.	10705.	12164.
45	1863.	2794.	3726.	4192.	4658.	5589.	6987.	8384.	9316.	10248.	11645.
43	1716.	2575.	3433.	3862.	4291.	5150.	6437.	7725.	8583.	9442.	10729.
3/32	1630.	2445.	3261.	3668.	4076.	4891.	6114.	7337.	8153.	8968.	10191.
40	1559.	2338.	3118.	3507.	3897.	4677.	5846.	7015.	7795.	8574.	9744.
38	1505.	2257.	3010.	3386.	3763.	4515.	5644.	6773.	7526.	8279.	9408.
36	1434.	2151.	2869.	3227.	3586.	4303.	5379.	6455.	7173.	7890.	8966.
35	1388.	2083.	2777.	3125.	3472.	4166.	5208.	6250.	6944.	7639.	8681.
33	1352.	2028.	2704.	3042.	3380.	4056.	5070.	6084.	6760.	7436.	8450.
31	1273.	1909.	2546.	2864.	3183.	3819.	4774.	5729.	6366.	7002.	7957.
30	1189.	1783.	2378.	2675.	2972.	3567.	4458.	5350.	5945.	6539.	7431.
28	1087.	1631.	2174.	2446.	2718.	3262.	4078.	4893.	5437.	5981.	6796.
27	1061.	1591.	2122.	2387.	2652.	3183.	3978.	4774.	5305.	5835.	6631.
25	1022.	1533.	2044.	2299.	2555.	3066.	3832.	4599.	5110.	5621.	6387.
23	992.	1488.	1984.	2232.	2480.	2976.	3720.	4464.	4960.	5456.	6200.
22	973.	1459.	1946.	2189.	2432.	2919.	3649.	4379.	4865.	5352.	6082.

CUTTING SPEEDS IN FEET PER MINUTE

Drill Size	40	60	80	90	100	120	150	180	200	220	250
					Revolutions per Minute						
20	949.	1423.	1898.	2135.	2372.	2847.	3558.	4270.	4745.	5219.	5931.
18	901.	1352.	1802.	2028.	2253.	2704.	3380.	4056.	4507.	4957.	5633.
17	883.	1324.	1766.	1987.	2207.	2649.	3311.	3974.	4415.	4857.	5519.
15	848.	1273.	1697.	1909.	2122.	2546.	3183.	3819.	4244.	4668.	5305.
13	825.	1238.	1651.	1858.	2064.	2477.	3097.	3716.	4129.	4542.	5161.
12	808.	1212.	1616.	1818.	2021.	2425.	3031.	3637.	4042.	4446.	5052.
10	789.	1184.	1579.	1776.	1974.	2368.	2961.	3553.	3948.	4342.	4935.
8	767.	1151.	1535.	1727.	1919.	2303.	2879.	3455.	3838.	4222.	4798.
13/64	752.	1128.	1504.	1692.	1880.	2256.	2821.	3385.	3761.	4137.	4701.
5	743.	1115.	1487.	1672.	1858.	2230.	2788.	3345.	3717.	4089.	4646.
3	717.	1075.	1434.	1613.	1793.	2151.	2689.	3227.	3586.	3945.	4483.
2	691.	1037.	1382.	1555.	1728.	2074.	2592.	3111.	3456.	3802.	4320.
A	652.	979.	1305.	1469.	1632.	1958.	2448.	2938.	3264.	3591.	4080.
B	641.	962.	1283.	1444.	1604.	1925.	2407.	2888.	3209.	3530.	4012.
D	621.	931.	1242.	1397.	1552.	1863.	2329.	2794.	3105.	3416.	3881.
F	594.	891.	1189.	1337.	1486.	1783.	2229.	2675.	2972.	3269.	3715.
17/64	575.	862.	1150.	1294.	1438.	1725.	2157.	2588.	2876.	3163.	3595.
I	561.	842.	1123.	1263.	1404.	1685.	2106.	2527.	2808.	3089.	3510.
K	543.	815.	1087.	1223.	1359.	1631.	2039.	2446.	2718.	2990.	3398.
L	526.	790.	1053.	1185.	1317.	1580.	1975.	2370.	2634.	2897.	3292.
19/64	514.	771.	1029.	1157.	1286.	1543.	1929.	2315.	2573.	2830.	3216.
5/16	488.	733.	977.	1100.	1222.	1466.	1833.	2200.	2444.	2689.	3055.
P	473.	709.	946.	1064.	1182.	1419.	1773.	2128.	2365.	2601.	2956.
Q	460.	690.	920.	1035.	1150.	1380.	1725.	2070.	2301.	2531.	2876.
11/32	444.	666.	888.	999.	1111.	1333.	1666.	1999.	2222.	2444.	2777.
T	426.	640.	853.	960.	1066.	1280.	1600.	1920.	2133.	2347.	2667.
U	415.	622.	830.	934.	1037.	1245.	1556.	1868.	2075.	2283.	2594.
3/8	407.	611.	814.	916.	1018.	1222.	1527.	1833.	2037.	2240.	2546.
W	395.	593.	791.	890.	989.	1187.	1484.	1781.	1979.	2177.	2473.
X	384.	577.	769.	865.	962.	1154.	1443.	1731.	1924.	2116.	2405.
13/32	376.	564.	752.	846.	940.	1128.	1410.	1692.	1880.	2068.	2350.
27/64	362.	543.	724.	814.	905.	1086.	1358.	1629.	1810.	1991.	2263.
19/64	337.	505.	674.	758.	843.	1011.	1264.	1517.	1686.	1854.	2107.
31/64	315.	473.	630.	709.	788.	946.	1182.	1419.	1577.	1734.	1971.
1/2	305.	458.	611.	687.	763.	916.	1145.	1375.	1527.	1680.	1909.

Printing Inks Used for Silk-Screen Printing

Paints (printing inks)	Type	Screen Type	Stencil Type	Surface Type	Air-drying Time	Baking Time
Naz-Dar BE and BEF series	Epoxy	#12, #14, #16xx silk	Water soluble	Glass, ceramic, metal, thermosetting plastics	Not recommended	7 min. at 350°F 4 min. at 400°F
Naz-Dar ER and ERF series	Epoxy	#12, #14, #16xx silk	Lacquer proof	Ceramic, metal, phenolic polyester, melamine, silicone	7 to 10 days	7 min. at 300°F 4 min. at 350°F
Wornow 129-000 series	Epoxy	No coarser than #14xx	Lacquer proof	Glass	Not recommended	10 min. at 350°F 5 min. at 375°F
Wornow 50-000 series	Epoxy	No coarser than #14xx	Lacquer proof	Glass, metal, and thermosetting plastics	Not recommended	10 min. at 350°F 5 min. at 375°F
Naz-Dar 59-000 series	Flexible synthetic enamel	#14 silk	Any type stencil	Anodized aluminum and glass	6 hours	30 min. at 180°F
Naz-Dar GV series	Vinyl	#12 silk	Water soluble	Vinyl and cellulose	20 to 40 minutes	Not recommended
Naz-Dar 70-000 series	Lacquer	#12, #14, #16xx silk	Water soluble	Styrene, acetate, rigid vinyl, cellulose	20 to 40 minutes	Not recommended
Candoc SGV-700 series	Vinyl	#12xx silk	Water soluble	Flexible and rigid vinyl	10 to 15 minutes	Not recommended
Naz-Dar IL series	Lacquer	#12, #14, #16xx silk	Lacquer proof	Rigid plastics and lacquer-coated surface	20 to 40 min. or less with force-dry hot air	Not recommended

571

Machine Screw Sizes and Decimal Equivalents

Gauge No. / Fraction	Diameters		Threads per Inch	
	Maximum† Diameter (G)	Minimum† Diameter (T)	UNC	UNF
0	0.0600	0.0438	————	80
1	0.0730	0.0527 / 0.0550	64	72
2*	0.0860	0.0628 / 0.0657	56	64
3	0.0990	0.0719 / 0.0758	48	56
4*	0.1120	0.0795 / 0.0849	40	48
5, 1/8	0.1250	0.0925 / 0.0955	40	44
6*	0.1380	0.0974 / 0.1055	32	40
8*	0.1640	0.1234 / 0.1279	32	36
10*	0.1900	0.1359 / 0.1494	24	32
12*	0.2160	0.1619 / 0.1696	24	28
1/4	0.2500	0.1850 / 0.2036	20	28
5/16	0.3125	0.2403 / 0.2584	18	24
3/8	0.3750	0.2938 / 0.3209	16	24
7/16	0.4375	0.3447 / 0.3726	14	20
1/2	0.5000	0.4001 / 0.4351	13	20
9/16	0.5625	0.4542 / 0.4903	12	18
5/8	0.6250	0.5069 / 0.5528	11	18
3/4	0.7500	0.6201 / 0.6688	10	16
7/8	0.8750	0.7307 / 0.7822	9	14
1"	1.0000	0.8376 / 0.9072	8	14

* Sizes most frequently used in electronics.
† Diameters in inches.

Clearance Hole Sizes and Hole Tolerance Requirements for Machine Screws

Screw Size	Clearance Drill	*Decimal Equivalent of Drill	*Hole Tolerance in Inches
0–80	51	0.0670	0.0070
1–64 1–72	5/64	0.0781	0.0081
2–56 2–64	42	0.0935	0.0075
3–48 3–56	37	0.1040	0.0050
4–40 4–48	31	0.1200	0.0080
5–40 5–44	29	0.1360	0.0110
6–32 6–40	27	0.1440	0.0060
8–32 8–36	17	0.1730	0.0090
10–24 10–32	8	0.1990	0.0090
12–24 12–28	1	0.2280	0.0120
1/4–20 1/4–28	G	0.2610	0.0110
5/16–18 5/16–24	21/64	0.3281	0.0156
3/8–16 3/8–24	25/64	0.3906	0.0156
7/16–14 7/16–20	29/64	0.4531	0.0156
1/2–13 1/2–20	33/64	0.5156	0.0156
9/16–12 9/16–18	37/64	0.5781	0.0156
5/8–11 5/8–18	41/64	0.6406	0.0156
3/4–10 3/4–16	25/32	0.7812	0.0312
7/8–9 7/8–14	29/32	0.9062	0.0312

*All dimensions in inches.

Dimensional Information for Hexagonal Machine Screw Nuts

Machine Screw Nut Size	Width Across Flats in Inches (W)	Thickness in Inches (T)
0−80	5/32	3/64
1−64 1−72	5/32	3/64
2−56	5/32	1/16
2−56 2−64	3/16	1/16
3−48 3−56	3/16	1/16
4−40	3/16	1/16
4−40 4−48	1/4	3/32
5−40	1/4	3/32
5−40 5−44	5/16	7/64
6−32	1/4	3/32
6−32 6−40	5/16	7/64
8−32	1/4 5/16 11/32	3/32 7/64 1/8
8−36	11/32	1/8
10−32	5/16 11/32 3/8	1/8 1/8 1/8
10−24	3/8	1/8
12−24	7/16	5/32
1/4−20 1/4−28	7/16	3/16
5/16−18 5/16−24	9/16	7/32
3/8−16 3/8−24	5/8	1/4

Screw Size	Flat			Internal			Lock Washers External			Split	
	O.D.	I.D.	Thick.	O.D	I.D.	Thick.	O.D.	I.D.	Thick.	W	T
0	5/32	1/16	0.020	–	–	–	–	–	–	0.022	0.022
1	3/16	5/64	0.016	–	–	–	–	–	–	0.022	0.022
										0.030	0.015
										0.030	0.020
	7/32	3/32	0.018							0.030	0.015
2	1/4	3/32	0.016	0.200	0.095	0.015	0.275	0.095	0.019		
	1/4	3/32	0.031							0.035	0.020
	7/32	7/64	0.022							0.035	0.020
3	1/4	7/64	0.016	0.225	0.109	0.015	–	–	–		
	1/4	7/64	0.031							0.040	0.025
	1/4	1/8	0.017							0.035	0.020
	1/4	1/8	0.028								
	5/16	1/8	0.016							0.040	0.025
4	5/16	1/8	0.025	0.270	0.123	0.019	0.260	0.123	0.019		
	5/16	1/8	0.031							0.047	0.031
	3/8	1/8	0.036								
5	9/32	9/64	0.025	0.280	0.136	0.019	0.280	0.136	0.019	0.040	0.025
										0.047	0.031
	5/16	5/32	0.027							0.040	0.025
	11/32	11/64	0.015								
	3/8	5/32	0.016							0.047	0.031
6	3/8	5/32	0.031	0.295	0.150	0.021	0.320	0.150	0.022		
	3/8	5/32	0.036								
	7/16	5/32	0.036							0.055	0.040
	3/8	11/64	0.016							0.047	0.031
	3/8	11/64	0.031							0.055	0.040
8	3/8	3/16	0.036	0.340	0.176	0.023	0.381	0.176	0.023		
	1/2	3/16	0.036							0.062	0.047
	3/8	13/64	0.047							0.055	0.040
	13/32	13/64	0.036								
	7/16	13/64	0.031							0.062	0.047
10	7/16	13/64	0.062	0.381	0.204	0.025	0.410	0.204	0.025		
	7/16	7/32	0.036								
	1/2	7/32	0.062							0.070	0.056
	9/16	13/64	0.036								
	9/16	1/4	0.051								
12	1/2	15/64	0.040	0.410	0.231	0.025	0.475	0.231	0.025	0.062	0.047
										0.070	0.056
	1/2	17/64	0.031							0.107	0.047
	1/2	17/64	0.062								
	1/2	9/32	0.056							0.109	0.062
	5/8	9/32	0.051								
1/4	5/8	9/32	0.062	0.478	0.267	0.028	0.510	0.267	0.028	0.110	0.077
	11/16	17/64	0.050								
	3/4	9/32	0.056								
	3/4	5/16	0.051								

Dimensional Information for Metal Washers

| Screw Size | Flat | | | Lock Washers | | | | | | Split | |
| | | | | Internal | | | External | | | | |
	O.D.	I.D.	Thick	O.D.	I.D.	Thick.	O.D.	I.D.	Thick.	W	T
5/16	9/16	21/64	0.031								
	9/16	21/64	0.062							0.117	0.056
	11/16	11/32	0.051								
	11/16	11/32	0.062	0.610	0.332	0.034	0.610	0.332	0.033	0.125	0.078
	5/8	11/32	0.056								
	3/4	11/32	0.050							0.130	0.097
	7/8	3/8	0.064								
3/8	5/8	25/64	0.031								
	5/8	25/64	0.062							0.136	0.070
	3/4	13/32	0.056								
	13/16	13/32	0.062	0.692	0.398	0.040	0.694	0.398	0.040		
	13/16	13/32	0.051								
	1	7/16	0.064							0.141	0.094

All dimensions in inches.

Tap Drill and Tap Requirements for Machine Screws

Screw Size	Tap Drill	*Decimal Equivalent of Drill	Tap Size
0 – 80	3/64	0.0469	0 – 80
1 – 64	53	0.0595	1 – 64
1 – 72			1 – 72
2 – 56	50	0.0700	2 – 56
2 – 64			2 – 64
3 – 48	47	0.0785	3 – 48
3 – 56	45	0.0820	3 – 56
4 – 40	43	0.0890	4 – 40
4 – 48	42	0.0935	4 – 48
5 – 40	38	0.1015	5 – 40
5 – 44	37	0.1040	5 – 44
6 – 32	36	0.1065	6 – 32
6 – 40	33	0.1130	6 – 40
8 – 32	29	0.1360	8 – 32
8 – 36			8 – 36
10 – 24	25	0.1495	10 – 24
10 – 32	21	0.1590	10 – 32
12 – 24	16	0.1770	12 – 24
12 - 28	14	0.1820	12 – 28
1/4 – 20	7	0.2010	1/4 – 20
1/4 – 28	3	0.2130	1/4 – 28
5/16 – 18	F	0.2570	5/16 – 18
5/16 – 24	I	0.2720	5/16 – 24
3/8 – 16	5/16	0.3125	3/8 – 16
3/8 – 24	Q	0.3320	3/8 – 24
7/16 – 14	U	0.3680	7/16 – 14
7/16 – 20	25/64	0.3906	7/16 – 20
1/2 – 13	27/64	0.4219	1/2 – 13
1/2 – 20	29/64	0.4531	1/2 – 20
9/16 – 12	31/64	0.4844	9/16 – 12
9/16 – 18	33/64	0.5156	9/16 – 18
5/8 – 11	17/32	0.5312	5/8 – 11
5/8 – 18	37/64	0.5781	5/8 – 18
3/4 – 10	21/32	0.6562	3/4 – 10
3/4 – 16	11/16	0.6871	3/4 – 16
7/8 – 9	49/64	0.7656	7/8 – 9
7/8 – 14	13/16	0.8125	7/8 – 14
1″ – 8	7/8	0.8750	1″ – 8
1″ – 14	15/16	0.9375	1″ – 14

* Dimensions in inches.

Wire Table for Uninsulated Round Copper Conductors

AWG* Number	Diameter in Mils (approximate)	Area in Circular Mils	Maximum† Recommended Current Capacity in Amperes
00	365	133,080	
0	325	105,560	
1	289	83,690	
2	257	66,360	
3	229	52,620	
4	204	41,740	
5	182	33,090	
6	162	26,240	
7	144	20,820	
8	128	16,510	33.0
9	114	13,090	26.0
10	102	10,380	20.5
11	90.7	8,226	16.5
12	80.8	6,529	12.0
13	72.0	5,184	10.0
14	64.1	4,109	8.0
15	57.1	3,260	6.5
16	50.8	2,580	5.0
17	45.3	2,052	4.0
18	40.3	1,624	3.0
19	35.9	1,289	2.5
20	32.0	1,024	2.0
21	28.5	812	1.5
22	25.3	640	1.0
23	22.6	511	1.0
24	20.1	404	.80
25	17.9	320	.60
26	15.9	253	.50
27	14.2	202	.40
28	12.6	159	.30
29	11.3	128	.25
30	10.0	100	.20
31	8.9	79	
32	8.0	64	
33	7.1	50	
34	6.3	40	
35	5.6	31	
36	5.0	25	
37	4.5	20	

Wire Table for Uninsulated Round Copper Conductors

AWG* Number	Diameter in Mils (approximate)	Area in Circular Mils	Maximum† Recommended Current Capacity in Amperes
38	4.0	16	
39	3.5	12	
40	3.1	10	
41	2.8	8	
42	2.5	6	
43	2.2	5	
44	2.0	4	
45	1.8	3	
46	1.6	2.5	
47	1.4	2	
48	1.2	1.5	
49	1.1	1.2	
50	1.0	1.0	

* American Wire Gauge Standard (AWG).
† Current capacity based on 500 circular mils per ampere.

Characteristics and Applications of Heat-Shrinkable Tubing. Courtesy of Electronized Chemicals Corporation, "Copyright Chilton Company."

TYPICAL CHARACTERISTICS, PROPERTIES and

MATERIAL[1,2]	Oper. Temp. Range-Contin.	Vol. Resist.& Ohm-cm@ R.T. Nom. (ASTM D257)	Dielectric Strength V/Mil. Min. (ASTM D149)	Dielectric Constant (ASTM D150)	Tensile Strength PSI (ASTM D876)	Ultimate Elongation % (ASTM D876)	Flammability[4]	Copper[5] Corrosion	Water[7] Absorption (ASTM D570)	Specific Gravity (ASTM D792)	Abrasion Resist.	Flexibility @ Room Temp.
Polyvinyl Chloride (PVC-Vinyl)	−20°C to +105°C	10^{11}	500	7.0	2000 to 4000	200 to 300	Self Extinguishing	Fair	1.0%	1.3	Fair	Good
Irradiated Modified Polyvinyl Chloride	−20°C to +105°C	10^{11}	500	5.0	2000 to 4000	200 to 300	Self Extinguishing	Fair	1.0%	1.3	Good	Good
Irradiated Modified Polyolefin	−55°C to +135°C	10^{14}	600^3	2.5	1500 to 3000	200 to 400	Self Extinguishing (except clear)	Good	0.5%	1.35	Good	Good
Polyester (Mylar—DuPont Trade Name)	−60°C to +135°C	10^{15}	3	3.4	20,000 to 30,000	35 to 110	Self Extinguishing	Good[6]	0.5%	1.4	Good	**Poor**
Polyvinylidine Fluoride (Kynar—Pennsalt Trade Name)	−55°C to +175°C	10^{13}	250^3	5.5	5000 to 8000	100 to 300	Self Extinguishing	Excellent	0.5%	1.8	Excellent	Fair
Ethylene Fluorocarbon Copolymer (Tefzel—DuPont Trade Name) *Preliminary Data	−100°C to +175°C	10^{16}	400	2.6	6500	100 to 400	Non-burning	Excellent	0.1%	1.7	Excellent	Fair
Irradiated Modified Fluoropolymer (Irrefluor—Electronized Chemicals Trademark)	−55°C to +200°C	10^{10}	600	5.5	3000 to 5000	250 to 350	Non-burning	Excellent	0.5%	2.3	Excellent	Good
Fluorinated Ethylene Propylene (FEP)	−67°C to +200°C	10^{17}	600	2.1	2700 to 3100	250 to 350	Non-burning	Excellent	0.01%	2.1	Very Good	Fair
Polytetrafluorethylene (Teflon-(P) TFE—DuPont Trade Name)	−67°C to +250°C	10^{18}	600	2.1	2000 to 5000	200 to 400	Non-burning	Excellent	0.01%	2.2	Very Good	Fair
Flexible Modified Neoprene	−55°C to +90°C	10^{10}	150 to 600	9.0	1500 to 3000	200 to 300	**Self Extinguishing**	Fair	1.0%	1.3	Excellent	Excellent
Butyl	−55°C to +135°C	10^{17}	125	2.6	1200 to 1600	200 to 500	Not Flame Retarded	Good	0.5%	1.4	**Good**	Excellent
Flexible, Modified Silicone Elastomer	−75°C to +175°C	10^{11}	100 to 600	3.2	600 to 900	200 to 400	Self Extinguishing	Good	1.0%	1.3	Poor	Excellent
Flexible, Modified Fluorosilicone Elastomer	−50°C to +175°C	5×10^{11}	320	n. a..	600 to 900	200 to 500	Self Extinguishing	Good	1.0%	1.7	Poor to Good	Excellent
Flexible, Modified Fluoroelastomer (Viton—DuPont Trade Name)	−55°C to +200°C	10^{12}	200	n. a.	1900 to 2400	350	Self Extinguishing	Good	1.0%	1.7	Good	Good

FEATURES OF HEAT-SHRINKABLE PRODUCTS and MATERIALS

Weatherability	Fuel and Oil Resist.[9] (at R.T.)	Solvent Resistance[10] Hydro Carbons (at R.T.)			Resist. to Acids and Alkalis	Tubing Sizes[8] Expanded Dia. Inches I.D.	Wall Thick.[8] Available (Min./Max.)	Shrink Ratio	Longitudinal Change % Max.	Shrink Temp. (Min.)	AIEE Classification (Class)	Available Spec.	Relative Cost[11]
		Aliphatic	Aromatic	Chlorinated									
Excellent[12]	Very Good	Poor	Poor	Fair	Excellent	0.063 to 6.00	0.014 in. to 0.034 in.	1.6:1 2:1	−25%	235°F 113°C	105 A	MIL UL NASA	$ 10.85
Excellent[12]	Very Good	Good	Good	Good	Excellent	0.046 to 2.00	0.020 in. to 0.050 in.	2:1	+1% to −10%	347°F 175°C	105 A	MIL UL NASA	$ 14.70
Excellent[12]	Good	Excellent	Excellent	Excellent	Excellent	0.046 to 4.00	0.016 in. to 0.055 in.	2:1 3:1	±5%	250°F 121°C	130 B	MIL AMS UL NASA	$ 25.75
Poor	Excellent	Excellent	Excellent	Excellent	Good	0.063 to 4.00	0.0023 in.	1.6:1	±35%	175°F 80°C	130 B(F) (13)	MIL NASA	$ 5.00
Excellent	Excellent	Excellent	Excellent	Excellent	Excellent	0.046 to 1.50	0.010 in. to 0.020 in.	2:1	±10%	350°F 177°C	155 F	MIL AMS UL NASA	$ 53.60
Excellent	Excellent	Excellent	Excellent	Excellent	Excellent	0.125 to 1.50	0.010 to 0.030 in.	1.5:1	+10% to 20%	300°F 149°C	155 F	—	$ 64.30
Excellent	Excellent	Excellent	Excellent	Excellent	Excellent	0.046 to 1.00	0.010 in. to 0.023 in.	2:1	±10%	347°F 175°C	180 H	—	$ 59.00
Excellent	Excellent	Excellent	Excellent	Excellent	Excellent	0.23 to 1.00	0.008 in. to 0.035 in.	1.6:1	±15%	347°F 175°C	180 H	MIL UL	$ 59.40
Excellent	Excellent	Excellent	Excellent	Excellent	Excellent	0.030 to 0.470	0.006 in. to 0.050 in.	1.6:1 1.33:1 4:1	±20%	625°F 329°C	220 C	MIL UL NASA	$103.12
Excellent[12]	Good	Good	Poor	Excellent	Good	0.250 to 4.00	0.035 in. to 0.140 in.	1.8:1	±10%	350°F 177°C	90 O	MIL AMS NASA	$ 61.20
Very Good	Poor	Poor	Poor	Poor	Good	0.025 to 4.00	0.035 in. to 0.150 in.	2:1	±15%	250°F 121°C	130 B	None	$ 98.55
Excellent	Poor to Good	Good	Fair	Good	Good	0.250 to 2.00	0.020 in. to 0.110 in.	1.8:1	+3% to −10%	350°F 177°C	155 F	MIL NASA	$108.25
Excellent	Excellent	Excellent	Excellent	Excellent	Excellent	0.50 to 2.0	0.048 in. to 0.190 in.	2:1	+3% to −10%	302°F 150°C	155 F	MIL	n. a.
Excellent	Excellent	Excellent	Excellent	Good	Excellent	0.250 to 1.50	0.030 in. to 0.075 in.	2:1	±15%	347°F 175°C	180 H	MIL	$181.05

Calibration of Exposure Unit for Dry-Film Photo-Resist

Proper exposure time for a specific resist is determined empirically for each exposure unit with the use of a *step table* such as the Stouffer 21-step exposure chart shown in Fig. A. This step table consists of a sheet of plastic having 21 areas (windows) each having increasing optical density. Step 1 is essentially clear and totally transparent, whereas step 21 is the most opaque. For any given exposure time and light source, the amount of light striking the resist surface is a function of the density of the window on the step table. Use of this table to determine the correct exposure requires a knowledge of the resist manufacturer's recommendations. Dynachem recommends the following for Laminar A resist. After exposure and developing, the results of the resist exposed under a 21-step exposure table should be a "solid No. 6" with high-gloss green resist existing totally under that window. Under step 7, some resist will have been attacked during developing and the remaining resist under that window will be a dull-green color. Under step 8 and above to step 21, only clear copper will be visible.

Too long an exposure time will result in "high-gloss" resist occurring under a *higher* step number than 6 as a result of enough light having penetrated the more-opaque windows of the step table. This exposure time will result in resist brittleness. Too short an exposure time will result in "high-gloss" resist occurring under a *smaller* step number than 6 because of the opaqueness of the smaller step numbers being sufficient to block enough light from striking the resist. Incomplete resist polymerization results. This may appear, after developing, as a dull resist image or simply an incomplete image of the conductor pattern.

The use of a 21-step exposure table to determine the correct exposure time for a particular exposure unit will require that six small pc boards be cleaned, dried, preheated, laminated, and allowed to normalize to room temperature. (These processes are detailed in Secs. 17.2, 17.3, 17.4, and 17.5.) Each board is then exposed with the Stouffer exposure chart as the photo tool for a predetermined amount of time with this time designated on the board. The first board should be exposed for 30 seconds, the second for 1 minute, the third for $1\frac{1}{2}$ minutes, and so on, until all boards are exposed. After all boards have been normalized to room temperature for 10 minutes, they should be developed, rinsed, neutralized in a 5% sulfuric acid bath, rinsed, and dried. (Refer to Secs. 17.6 and 17.7.) Six properly processed samples are shown in Fig. B. The time required to obtain a "solid No. 6" on the step table is the proper exposure time for the specific resist and exposure unit employed.

STOUFFER
Graphic Arts

2
3
4
5
6
7
8
9
10
11
12
13
14
15
16
17
18
19
20
21

FIGURE A. Stouffer 21 step exposure chart.

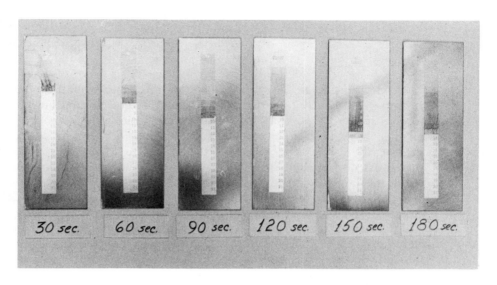

FIGURE B. Exposure time to obtain a solid #6 is approximately 60 seconds.

As discussed in Sec. 17.10, the tin–lead plating solution described is extremely susceptible to organic and metallic contaminations. It is therefore absolutely essential that all contaminations be removed from all objects that will contact the plating solution. New plating tanks must be thoroughly leached before being used for the first time. The contaminants that must be removed from these tanks are cements, plastic welding residues, mold-release residues, machine oils, airborn particles, and simple dirts and greases. A new tank may look clean but is probably the single largest source of contamination that will result in plating failures.

Leaching involves first exposing the tank to an extremely strong alkaline solution followed by a relatively mild acid solution. The alkaline solution is used to remove the contaminations, and the acid solution neutralizes the alkaline residue.

A 1000-watt quartz heater with temperature control is first positioned into the tank with all hardware removed from the tank. The tank is filled with tap water to within ½ inch of the top rim. The following are then added to the tank:

Trisodium phosphate (Na_3PO_4), 6 ounces per gallon

Sodium hydroxide (NaOH)–caustic soda, 3 ounces per gallon

Plastic gloves should be worn when handling these strongly alkaline compounds. The chemicals are stirred into the solution with a plastic paddle and the temperature is raised to 130°F (54°C). The tank is subjected to this temperature for 8 hours, after which the heater is shut off, allowed to cool, and removed from the tank. The alkaline solution is then discarded. The tank is then *twice* filled with cold tap water and emptied.

To neutralize the alkalinity remaining in the tank, it is then filled with a 10% by volume solution of fluoboric acid. Using a clean plastic paddle, the solution is slowly stirred. This solution is allowed to remain in the tank for 24 hours at room temperature before discarding and again the tank is twice filled with cold tap water and emptied. The use of deionized water as a final rinse is recommended, but cold tap water will suffice if deionized water is unavailable.

The tank is now properly leached and ready for the plating solution.

APPENDIX XVI

Conversions
Degrees Fahrenheit into Degrees Celsius

°F* → °C		°F → °C		°F → °C		°F → °C	
0	−17.8	140	60.0	280	137.8	420	215.6
5	−15.0	145	62.7	285	140.6	425	218.3
10	−12.2	150	65.6	290	143.3	430	221.1
15	− 9.4	155	68.3	295	146.1	435	223.9
20	− 6.7	160	71.1	300	148.9	440	226.7
25	− 3.9	165	73.9	305	151.7	445	229.4
30	− 1.1	170	76.7	310	154.4	450	232.2
35	1.7	175	79.4	315	157.2	455	235.0
40	4.4	180	82.2	320	160.0	460	237.8
45	7.2	185	85.0	325	162.8	465	240.6
50	10.0	190	87.8	330	165.6	470	243.3
55	12.8	195	90.6	335	168.3	475	246.1
60	15.6	200	93.3	340	171.1	480	248.9
65	18.3	205	96.1	345	173.9	485	251.7
70	21.1	210	98.9	350	176.7	490	254.4
75	23.9	215	101.7	355	179.4	495	257.2
80	26.7	220	104.4	360	182.2	500	260.0
85	29.4	225	107.2	365	185.0	505	262.8
90	32.2	230	110.0	370	187.8	510	265.6
95	35.0	235	112.8	375	190.6	515	268.3
100	37.8	240	115.6	380	193.3	520	271.1
105	40.6	245	118.3	385	196.1	525	273.9
110	43.3	250	121.1	390	198.9	530	276.7
115	46.1	255	123.9	395	201.7	535	279.4
120	48.9	260	126.7	400	204.4	540	282.2
125	51.7	265	129.4	405	207.2	545	285.0
130	54.4	270	132.2	410	210.0	550	287.8
135	57.2	275	135.0	415	212.8	555	290.6

*For each °F between those listed add 0.56°C. ⟶ Example: 74° F = 21.1°C + 4(0.56°C)

$$°C = \tfrac{5}{9}\,(°F - 32)$$
Celsius

$$°F = \tfrac{9}{5}\,°C + 32$$
Fahrenheit

74° F = 23.3°C

Weights and Measures

1 pound = 453.6 grams

1 ounce = 28.35 grams

1 gallon = 3.785 liters

1 quart = 946 milliliters

1 ounce = 29.57 milliliters

1 inch = 25.4 millimeters

1 foot = 304.8 millimeters

1 square inch = 645.2 square millimeters

DECIMALS TO MILLIMETERS				FRACTIONS TO DECIMALS TO MILLIMETERS					
Deci-mal	mm	Deci-mal	mm	Frac-tion	Deci-mal	mm	Frac-tion	Deci-mal	mm
0.001	0.0254	0.500	12.7000	1/64	0.0156	0.3969	33/64	0.5156	13.0969
0.002	0.0508	0.510	12.9540	1/32	0.0312	0.7938	17/32	0.5312	13.4938
0.003	0.0762	0.520	13.2080	3/64	0.0469	1.1906	35/64	0.5469	13.8906
0.004	0.1016	0.530	13.4620						
0.005	0.1270	0.540	13.7160						
0.006	0.1524	0.550	13.9700	1/16	0.0625	1.5875	9/16	0.5625	14.2875
0.007	0.1778	0.560	14.2240						
0.008	0.2032	0.570	14.4780						
0.009	0.2286	0.580	14.7320	5/64	0.0781	1.9844	37/64	0.5781	14.6844
		0.590	14.9860	3/32	0.0938	2.3812	19/32	0.5938	15.0812
0.010	0.2540								
0.020	0.5080			7/64	0.1094	2.7781	39/64	0.6094	15.4781
0.030	0.7620								
0.040	1.0160	0.600	15.2400	1/8	0.1250	3.1750	5/8	0.6250	15.8750
0.050	1.2700	0.610	15.4940						
0.060	1.5240	0.620	15.7480						
0.070	1.7780	0.630	16.0020	9/64	0.1406	3.5719	41/64	0.6406	16.2719
0.080	2.0320	0.640	16.2560	5/32	0.1562	3.9688	21/32	0.6562	16.6688
0.090	2.2860	0.650	16.5100						
		0.660	16.7640	11/64	0.1719	4.3656	43/64	0.6719	17.0656
0.100	2.5400	0.670	17.0180						
0.110	2.7940	0.680	17.2720	3/16	0.1875	4.7625	11/16	0.6875	17.4625
0.120	3.0480	0.690	17.5260						
0.130	3.3020								
0.140	3.5560								
0.150	3.8100			13/64	0.2031	5.1594	45/64	0.7031	17.8594
0.160	4.0640	0.700	17.7800	7/32	0.2188	5.5562	23/32	0.7188	18.2562
0.170	4.3180	0.710	18.0340	15/64	0.2344	5.9531	47/64	0.7344	18.6531
0.180	4.5720	0.720	18.2880						
0.190	4.8260	0.730	18.5420						
		0.740	18.7960	1/4	0.2500	6.3500	3/4	0.7500	19.0500
0.200	5.0800	0.750	19.0500						
0.210	5.3340	0.760	19.3040						
0.220	5.5880	0.770	19.5580	17/64	0.2656	6.7469	49/64	0.7656	19.4469
0.230	5.8420	0.780	19.8120	9/32	0.2812	7.1438	25/32	0.7812	19.8438
0.240	6.0960	0.790	20.0660	19/64	0.2969	7.5406	51/64	0.7969	20.2406
0.250	6.3500								
0.260	6.6040								
0.270	6.8580			5/16	0.3125	7.9375	13/16	0.8125	20.6375
0.280	7.1120	0.800	20.3200						
0.290	7.3660	0.810	20.5740						
		0.820	20.8280	21/64	0.3281	8.3344	53/64	0.8281	21.0344
0.300	7.6200	0.830	21.0820	11/32	0.3438	8.7312	27/32	0.8438	21.4312
0.310	7.8740	0.840	21.3360	23/64	0.3594	9.1281	55/64	0.8594	21.8281
0.320	8.1280	0.850	21.5900						
0.330	8.3820	0.860	21.8440						

DECIMALS TO MILLIMETERS				*FRACTIONS TO DECIMALS TO MILLIMETERS*					
Deci-mal	mm	Deci-mal	mm	Frac-tion	Deci-mal	mm	Frac-tion	Deci-mal	mm
0.340	8.6360	0.870	22.0980	3/8	0.3750	9.5250	7/8	0.8750	22.2250
0.350	8.8900	0.880	22.3520						
0.360	9.1440	0.890	22.6060						
0.370	9.3980			25/64	0.3906	9.9219	57/64	0.8906	22.6219
0.380	9.6520			13/32	0.4062	10.3188	29/32	0.9062	23.0188
0.390	9.9060			27/64	0.4219	10.7156	59/64	0.9219	23.4156
		0.900	22.8600						
0.400	10.1600	0.910	23.1140						
0.410	10.4140	0.920	23.3680	7/16	0.4375	11.1125	15/16	0.9375	23.8125
0.420	10.6680	0.930	23.6220						
0.430	10.9220	0.940	23.8760						
0.440	11.1760	0.950	24.1300	29/64	0.4531	11.5094	61/64	0.9531	24.2094
0.450	11.4300	0.960	24.3840	15/32	0.4688	11.9062	31/32	0.9688	24.6062
0.460	11.6840	0.970	24.6380	31/64	0.4844	12.3031	63/64	0.9844	25.0031
0.470	11.9380	0.980	24.8920						
0.480	12.1920	0.990	25.1460	1/2	0.5000	12.7000	1	1.0000	25.4000
0.490	12.4460	1.000	25.4000						

MILLIMETERS TO DECIMALS

mm	Deci-mal	mm	Deci-mal	mm	Deci-mal	mm	Deci-mal	mm	Deci-mal
0.01	.00039	0.41	.01614	0.81	.03189	21	.82677	61	2.40157
0.02	.00079	0.42	.01654	0.82	.03228	22	.86614	62	2.44094
0.03	.00118	0.43	.01693	0.83	.03268	23	.90551	63	2.48031
0.04	.00157	0.44	.01732	0.84	.03307	24	.94488	64	2.51969
0.05	.00197	0.45	.01772	0.85	.03346	25	.98425	65	2.55906
0.06	.00236	0.46	.01811	0.86	.03386	26	1.02362	66	2.59843
0.07	.00276	0.47	.01850	0.87	.03425	27	1.06299	67	2.63780
0.08	.00315	0.48	.01890	0.88	.03465	28	1.10236	68	2.67717
0.09	.00354	0.49	.01929	0.89	.03504	29	1.14173	69	2.71654
0.10	.00394	0.50	.01969	0.90	.03543	30	1.18110	70	2.75591
0.11	.00433	0.51	.02008	0.91	.03583	31	1.22047	71	2.79528
0.12	.00472	0.52	.02047	0.92	.03622	32	1.25984	72	2.83465
0.13	.00512	0.53	.02087	0.93	.03661	33	1.29921	73	2.87402
0.14	.00551	0.54	.02126	0.94	.03701	34	1.33858	74	2.91339
0.15	.00591	0.55	.02165	0.95	.03740	35	1.37795	75	2.95276
0.16	.00630	0.56	.02205	0.96	.03780	36	1.41732	76	2.99213
0.17	.00669	0.57	.02244	0.97	.03819	37	1.45669	77	3.03150
0.18	.00709	0.58	.02283	0.98	.03858	38	1.49606	78	3.07087
0.19	.00748	0.59	.02323	0.99	.03898	39	1.53543	79	3.11024
0.20	.00787	0.60	.02362	1.00	.03937	40	1.57480	80	3.14961
0.21	.00827	0.61	.02402	1	.03937	41	1.61417	81	3.18898
0.22	.00866	0.62	.02441	2	.07874	42	1.65354	82	3.22835
0.23	.00906	0.63	.02480	3	.11811	43	1.69291	83	3.26772
0.24	.00945	0.64	.02520	4	.15748	44	1.73228	84	3.30709
0.25	.00984	0.65	.02559	5	.19685	45	1.77165	85	3.34646
0.26	.01024	0.66	.02598	6	.23622	46	1.81102	86	3.38583
0.27	.01063	0.67	.02638	7	.27559	47	1.85039	87	3.42520
0.28	.01102	0.68	.02677	8	.31496	48	1.88976	88	3.46457
0.29	.01142	0.69	.02717	9	.35433	49	1.92913	89	3.50394
0.30	.01181	0.70	.02756	10	.39370	50	1.96850	90	3.54331
0.31	.01220	0.71	.02795	11	.43307	51	2.00787	91	3.58268
0.32	.01260	0.72	.02835	12	.47244	52	2.04724	92	3.62205
0.33	.01299	0.73	.02874	13	.51181	53	2.08661	93	3.66142
0.34	.01339	0.74	.02913	14	.55118	54	2.12598	94	3.70079
0.35	.01378	0.75	.02953	15	.59055	55	2.16535	95	3.74016
0.36	.01417	0.76	.02992	16	.62992	56	2.20472	96	3.77953
0.37	.01457	0.77	.03032	17	.66929	57	2.24409	97	3.81890
0.38	.01496	0.78	.03071	18	.70866	58	2.28346	98	3.85827
0.39	.01535	0.79	.03110	19	.74803	59	2.32283	99	3.89764
0.40	.01575	0.80	.03150	20	.78740	60	2.36220	100	3.93701

Resistance Color Code

Band colors	1st Band	2nd Band	3rd Band	4th Band
Black ⟶	0	0	0	
Brown ⟶	1	1	10	
Red ⟶	2	2	100	
Orange ⟶	3	3	1,000	
Yellow ⟶	4	4	10,000	
Green ⟶	5	5	100,000	
Blue ⟶	6	6	1,000,000	
Violet ⟶	7	7	—	
Gray ⟶	8	8	—	
White ⟶	9	9	—	
Gold ⟶	—	—	0.1	±5%
Silver ⟶	—	—	0.01	±10%
No color ⟶	—	—	—	±20%
	(A)	(B)	(C)	(D)

Arrows point to possible numerical values for each BAND COLOR depending upon position

Example:	R	=	(A)	before	(B)	X	(C)	±	(D)%
			Yellow		Violet		Orange		Silver
	R	=	4		7	X	1,000	±	10%

R = 47,000 Ω ± 10% or 47 KΩ ± 10%

Engineering notation
1,000 = K
10,000 = 10 K
100,000 = 100 K
1,000,000 = M

Bibliography

Chemical Milling with Kodak Photosensitive Resists. Eastman Kodak Company, Rochester, N.Y., 1971.

Coombs, C. F. *Printed Circuits Handbook.* McGraw-Hill Book Company, New York, 1979.

Coughlin, R. F. *Power Supplies.* Massachusetts Institute of Technology, Cambridge, Mass., 1978.

Coughlin, R. F. *Principles and Applications of Semiconductors and Circuits.* Prentice-Hall, Inc., Englewood Cliffs, N.J., 1971.

Coughlin, R. F., and F. F. Driscoll. *Operational Amplifiers and Linear Integrated Circuits.* Prentice-Hall, Inc., Englewood Cliffs, N.J., 1977.

Coughlin, R. F., and F. F. Driscoll. *Semiconductor Fundamentals.* Prentice-Hall, Inc., Englewood Cliffs, N.J., 1976.

The Design and Drafting of Printed Circuits. Bishop Graphics, Inc., Westlake Village, Calif., 1979.

DeForest, W. S. *Photoresist Materials and Processes.* McGraw-Hill Book Company, New York, 1975.

Driscoll, F. F. *Analysis of Electric Circuits.* Prentice-Hall, Inc., Englewood Cliffs, N.J., 1973.

Driscoll, F. F. *Operational Amplifiers.* Massachusetts Institute of Technology, Cambridge, Mass. 1978.

Driscoll, F. F., and F. Coughlin. *Solid State Devices and Applications.* Prentice-Hall, Inc., Englewood Cliffs, N.J., 1975.

Dynachem Technical Data. Dynachem Corporation, Santa Fe Springs, Calif., 1971.

Einarson, N. S. *Printed Circuit Technology.* Printed Circuit Technology, Burlington, Mass., 1977.

Harper, C. A. *Handbook of Electronic Packaging.* McGraw-Hill Book Compnay, New York, 1969.

An Introduction to Photofabrication Using Kodak Photosensitive Resists. Kodak Company, Rochester, N.Y., 1971.

Kirshner, C., and K. M. Stone. *Electronics Drafting Workbook.* McGraw-Hill Book Company, New York, 1978.

Lambert Screen Process Supplies Catalog. Lambert Company, Inc., Boston, Mass., 1970.

Low-cost Plastic Power Audio Output. Application Note CA-116A, Texas Instruments, Dallas, Tex., 1969.

Lowenheim, F. A. *Electroplating Fundamentals of Surface Finishing.* McGraw-Hill Book Company, New York, 1978.

McWane, John W., R. J. Duffy, and R. F. Coughlin. *DC Current Voltage and Resistance.* Massachusetts Institute of Technology, Cambridge, Mass., 1976.

Metals Finishing Guidebook and Directory. Metals and Plastics Publications, Inc., Westwood, N.J., 1971.

MOS/LSI Databook. National Semiconductor Corporation, Santa Clara, Calif., 1977.

Patton, W. J. *Materials in Industry.* Prentice-Hall, Inc., Englewood Cliffs, N.J., 1968.

Prise, W. J. *Electronic Circuit Packaging.* Charles E. Merrill Books, Inc., Columbus, Ohio, 1967.

Raskhodoff, N. M. *Electronic Drafting and Design,* 3rd ed. Prentice-Hall, Inc., Englewood Cliffs, N.J., 1977.

RCA Linear Integrated Circuits. Technical Series IC-42, Radio Corporation of America, Harrison, N.J., 1970.

RCA Transistor, Thyristor and Diode Manual. Technical Series SC-14, Radio Corporation of America, Harrison, N.J., 1969.

Reliable Electrical Connections. Quality Division, George C. Marshall Space Flight Center, Huntsville, Ala., 1961.

SCR Manual, 4th ed. General Electric Company, Auburn, N.Y., 1967.

Soldering Electrical Connections. National Aeronautics and Space Administration, Washington, D.C., 1967.

Teeling, John. *Mounting and Heat Sinking Uniwatt Plastic Transistors.* AN-472, Motorola Semiconductor Products, Inc., Phoenix, Ariz., 1969.

Teeling, John. *Using Uniwatt Transistors in High-Fidelity Amplifiers,* AN-428A, Motorola Semiconductor Products, Inc., Phoenix, Ariz., 1968.

Wheeler, G. J. *Electronic Assembly and Fabrication.* Reston Publishing Company, Inc., Reston, Va., 1976.

Index

Hydraulic power-driven shears, 61

T